Secure Cyber–Physical Systems for Smart Cities

Riaz Ahmed Shaikh
King Abdulaziz University, Saudi Arabia

A volume in the Advances
in Computer and Electrical
Engineering (ACEE) Book Series

Published in the United States of America by
    IGI Global
    Engineering Science Reference (an imprint of IGI Global)
    701 E. Chocolate Avenue
    Hershey PA, USA 17033
    Tel: 717-533-8845
    Fax: 717-533-8661
    E-mail: cust@igi-global.com
    Web site: http://www.igi-global.com

Library of Congress Cataloging-in-Publication Data

Names: Shaikh, Riaz Ahmed, 1981- editor.
Title: Secure cyber-physical systems for smart cities / Riaz Ahmed Shaikh,
    editor.
Description: Hershey, PA : Engineering Science Reference, an imprint of IGI
    Global, [2019] | Includes bibliographical references and index.
Identifiers: LCCN 2018019091| ISBN 9781522571896 (hardcover) | ISBN
    9781522571902 (ebook)
Subjects: LCSH: Municipal engineering. | Smart cities. | Cooperating objects
    (Computer systems)
Classification: LCC TD159.4 .S43 2019 | DDC 363.60285/58--dc23 LC record available at https://
lccn.loc.gov/2018019091

This book is published in the IGI Global book series Advances in Computer and Electrical
Engineering (ACEE) (ISSN: 2327-039X; eISSN: 2327-0403)

British Cataloguing in Publication Data
A Cataloguing in Publication record for this book is available from the British Library.

All work contributed to this book is new, previously-unpublished material.
The views expressed in this book are those of the authors, but not necessarily of the publisher.

For electronic access to this publication, please contact: eresources@igi-global.com.

Advances in Computer and Electrical Engineering (ACEE) Book Series

ISSN:2327-039X
EISSN:2327-0403

Editor-in-Chief: Srikanta Patnaik, SOA University, India

MISSION

The fields of computer engineering and electrical engineering encompass a broad range of interdisciplinary topics allowing for expansive research developments across multiple fields. Research in these areas continues to develop and become increasingly important as computer and electrical systems have become an integral part of everyday life.

The **Advances in Computer and Electrical Engineering (ACEE) Book Series** aims to publish research on diverse topics pertaining to computer engineering and electrical engineering. **ACEE** encourages scholarly discourse on the latest applications, tools, and methodologies being implemented in the field for the design and development of computer and electrical systems.

COVERAGE

- VLSI Fabrication
- Computer Architecture
- Computer Hardware
- Chip Design
- Analog Electronics
- Optical Electronics
- Electrical Power Conversion
- Applied Electromagnetics
- Sensor Technologies
- Circuit Analysis

IGI Global is currently accepting manuscripts for publication within this series. To submit a proposal for a volume in this series, please contact our Acquisition Editors at Acquisitions@igi-global.com or visit: http://www.igi-global.com/publish/.

Titles in this Series

For a list of additional titles in this series, please visit:
https://www.igi-global.com/book-series/advances-computer-electrical-engineering/73675

Multi-Objective Stochastic Programming in Fuzzy Enironments
Animesh Biswas (University of Kalyani, India) and Arnab Kumar De (Government College of Engineering and Textile Technology Serampore, India)
Engineering Science Reference • ©2019 • 420pp • H/C (ISBN: 9781522583011) • US $215.00

Renewable Energy and Power Supply Challenges for Rural Regions
Valeriy Kharchenko (Federal Scientific Agroengineering Center VIM, Russia) and Pandian Vasant (Universiti Teknologi PETRONAS, Malaysia)
Engineering Science Reference • ©2019 • 432pp • H/C (ISBN: 9781522591795) • US $205.00

Handbook of Research on Smart Power System Operation and Control
Hassan Haes Alhelou (Tishreen University, Syria) and Ghassan Hayek (Tishreen University, Syria)
Engineering Science Reference • ©2019 • 489pp • H/C (ISBN: 9781522580300) • US $265.00

Cases on Modern Computer Systems in Aviation
Tetiana Shmelova (National Aviation University, Ukraine) Yuliya Sikirda (National Aviation University, Ukraine) Nina Rizun (Gdansk University of Technology, Poland) and Dmytro Kucherov (National Aviation University, Ukraine)
Engineering Science Reference • ©2019 • 488pp • H/C (ISBN: 9781522575887) • US $225.00

Harnessing the Internet of Everything (IoE) for Accelerated Innovation Opportunities
Pedro J.S. Cardoso (University of Algarve, Portugal) Jânio Monteiro (University of Algarve, Portugal) Jorge Semião (University of Algarve, Portugal) and João M.F. Rodrigues (University of Algarve, Portugal)
Engineering Science Reference • ©2019 • 383pp • H/C (ISBN: 9781522573326) • US $225.00

For an entire list of titles in this series, please visit:
https://www.igi-global.com/book-series/advances-computer-electrical-engineering/73675

701 East Chocolate Avenue, Hershey, PA 17033, USA
Tel: 717-533-8845 x100 • Fax: 717-533-8661
E-Mail: cust@igi-global.com • www.igi-global.com

Table of Contents

Detailed Table of Contents

Chapter 1

 Vijey Thayananthan, King Abdulaziz University, Saudi Arabia
 Javad Yazdani, University of Central Lancashire, UK

The main aim of this strategic research proposal is to develop a model of secure transportation system using efficient CPS which not only reduce the unnecessary accident rates but also increase safety system that enhances the livability of smart cities and Industry 4.0. Although the main focus is efficient security solutions, dynamic and intelligent approaches of the future security solutions will be able to detect the evolving threats and cyberattacks during the data or signal transmission between the users and service providers.

Chapter 2

 Mamata Rath, Birla Global University, India

Smart traffic administration has been a challenge for the engineers of smart applications in smart urban areas. Regardless of many smart applications such difficulties have not yet been completely comprehended because of an assortment of unpredicted traffic situations in various areas and also the necessity of progressive choice taking parameters. Because of street clog in traffic focuses, individuals in significant metropolitan urban communities are confronting issues amid heading out from one place to another. It makes unforeseen postponements amid travel, increased odds of mishaps, pointless fuel utilization, and unhygienic conditions because of contamination additionally corrupts the wellbeing of general individuals in a typical city situation. This chapter focuses on modern traffic systems and frameworks for smart cities and explains how various challenges are met with an improved mechanism.

Ubiquitous use of wireless technology and ad-hoc networks have paved the way for intelligent transportation systems also known as vehicular ad-hoc networks (VANETs). Several trust-based frameworks have been proposed to counter the challenges posed by such fast mobile networks. However, the dynamic nature of VANETs make it difficult to maintain security and reliability solely based on trust within peers. Decision-making upon collaborative communications is critical to functioning of VANETs in safe, secured, and reliable manner. Decision taken over malicious or wrong information could lead to serious consequences. Hence, risk management within paradigm of trust becomes an important factor to be considered. In this chapter, a survey of the existing works having incorporated risk factor in their trust models has been explored to give an overview of approaches utilized. The parameters chosen in these models are analyzed and categorized based on the approaches modeled. Finally, future research directions will be presented.

Chapter 4

Interfacing the smart cities with cyber-physical systems (CPSs) improves cyber infrastructures while introducing security vulnerabilities that may lead to severe problems such as system failure, privacy violation, and/or issues related to data integrity if security and privacy are not addressed properly. In order for the CPSs of smart cities to be designed with proactive intelligence against such vulnerabilities, anomaly detection approaches need to be employed. This chapter will provide a brief overview of the security vulnerabilities in CPSs of smart cities. Following a thorough discussion on the applicability of conventional anomaly detection schemes in CPSs of smart cities, possible adoption of distributed anomaly detection systems by CPSs of smart cities will be discussed along with a comprehensive survey of the state of the art. The chapter will discuss challenges in tailoring appropriate anomaly detection schemes for CPSs of smart cities and provide insights into future directions for the researchers working in this field.

Chapter 5

The internet of things (IoT) is expected to influence both architecture and
infrastructure of current and future smart cities vision. Thus, the requirement and
effectiveness of making cities smarter demands suitable provision of secure and
efficient communication networks between IoT networking devices. Trust-based
routing protocols play an important role in IoT for secure information exchange
and communications between IoT networking elements. Thus, this chapter presents
the foundation of trust-based protocols from social science to IoT for secure smart
city environments. The chapter outlines and discusses the key ideas, notions, and
theories that may help the reader to understand the current status and the possible
future trends of trust-based protocols in IoT networks for smart cities. The chapter
also discusses the implications, requirements, and future research challenges of
trust-based protocols in IoT for smart cities.

Chapter 6

Smart cities are established on some smart components such as smart governances,
smart economy, science and technology, smart politics, smart transportation, and
smart life. Each and every smart object is interconnected through the internet,
challenging the security and privacy of citizen's sensitive information. A secure
framework for smart cities is the only solution for better and smart living. This can
be achieved through IoT infrastructure and cloud computing. The combination of
IoT and Cloud also increases the storage capacity and computational power and
make services pervasive, cost-effective, and accessed from anywhere and any device.
This chapter will discuss security issues and challenges of smart city along with
cyber security framework and architecture of smart cities for smart infrastructures
and smart applications. It also presents a general study about security mechanism
for smart city applications and security protection methodology using IOT service
to stand against cyber-attacks.

Chapter 7

Smart grids are conceived to ensure smarter generation, transmission, distribution, and consumption of electricity. It integrates the traditional electricity grid with information and communication technology. This enables a two-way communication among the smart grid entities, which translates to exchange of information about fine-grained user energy consumption between the smart grid entities. However, the flow of user energy consumption data may lead to the violation of user privacy. Inference on such data can expose the daily habits and types of appliances of users. Thus, several privacy preservation schemes have been proposed in the literature to ensure the privacy and security requirements of smart grid users. This chapter provides a review of some privacy preservation schemes. The schemes operational procedure, strengths, and weaknesses are discussed. A taxonomy, comparison table, and comparative analysis are also presented. The comparative analysis gives an insight on open research issues in privacy preservation schemes.

Research questions remain to be answered in terms of discovering how security could be provided for different resources, such as data, devices, and networks. Most organizations compromise their security measures due to high budgets despite its primary importance in today's highly dependent cyber world and as such there are always some loopholes in security systems, which cybercriminals take advantage of. In this chapter, the authors have completed an analysis of data obtained from 31 peer-reviewed scientific research studies (2009-2017) describing cybersecurity issues and solutions. The results demonstrated that the majority of applications in this area are from the government and the public sector (17%) whereas transportation and other areas have a minor percentage (6%). This study determined that the government sector is the main application area in cybersecurity and is more susceptible to cyber-attacks whereas the wireless sensor network and healthcare areas are less exposed to attack.

Chapter 9
Jayapandian N., Christ University, India

The main objective of this chapter is to discuss various security and privacy issues in smart cities. The development of smart cities involves both the private and public sectors. The theoretical background is also discussed in future growth of smart city devices. Thus, the literature survey part discusses different smart devices and their working principle is elaborated. Cyber security and internet security play a major role in smart cities. The primary solution of smart city security issues is to find some encryption methods. The symmetric and asymmetric encryption algorithm is analyzed and given some comparative statement. The final section discusses some possible ways to solve smart city security issues. This chapter showcases the security issues and solutions for smart city devices.

Preface

Recently, the interest of various stakeholders (government, industry, academia, and citizens) in building smart cities has grown exponentially. Various, governments have a plan to initiate more than 1000 smart cities projects by 2025. One of the key challenges for the success of any smart city project is the assurance of smart security and privacy of the citizens. The concept of the smart city can be realized using a wide range of interconnected cyber-physical systems such as ZigBee, RFID, WiFi, Bluetooth, VANETs etc. With the help of these heterogeneous systems, data is processed and exchanged between multiple stakeholders. This phenomenon introduces large and complex attack surface, which cannot be handled with traditional security solutions.

According to Frost & Sullivan, the global smart cities market is projected to reach $1.56 trillion by 2020. Security in a smart city is one of the major challenging issue as discussed recently in various major conferences like RSA, ACM CCS, IEEE Security and Privacy etc. Currently, there are not many books available in the market which discusses the privacy and security issues of a smart city in a comprehensive manner. This book tries to fill this gap by providing recent advances in smart city security and privacy. It will help the researchers and engineers to design and implement a safe, secure and reliable smart city project.

This edited book, *Secure Cyber-Physical Systems for Smart Cities,* aims to present the discussion on recent security frameworks, solutions, and challenges of a smart city environment. For this book, 25 chapter proposals were submitted from the researchers and students. The nine chapters were finally accepted for the inclusion in this book. The acceptance rate was 36%. The contributed chapters in this book cover a broad range of security topics related to Cyber-Physical Systems, including:

- Secure Intelligent transportation frameworks
- Smart traffic management
- Intrusion Detection

- Risk-based decision systems
- Secure routing protocols for IoT
- Secure Smart Grid Environment
- Cyber Security frameworks and threats

Chapter 1 presents a theoretical security model for cyber-physical systems. This model specifically developed for the automotive system (driverless cars) used in the industry 4.0. The escalation of new digital *industrial* technology enables researchers and engineers to design and implement a successful smart city project.

Chapter 2 presents an overview of existing smart traffic management frameworks that incorporates security and safety features. The qualitative comparison of the existing smart urban traffic control systems and frameworks is presented. Also, research and implementation challenges in building urban traffic control systems are also discussed in this chapter.

Chapter 3 presents an overview of existing risk-based decision methods of vehicular ad ho networks (VANETs). In VANETs, vehicles collaborate with each other for the purpose of achieving better traffic efficiency, and safety. However, malicious vehicles may forward bogus or incorrect information, which may lead to some serious problems like road accidents, traffic congestions etc. Therefore, it is important to incorporate risk in the decision-making process. This chapter presents a taxonomy and qualitative comparison of the various risk management schemes of vehicular networks.

Chapter 4 discusses the vulnerabilities of various cyber-physical systems (CPS) used in a smart city environment. Since CPS needs to be protected against cyber-attacks, therefore, robust algorithms are needed for the efficient intrusion detection system (IDS). This chapter presents the classification of IDSs according to the source of audit data and detection methodologies. With respect to these classification methods, overview and problems of existing schemes are presented.

Chapter 5 presents an overview of the existing trust-based routing schemes used in IoT-enabled smart environments. Incorporating trust management concept in routing protocols gives assurance that the data will reach the destination by passing through reliable, non-faulty, and non-malicious intermediate nodes. This chapter highlights the requirements, challenges, and implication of trust-based routing protocols.

Chapter 6 presents the cybersecurity frameworks of smart cities. It discusses the various technologies and techniques (e.g. blockchain, cryptography, biometrics etc.) that are used in building cybersecurity solutions. Various domains of secure smart city frameworks are also highlighted in this chapter.

Chapter 7 presents the privacy-preserving schemes used in the smart grid. The need for privacy preservation in the smart grid environment is essential as knowledge of such information exchanged between the smart grand entities can cause unforeseen damage. This chapter presents a brief background of the smart grid, a review of some proposed privacy preservation schemes along with a hierarchical taxonomy and comparison table of the schemes. Finally, a comparative analysis which offers open research challenges is presented.

Chapter 8 discusses the cybersecurity and privacy issues, threats and challenges of various technologies used in cyber-physical systems. It also identifies the application sectors which are more vulnerable to security threats and needs more protection. This chapter also presents solutions that can be used to increase the privacy and security of cyber-physical systems.

Chapter 9 focuses on the security issues related to smart devices. Smart devices like mobile phones, watches, GPS devices, smart TVs, and other IoT-connected gadgets are an essential part of our daily life. Unfortunately, most of the smart devices are insecure due to lack of clear standards. Therefore, the privacy and security of these devices is critical. This chapter discusses the various methodologies that can be used to secure smart devices.

The book addresses researchers, academicians, and industry professionals who are working in the domain of Secure Cyber-Physical Systems for Smart Cities. It provides insight into state-of-the-art research to various graduate-level academic departments e.g. electrical and computer engineering, computer science etc.

Riaz Ahmed Shaikh
King Abdulaziz University, Saudi Arabia
30 January 2019

Acknowledgment

I would like to convey my gratitude to all contributors including the chapters' authors, and reviewers. Special thanks to the following people who participated in the reviewing process: Al-Sakib Khan Pathan, M. Mustafa Monowar, Vijey Thayananthan, Syed Raheel Hassan, Muhammad Aminu Lawal, Mamata Rath, Ismail Butun, Jayapandian N., Thenmozhi S., and Aminu Usman.

My special thanks to IGI development team for their assistance in formatting, designing, and marketing of the book. Special thanks to Amanda Fanton, Josephine Dadeboe, and Maria Rohde, the ever-patient Assistant Development Editors.

Chapter 1
Secure Cyber–Physical Systems for Improving Transportation Facilities in Smart Cities and Industry 4.0

Vijey Thayananthan
King Abdulaziz University, Saudi Arabia

Javad Yazdani
University of Central Lancashire, UK

ABSTRACT

The main aim of this strategic research proposal is to develop a model of secure transportation system using efficient CPS which not only reduce the unnecessary accident rates but also increase safety system that enhances the livability of smart cities and Industry 4.0. Although the main focus is efficient security solutions, dynamic and intelligent approaches of the future security solutions will be able to detect the evolving threats and cyberattacks during the data or signal transmission between the users and service providers.

DOI: 10.4018/978-1-5225-7189-6.ch001

INTRODUCTION

Minimization of traffic problems, optimization of accuracy measurements (reception of vehicular communication) and automation facilities of the future driverless vehicles can be anticipated through real-time communications between advanced cyber-physical system (CPS) sensors and devices. Although an efficient transportation system and vehicular communication technologies are essential to improve the smart facilities such as minimum energy consumption with maximum security, smart transportation system depends on the CPS which is the revolution of industry 4.0. However, cyber-physical infrastructure is the major driving force of the future smart transportation system like a traffic monitoring system with all necessary security facilities. Transportation industries make sure that security solutions designed, developed and implemented for future evolving threats should protect the complete transportation system. Although 4G and basic 5G dominate the vehicular communication and main infrastructure of the legacy transportation system, 5G introduces many novel approaches and features to improve the traffic monitoring facilities through the intelligent transport systems.

The main aim of this strategic research proposal is to develop a model of secure transportation system using efficient CPS which not only reduce the unnecessary accident rates but also increase safety system that enhances the livability of smart cities and industry 4.0. Although the main focus is efficient security solutions, dynamic and intelligent approaches of the future security solutions will be able to detect the evolving threats and cyber attacks during the data or signal transmission between the users and service providers.

Based on the literature review and existing data related to the accidental rate, locations, etc., efficient CPS can be modified to combat the potential attacks. Employing CPS will not only allow us to improve the future transportation system but also will help us to develop an appropriate theoretical model with minimum energy consumption and maximum security. Further, CPS will allow us to enhance the security features according to the evolving technologies. Here, the appropriate technology of industry 4.0 revolutions and necessary security policies based on blockchain (BC) will be employed to improve the transportation services. As a specific methodology, a theoretical model designed and developed for securing automotive service allowed me to analyze the various security issues focused on the transportation system. Despite many unsecured traffic services between the users and service providers, the intelligent system considered in the theoretical model detects the malicious messages and information from the transportation services quickly and efficiently.

According to the research idea of securing future transportation system, I hope that the proposed model will provide the new security solutions with minimum energy consumption and maximum security solutions. Although selected topology of block structure based on the size of the specific road is considered with BC, the study of early results confirm that BC enhances the security solutions in the future transportation system used in the industry 4.0.

This CPS based research will be leading us to deliver a practical security solution based on industry 4.0 for the future transport systems in smart cities. Here, evolving threats created from the dynamic environments must be identified, detected and resolved with minimum cost and maximum energy efficiency.

Improving secure automotive environments in favorite cities is one of the 2030 visions. In the current industrial revolution, governments of the popular countries plan to execute the driverless vehicles soon. In this environments, automotive services will be improved with the efficient use of security technologies based on industry 4.0. Despite many solutions, efficient security solutions will reduce accidental rate. Regarding the driverless vehicles, the accidental rate may be slightly lower than existing transport systems, but cybersecurity will create the different types of accidents without any warning or pre-knowledge. Here, passengers (users) should be able to solve the cybersecurity problems as quick as possible. In this situation, CPS will help us to provide the necessary solutions and immediate actions. Instead of improving technological capabilities, security issues should be applied to protect the current automotive system as well as the future technologies used within the industry 4.0. The accident may happen in many different ways such as drivers' attitude, conditions of the roads, etc. The automotive system can monitor the drivers' attitudes, but it has to monitor the passengers' attitude and behavior when passengers are using driverless vehicles. Infotainments created through the vehicular communication are for useful purposes such as shortest path during the traffic situation. However, the wrong information sent as infotainment creates many unnecessary accidents and problems. Environmental conditions sent through the sensors provide the security solutions which enhance the monitoring facilities in CPS.

Purpose of the chapter: The main purpose of this chapter is to develop a theoretical model of secure transportation system using efficient CPS which not only reduce the unnecessary accident rates but also increase safety system that enhances the livability of smart cities and industry 4.0. Although the main focus is efficient security solutions, this chapter proposes the dynamic and intelligent approaches of the future security solutions to detect the evolving threats and cyber attacks during the data or signal transmission between the users and transportation service providers.

Future wireless systems such as the 5th, generation (5G) of wireless and mobile networks support to design the automotive services. Here, the infrastructure of 5G is being developed to reduce the cyber threats which create not only the unnecessary accident but damage the transport network. Despite the basic level of the infrastructure in the automotive systems, CPS in an automotive system enhances monitoring facilities. Also, 5G introduces many novel approaches to improve the traffic monitoring facilities through the intelligent transport systems Menouar et al. (2017), Camacho et al. (2018). Vehicular Communication Functions (VCF) in public transport system needs to be handled securely through the appropriate security solutions. Here, a combination of IoT based 5G and blockchain (BC) scheme provides us to design the secure transport.

Problem statement: Minimization of traffic problems depend on many factors including security issues. To improve the transportation and traffic problems, all services used within the transport industry should be monitored securely. Despite many monitoring facilities and security issues, optimization of accurate measurements and automation facilities of the future driverless vehicles can be considered with real-time communications between the advanced cyber-physical system (CPS) and sensors or devices used in the vehicles, industry 4.0, smart cities, etc. Here, passengers (users) face the cybersecurity problems created by inefficient monitoring and security solutions. In this situation, CPS will help us to provide the necessary solutions and immediate actions. Further, the wrong information creates many unnecessary accidents and problems when infotainment considered as an example of the automotive services shows the unsecure messages. Real-time and reliable integration of driverless vehicles with the smart grid could solve security problems related to demand response, cost and time of charging. The security through the layers, sensors network and SDN within the vehicular communication is one of the key problems in modern infrastructure. For instance, lower and upper layer attacks usually affect many communication services between the components used in this modern infrastructure.

Although future mobile communication influenced with automotive systems, security challenges, Radio Access Network (RAN), IoT, etc. can be considered to design the secure infrastructure Sfar et al. (2018), Samaaila et al. (2017) for securing industry 4.0. Further, technological advances such as machine to machine communication, non-orthogonal modulation, ultra-dense cells, SDN, etc. are also applicable to control the traffic services. In this research, the following strategic approaches are important and motivating to design the efficient security solution. Security issues of the CPS are the focal concern of automobile enterprises which are evolving within the smart industry 4.0. Although the application layer is responsible for setting many different automotive applications, it provides the security management through the existing security protocols. CPS helps service providers to manage the access control. Therefore, we secure access control which protects the application layer. Despite many security algorithms, security issues of access control should be considered to improve the automotive management services. Despite these facts, threats and motivations are increasing within the automotive services because most of the vehicles are wirelessly linked. This means that cyber attacks such as Internet-borne malware can create the greatest accident risks and financial loss within the automotive organizations.

The rest of the paper is organized as follows. Section 2 presents the quick state of the art. Section 3 focuses on the literature review, and related work includes existing security issues of CPS, SDN security solutions and IoT issues based on 5G networks. In section 4, this paper provides security issues of CPS in an automotive system and the proposed theoretical model of the secure automotive system. Section 5 explains the details of securing automotive services with results and analysis. In Section 6; overall conclusions are written based on the theoretical analysis and results.

STATE OF ART

Regarding the security issues of CPS, medical communication of ambulance, smart traffic signal systems, advanced automotive system, etc. are the examples of the applications in different sectors used in the industry 4.0. Here, many sectors are fighting to solve the potential cybersecurity problems Wang et al. (2015), O'Donovan et al. (2018). More specifically, the cybersecurity of CPS measures for remotely controlling applications, medical devices, machines used in the transportation system, etc. and the information sent to them is limited.

The idea behind a CPS is to connect the virtual and physical industry 4.0 environments where sensors used in the automotive system are able to gather all transportation data and provide useful feedback to a driverless vehicle or all physical entities within the automotive system. The sensor data gathered from the physical machine is transmitted through IoT and allows a remote operator to monitor and control the equipment based on the data received. Therefore, the security of controlling who can access the remotely controlled machines directly affects the security of the CPS. A CPS allows the communication between machines and humans. These systems can gather and process data and use interfaces to communicate with humans.

Figure 1. The automotive system in industry 4.0

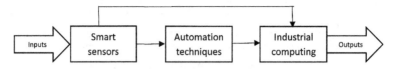

As shown in Figure 1, secure communication channels, smart sensors, and industrial PC maintain the current automotive system which is one of the examples in industry 4.0 and its smart warehouse Liu et al. (2018). However, insecure automotive services and systems affect the economy of the industry 4.0 and damage the transportation infrastructures. The automation system in industry 4.0 needs smart sensors to enhance the smart facilities such as energy saving when the specific application is involved in the automation processing. Here, automation system depends on the standard components; they are such as industrial computing, automation techniques, smart sensors, etc. During the automation processing, these components should be able to provide the smart facilities which improve the industry 4.0. Although many challenges and possible implementations are urgent to consider in our contributions, we have focused on security issues of CPS in an automotive system and few points related to securing the current automotive systems through the following contributions.

- Studying and investigating the existing security issues of CPS for protecting the automotive system of the IoT based 5G infrastructure in industry 4.0.
- Based on the study and investigation of current CPS, necessary details related to SDN and BC have been considered for enhancing cybersecurity solutions.
- Designing possible security protocols using SDN and BC.
- Developing the theoretical model based on the above three contributions and evaluation mechanism

As shown in Figure 2, industry 4.0 influences with CPS which connects many services deployed within the smart automotive system. Although smart lights on the smart streets provide the public safety, some lights do not get any maintenance which is one of the security issues.

State of the art in industry 4.0 based automotive system is the revolution of the existing automobile system. Despite many energy-efficient technologies, the automotive system needs maximum security with a minimum cost which depends on the energy management. The future automotive system expects to have dynamic power while driverless vehicles are moving. In this state of the art, CPS plays an important role in connecting physical nodes known as driverless vehicles and energy

Figure 2. The smart automotive system in industry 4.0

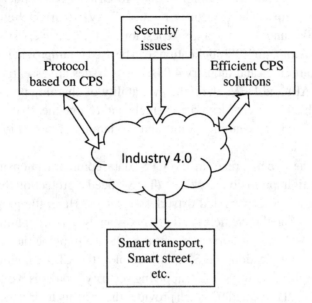

source which may be in remote locations. Here, wireless charging units considered as virtual energy sources may be employed.

Cyber and Physical Threats in CPS

Attacks and threats are increasing with cyber-physical interactions which are physical to cyber and cyber to physical. In physical to the cyber case, wireless sensor data attacks, threats influenced with the memory, etc. are still happening while monitoring the security in CPS where physical objects are dealing with cyber objects. In cyber to the physical case, vehicle charger is wirelessly attacked. Also, other attacks such as device coordination are involved between the devices when cyber objects control the security of physical objects. Recent articles and news show that the cyber and physical attacks are increasing the number of automation functions involved in the modern infrastructure of the VANET.

According to Al-Anbagi and Mouftah (2016), driverless electric vehicles (DEVs) is better for the green environment because it does not emit the carbon which is also another physical threat. One of the main limitations of DEVs is their safety regulations between the resources which enable to establish the secure communication. Real-time and reliable integration of DEVs with the smart grid could solve problems related to demand response, cost and time of charging. In CPS, Wireless Access in

Vehicular Environments (WAVE) for Vehicle-to-Grid (V2G) applications provide necessary power with available security. Keeping WAVE with which charging allows through the online and dynamic approaches, for the future driverless will be one of the challenges. Although V2G applications can be used for improving transportation facilities in smart cities and industry 4.0, security issues are essential between the Access Point (AP) and DEV. Here, the availability of charging points and virtual locations which allow users (driverless vehicles) to charge their battery wirelessly. In this case, maintaining charging points and availability of virtual locations is one of the security issues.

Regarding the cyber threats in the CPS, data communication influenced by traffic information provision services (TIPS) needs protection for delivering proper information to drivers and driverless vehicles. Here, the progress of data communication related to vehicles and drivers is analyzed through the data center and 3rd parties with the authorization of the TIPS. To improve the security issues in vehicular communication, TIPS should employ the CPS which enhances the security in the following ways. First, applying security protocols based on CPS for protecting the spiral information which provides the security to drivers and vehicles. Secondly, intelligent information processing which helps us to analyze the security issues in the data center and 3rd parties. Finally, information should be extracted through the secure interface which may be either human to machine or machine to machine Nawa et al. (2014).

Useful SDN Security Challenges and Solutions

There are many SDN-enabled security challenges and solutions implemented in wireless network applications. Following issues illustrate the security improvements in future wireless infrastructures employed in VANET.

To address the issue of internal and higher layer security which includes network security, Software-Defined Network (SDN) technology has been introduced to vehicular communication Ma et al. (2016). The SDN not only improves the VANET performance but also deals with the attacks on the wireless sensors network involved between the driverless vehicles. To prevent man-in-the-middle and denial of service attacks, SDN technology is proposed and studied with a Bayes-based algorithm.

According to Wu et al. (2016), in smart cities, driverless vehicles will be one of the major challenges which need many factors including security. In the modern infrastructure of the driverless vehicles, wireless sensor networks (WSNs) play an important role to collect all types of data including attacks. The most of the nodes used in WSN allow us to implement secure channels which provide secure transportation communication services through the SDN. The security through the WSN and SDN within the vehicular communication is one of the key issues in modern infrastructure. In most of the resources used in WSNs, lower and upper layer attacks usually affect many communication services between the components used in this modern infrastructure. Therefore, to resolve this problem, authors have proposed a hierarchical framework based on chance discovery and usage control (UCON) technologies to improve the security of VANETs. The features of the continuous decision and dynamic attributes in UCON can address ongoing attacks using advanced persistent threat detection. In addition, they have used a dynamic adaptive chance discovery mechanism to detect unknown attacks. To design and implement a system using the mechanism described above, a unified framework is proposed in which low-level attack detection with simple rules is performed in sensors, and high-level attack detection with complex rules is performed in sinks and at the base station. Moreover, SDN and network function virtualization (NFV) technologies are used to perform attack mitigation when either low level or high-level attacks are detected. This approach may be useful in all layers of the vehicular communication between the driverless vehicles.

The following examples of SDN security challenges may be added to improve the industry 4.0. They are automation processing when the system handles the confidential big data, secure communication of driverless vehicles, decision making when securing medical information, etc.

Blockchain

There are many applications of the blockchain, which improve the security issues in the network could be deployed in the industry 4.0. As shown in Figure 3, IoT forwarding devices from the transportation environment send the necessary information to distributed BC network through the controllers.

Figure 3. Example of BC for IoT devices

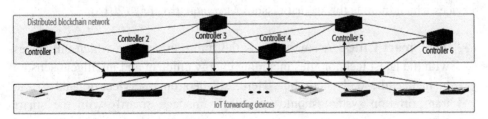

Although BC has 2 types of structures (public and permission), employing the permission BC technology allows the researchers to handle the information securely and efficiently. Some security policies based on BC and permission BC framework are considered to maintain the transportation services involved with the vehicle-related data. In the BC, the blocks are linked to each other, managed efficiently with security policies and secured cryptographically with the efficient security algorithms, policies, and protocols. The public BC allows all users to participate either as client or server which provides validating security policies. The permission BC allows the granted users to participate with the strict membership policies and to handle the transactions securely.

In the BC network, the static policy can be defined that any policy influenced by the transportation services waits for the limited confirmations from the blocks considered as security domains. Dynamic policy can be defined that any policy depends on both time and space. So, confirmations may be varied with the transportations services which influence the internal and external conditions.

Although the blockchain approach is new to industry 4.0, few examples allow us to enhance the security facilities which provide the cost efficiency in the future industries. Here examples are key management via BC, energy management which depends on the way of using secure smart devices, handing multiple decisions making simultaneously when multiple sensors employed in the smart cities, etc.

LITERATURE REVIEW AND RELATED WORK

According to Chatterjee et al. (2017), industry 4.0 in India needs secure network grids which allow the business communities to manage their business securely and peacefully. Despite the secure management and overall performance of Modern Network Grids, the security issues of CPS in industry 4.0 based automotive system needs dynamic security solutions. The IoT emerges with many devices connected to the Internet which provides the quick access to maintain the transport services. The everyday automotive activities are to encompass everything from IoT based services which include the security issues Mozzaquatro et al. (2015).

- **The Smart Cities:** A smart city is the future urban area which influences with the latest technologies including energy efficient and intelligent sensors to handle the daily activities. Here, all electronic services related to the transportation system should be able to manage smartly with the smart processing such as autonomous, intelligent, quick decision making, etc.

- **Industry 4.0:** The initial focus of Industry 4.0 revolutions started in the manufacturing industries to reduce the working loads, simplify the industrial tasks, increase the efficiency, etc. For instance, supply chain management, manufacturing, and production can be used as applications of industry 4.0. With Industry 4.0, smart factories of transportation industries can become smart in most of the tasks without human supervision and reduce the overall cost.
- **Transportation Facilities and Their Relationships:** Here, all services related to improving the transport users are the facilities they are navigation, intelligent traffic management, etc. Transportation system depends on smart management such as intelligent transport system with the efficient security solutions. Transportation facilities and their relationships can be improved when we employ the industry 4.0 approaches which depend on the CPS functions and smart technologies.

Classification of the SDN is shown in Figure 4. In this paper, all SDN security issues provide security solutions for CPS and allow researchers to monitor and control the automotive services dynamically and securely.

Figure 4. Classification of the SDNs security solutions
Akhunzada et al. (2016)

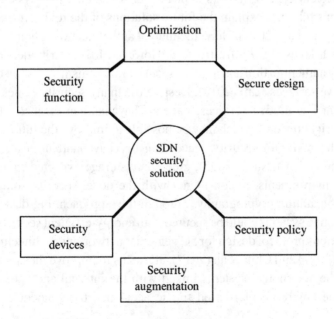

Following terms used in Figure 4 are explained to improve the transportation facilities we developers use the SDN security solutions.

- **Security Functions:** Implementation of security functions based on SDN allow us to enhance the security issues.
- **Optimization:** Employing SDN optimize not only the security management issues but also fault tolerance.
- **Secure Designs:** All designs need secure protocols and algorithms which depend on the secure infrastructure of SDNs in CPS
- **Security Devices:** Transportation can be enhanced through the efficient way of using sensors and devices which allow us to improve the monitoring facilities
- **Security Augmentation:** When organizations use the interactive tools based on SDN security solutions, transportation system can augment its content management systems more securely.
- **Security Policy:** All systems depend on the efficient management regulated through the updated policies used for improving future VANET.

In Tewari and Gupta (2017), ultra-lightweight mutual authentication approach has been considered. Here, two bitwise operations adopted for authentication purpose which ensures low computational and storage cost. This approach resists against a different type of attacks appeared in the automotive system. Different authentications schemes explain the future solutions of the reply attacks in the IoT environment Jan et al. (2017). Here, the mutual authentication scheme may help us to secure the IoT based 5G infrastructure. However, IoT security depends on end-point identity authentication and another security mechanism such as access control for automotive services Kim (2017). Despite the many security issues considered in the current automotive services, papers Thayananthan et al. (2015) provide relevant security concepts which may be possible to improve the future automotive services with maximum security. According to Thayananthan et al. (2017) and Thayananthan and Albeshri, (2015), green data storage expected to use in future automotive environments, is considered with the novel security solutions based on Li-Fi and quantum cryptography. Through this approach, big data used in the smart automotive systems could be secured with energy-efficient security protocols. Communications protocol design for 5G vehicular networking architecture Katsaros and Dianati (2017) and Chiti et al. (2017), allows us to improve the communication services of the automotive systems linked with the internal and external signals. Regarding the environmental-based traffic accidents, trust model of improving secure communication between the vehicles offers many facilities in the automotive services. These specific improvements depend on the vehicle contexts, drivers' attitude, etc. reduce the unnecessary accidents Thayananthan and Shaikh (2016) and Hasrouny et al. (2017).

According to Jimenez et al. (2015), SDN has become a new way to make a secure access mechanism and dynamic topologies in VANET. They have great potential in both the creation of network protocol which deals with security and resolution of vehicular communication.

According to Namal et al. (2016), the vehicular communication network is still adding new resources which are the extension of legacy. In VANET, legacy system will have to be modernized for driverless vehicles with flexible network management which includes the strong security in all conditions. Software Defined Radio (SDR) systems dominate the vehicular communication and VANET in many ways such as programmable communication services of the transportation system. Further, SDR and its new extended versions of cognitive and smart radio allow transports service providers to improve the security services. Managing network complexity, characterizing the communication channel and management inefficiencies in VANET need secure technologies, infrastructure, and management. Open Flow is the first standard interface that enables SDN for enhancing security issues in vehicular communication. It can be used for all devices employed in VANET and for improving automation and key management.

Regarding the potential security threats, the SDN has some functionalities which collect the security levels of the traditional wired and wireless network status between the driverless vehicles. In this case, SDN plays an important role to detect some attacks such as DDoS which depends on the traffic pattern and provides a programmable approach to control the traffic flow.

It happens between the physical systems where channels connect the transmitter and receiver of the vehicles. According to Kumar and Dutta (2016), intrusion attacks are defined as an attempt to compromise the confidentiality, integrity, and availability of the vehicular communication system, or bypass the security mechanism. The IDS (intrusion detection scheme) can be used to prevent the intrusion attacks appeared in the physical layer of the vehicular communication network where various levels of attacks are expected under different channel conditions. Further, an overview of the IDS taxonomy and another attempt are considered to compare different intrusion detection techniques which are useful for different channel conditions, their operational strengths, and limitations. The IDS can provide a partial security solution to most of the intrusion attacks appeared in the CPS of modern infrastructure implemented for driverless vehicles. In driverless vehicles, controller area network (CAN) Song et al. (2016) needs to be secured using efficient IDS.

According to Eze et al. (2016), wireless access technologies support the security of the driverless vehicles because vehicle-to-vehicle (V2V) and vehicle-to-infrastructure (V2I) are the main communication systems in the vehicular network. Improving these security issues within these communication systems protect not only the vehicular network but also enhance the road safety and traffic efficiency which are part of the development programs in the intelligent transport system (ITS). In VANET, many communication applications need better protection because malicious traffic information causes unnecessary vehicle collision Raut and Malik (2014).

Case study explained in Dorri et al. (2017) encouraged us to develop the theoretical model and security solutions to improve the transportation services in smart cities. Recent applications based on the case studies create more powerful solutions with the latest technologies.

- **Recent Applications:** Securing driverless vehicles in the smart cities, Wireless charging while vehicles are moving, the Dynamic approach of energy saving when the system employs the CPS with SDN and BC technologies, etc.
- **Case Studies:** Concepts of SDN based security solutions and BC policies are examples of the recent case studies used to improve the transportation facilities.

According to the Dorri et al. (2017) and Cebe et al. (2018), traffic flow monitored during the transportation services with BC, and existing security solutions provides following observations. They are energy consumptions are increasing when we use the BC-based security solutions because BC enhances the capacity of the security solutions. Although energy consumption and security are a trade-off, IoT based 5G design will allow researchers to find the compromised solutions. These 2 articles provide the necessary latest and relevant information for improving the transportation system.

Articles published on 2018 provide the necessary details and possible latest theoretical information which are not only relevant to this topic used in this chapter but also explain the appropriate solution using the latest technology for industry 4.0

SECURITY ISSUES OF CPS IN AN AUTOMOTIVE SYSTEM

The IoT based 5G schemes are growing among the services used within the automotive and industry 4.0. This scheme needs secure infrastructure which allows transport service provider to monitor and solve the automotive facilities quickly and dynamically. The CPS technology innovation Centre within the industry 4.0 will offer each program the possibility to develop applications related to smart systems Centea et al. (2018), as follows:

- **Automation:** IoT, smart automobile factory, establishing automation, electronics for the automotive system
- **Automotive:** Vehicle to vehicle technology (V2V), a vehicle for everything technology (V2X), electric vehicles, autonomous and connected vehicles
- **Manufacturing:** Additive manufacturing for smart industry 4.0, smart automobile factory
- **Infrastructure:** Smart transportation, smart traffic light system, and smart communication structures
- **Software:** Vehicular networking and infrastructure, Big Data of automotive schemes, Data Analytics, security
- **Energy:** Alternative energy, smart grid, charging facilities wirelessly and dynamically

The secure automotive system considered has been employed to future IoT based 5G infrastructure which allows improving the vehicular communications.

Theoretical Model of Automotive System

As shown in Figure 5, this paper proposes the theoretical model which represents the simple illustration of the secure automotive system. In this model, each layer accelerates the security solutions which secure the facilities and features used in the automotive system. Layers of the smart automotive and monitoring devices increase the security levels through secure IoT based 5G infrastructure.

The automotive services and IoT application support layer provides secure links between the application layer and the 5G network layer. Secure links support the information exchange between multiple subnetworks within the IoT based 5G infrastructure. In order to enhance the security issues of CPS, IoT needs to be secured from possible threats; they are eavesdropping, impersonation, relay, replay attacks, etc. In this theoretical model, Integration of SDN with Internet and IoT will help us to create navel security protocol which provides better solutions for cybersecurity issues of access control mechanisms. Future design of SDN is an intelligent networking paradigm which improves the security of the legacy system employed in V2V (vehicular-to-vehicular) network and V2I (vehicular-to-infrastructure).

Figure 5. A theoretical model for securing automotive service

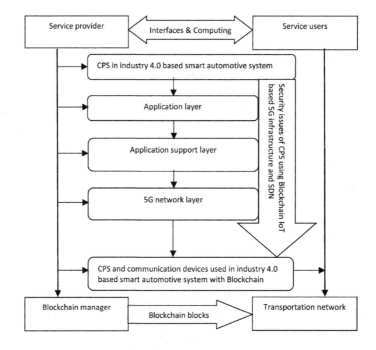

- **Service Providers:** Transportation handler who provides the necessary services to transport system within the 5G environment, improves the transportation facilities in the smart cities
- **Service Users:** People and all communication devices within the transport system are the regular users expected to have good services.
- **Blockchain Manager:** Here, the service provider expects to deliver correct security solutions depended on the BC policies and protocols from the BC manager.
- **Transportation Network:** Although it depends on the appropriate security solutions, managing service provider's priorities and users' instance protections should be able to handle dynamically and intelligently through the BC IoT, SDN, etc.
- **Layers:** Here, IoT based industry 4.0 applications can be considered through the application layer. All intelligent and dynamic actions need extra features such as autonomous services to support the security enhancements through the application support layer. The 5G network layer enhances the speed of the automotive system and 5G features used in the transportation services.

Infrastructure and Future Automotive System

Automotive systems in industry 4.0 should have efficient V2V and V2I. Despite many Internet connections within the automotive systems, efficient design of IoT based 5G infrastructure will connect and handle millions of users through the 5G network. Here, communications between V2V and V2I face many challenges; they are secure traffic communication, safety regulations, handling emergency situations and finding secure parking locations. Thus, infotainment is considered as an example of the automotive services, which is one of the problems in V2V and V2I. Further, system details can allow us to implement the IoT based 5G infrastructure for future automotive.

The future automotive system in industry 4.0 depends on the cost management which directly influences with maximum security and minimum energy consumptions. When driverless vehicles use more features, users have to pay according to the facilities of automotive services they use during the journey. For instance, the performance of driverless vehicles is one of the measurements. Although BC IoT based 5G services provide excellent security facilities, vehicles should be able to handle all features without any warnings. Especially electric vehicles should have alternative or temporary electrical source to detect the basic security issues.

Security policies can be defined from which evolving transportations services involved with the users' preferences and technologies. For instance, the gap between the 2 driverless vehicles should be within the limits which depend on the policies. Thus, we can derive many different policies influenced by the lengths of time T which depends on the 1second gap or 2 seconds gap using internal and external parameters. Here, we focus on the internal parameters which depend on the infrastructures involved with the CPS, 5G, IoT, etc. Although external parameters such as vehicle context are involved for securing transportation services, policies involved with the internal parameters such as technologies control not only the accidents but also improve the reliabilities of the electronic equipment used in the transportation services. Based on the topology of security domains Lei et al. (2017), T and cost values depend on the processing time of handover considered within either same or different security domains known as blocks of the topology. When time is elapsed, an attacker can easily follow some of the blocks (i = 1to N) used in the BC network which employs the following formula based on the Poisson distribution to calculate the probability (P) that the attacker still intends to damage the exact blocks.

$$P_i = \sum_{i=1}^{N} C_i + N\frac{v^i}{e^v i!}$$

where $v = \alpha \lambda T$ and λ represents the rate at which block of BC network are used when transportation service incorporate with the dynamic policies and efficient protocols. Here, C is the probability that the attacker is expecting to damage the blocks. In the transportation services, the one attacker with α fraction of the total blocks used in the BC network. In this security issues, the attacker can also generate the blocks at a rate of $\alpha\lambda$ and confuse the BC network with the additional blocks. Using above formula, our theoretical model can be integrated with the mathematical concepts which allow future researchers to improve the multiple security problems within the transportation services.

According to the Lei et al. (2017), as an assumed topology, BC network takes 9 blocks for mapping of 4 security domains which influence the transportation services. Different topologies may have more blocks with which many services could be between the mobile users and transportation system expected to monitor the security of selected domains. Dynamic policies can be used to secure the blocks which influence the processing time depending on the design parameters such as the size of blocks.

ANALYSIS OF PROPOSED SOLUTIONS FOR SECURING AUTOMOTIVE SERVICES

Although many security challenges considered in the current automotive services, few of the security solutions are successfully implemented within the automotive services. Security Solutions in Automotive Services. Despite many solutions implemented in the automotive services, employing secure IoT and 5G based infrastructure used within the automotive systems will be an efficient method to improve the current and future security problems. Evolving threats and cyber attacks are the potential security issues which need dynamic solutions during the manuring. However, security solution should increase the reliability and robustness of the future automotive system. The proposed theoretical model defined in section 4 provides the proposed solutions to improve the services used in the transportation systems which include the automation services considered to facilitate the smart cities within the industry 4.0 environments.

Operational Scenarios and Indicative Results

This section describes the showcases of estimating vehicle traffic, followed by indicative results for managing the traffic during the different conditions such as weather. For instance, In order to determine the security issues of CPS in an automotive system, the temperature policies of the road should be fixed to room temperature.

Scenario 1: Hot (Above 30 degree Celsius) In order to determine the security issues of CPS during the hot temperature of the road, physical protections are the most important approach based on cybersecurity solution.

Scenario 2: Icy (Below 0 degree Celsius) It depends on the decreasing temperatures and the thickness level of the snow. Again cybersecurity solution can provide the necessary physical protections through the remote monitoring.

Challenges of Possible Solutions via IoT Based 5G Services

Due to the insecure automotive services and complexity of V2V communication networks, the accident rate was increasing. Further, traditional security mechanisms such as authentication may not be possible in the future IoT based 5G services. However, authentication used in the theoretical model allows us to improve the security issues in V2Is influenced with IoT. Table I shows the security challenges of the IoT based 5G infrastructure for different approaches to automotive.

Regarding the minimum energy consumption, BC's autonomy in industry 4.0 is emerging with IoT based 5G, machine-to-machine (M2M), etc. and minimizing the complexity of the security designs and architectures. When we deploy such an optimized design within the 5G environment, we can easily reduce the energy consumption. In the future solution, transportation systems will use BC to improve autonomous systems which enhance the secure automotive services. Before we finalize features of the model, it is expected that advances in artificial intelligence (AI) will allow our possible solutions based on IoT to minimize the complex situation of the services. All cases, updated dynamic policies of BC enhance the security solutions with the reduced waiting time depending on the properties and parameters of the BC network.

Table 1. Security challenges and solutions for improving smart automotive system

Security Challenges of CPS	Possible Solutions to Policies and Protocols	Expected Improvements
Challenge 1	Cybersecurity of all transportation services with BC – IoT security mechanism	More than 10%
Challenge 2	Cybersecurity of all transportation services with BC – IoT based 5G security mechanism	More than 15%
Challenge 3	Cybersecurity of all transportation services with BC – IoT based 5G- SDN security mechanism	More than 20%

Challenges mention in Table 1 show the following relevant analysis depended on the technologies and policies employed. For instance, assume that cybersecurity depends on the reduced waiting time.

Challenge 1: When we use only BC-IoT infrastructure, waiting time is reduced to 10%.

Challenge 2: When we add the 5G infrastructure with challenge 1, waiting time is reduced more (almost 15%).

Challenge 3: When SDN was integrated with the challenge 2, waiting time reached even better than above challenges.

Energy Analysis With Different Structures

Based on the article Lei et al. (2017), three different schemes are compared for analyzing the time and cost when two different security domains are considered. Generally speaking, energy is directly proportional to time, which means that the BC approach is better than other approaches. Through the proposed theoretical model, we can increase the processing time to improve security solutions which allow users to reduce the overall cost as well as energy. As shown in Figure 6, we can compare the key transmission time with our proposed model where new technology and policies are introduced.

Figure 6. Comparison between structures and schemes (a) Within the same security (b) Different security

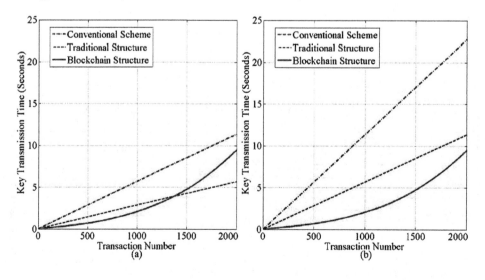

Although we have not analyzed the results through the separate experiments, we can deduce to achieve the goal of this research through their mathematical steps and approaches.

To analyze the security issues, the key transmission time is measured with a total transaction number recorded in all 3 approaches. Comparing the key transmission time of 3 different approaches not only allow us to improve the transportation facilities through low energy cost and time but also enhance the secure transportation services. According to the experiment Lei et al. (2017), the total transaction number is recorded using 3 different approaches in 2 following different situations.

1.   Key transmission time is measured between two security domains which mean different locations use same security approaches when appropriate security solution is deployed in the transportation service (Figure 6a).
2.   Key transmission time is measured between two security domains which mean different locations use different security approaches when separate security solution is deployed in the transportation service (Figure 6b).

Here, the transaction number can be assumed as transportation services. To analyze the energy through the 3 different security approaches, we need maximum secure services within the minimum key transmission time.

From the table 2, improvements can be achieved through the BC which is dominating in both situations to analyze the transportation services. Hence, efficient transportation and vehicular communication services are essential to improve the smart facilities such as minimum energy consumption with maximum security. In this analysis, the minimum key transmission time improves the smart transportation services depended on the CPS which is one of the key technologies in the revolution of industry 4.0.

Table 2. Security and energy analysis for transportation services

Transportation Services	Time in Second (s) for Conventional	Time (s) for Traditional	Time (s) for BC	Security Solutions With Minimum Energy in 2 Situations (i) and (ii)
500 (i)	3.1	2.1	1	BC can be used for both situations when services are less than 1500
500 (ii)	5.8	2.9	1	
1000 (i)	6	3	2	When services are between 500 and 1500, BC is much better
1000 (ii)	11	6	2	
1500 (i)	8.5	4.5	4.9	Here, traditional is better than BC
1500 (ii)	17	8.5	4.9	Again, BC provides improvement

CONCLUSION AND FUTURE WORK

The IoT based 5G concept in the automotive system will enhance the security solutions of the industry 4.0 when CPS influences are in the future driverless vehicles. This paper introduces the theoretical model that addresses the security issues of CPS using BC IoT based on 5G architecture. This model will not only the increases in the safety of the driverless vehicles but also it provides the systematic behaviors and overall secure environments in industry 4.0. In this research, we have studied the current and future security problems of CPS within the automotive system. Further, the theoretical model employs the BC and SDN to improve the security issues with minimum energy consumptions. Hence, this research will be leading us to implement an effective security solution for the smart automotive system used in the industry 4.0. The dynamic policies which we can develop for our theoretical model reduce the expected waiting time in the selected topology of the BC network. In future works, we can simulate the proposed theoretical model and analyze the results with different topologies, scenarios and environmental conditions of the automotive systems.

REFERENCES

Akhunzada, A., Gani, A., Anuar, N. B., Abdelaziz, A., Khan, M. K., Hayat, A., & Khan, S. U. (2016). Secure and dependable software defined networks. *Journal of Network and Computer Applications, 61*, 199–221. doi:10.1016/j.jnca.2015.11.012

Al-Anbagi, I., & Mouftah, H. T. (2016). WAVE 4 V2G: Wireless access in vehicular environments for vehicle-to-grid applications. *Vehicular Communications, 3*, 31–42. doi:10.1016/j.vehcom.2015.12.002

Camacho, F., Cárdenas, C., & Muñoz, D. (2018). Emerging technologies and research challenges for intelligent transportation systems: 5G, HetNets, and SDN. *International Journal on Interactive Design and Manufacturing, 12*(1), 327–335. doi:10.100712008-017-0391-2

Cebe, M., Erdin, E., Akkaya, K., Aksu, H., & Uluagac, S. (2018). *Block4Forensic: An Integrated Lightweight Blockchain Framework for Forensics Applications of Connected Vehicles.* arXiv preprint arXiv:1802.00561

Centea, D., Singh, I., & Elbestawi, M. (2018). Framework for the Development of a Cyber-Physical Systems Learning Centre. In *Online Engineering & Internet of Things* (pp. 919–930). Cham: Springer. doi:10.1007/978-3-319-64352-6_86

Chatterjee, S., Kar, A. K., & Gupta, M. P. (2017). Critical success factors to establish 5G network in smart cities: Inputs for security and privacy. *Journal of Global Information Management, 25*(2), 15–37. doi:10.4018/JGIM.2017040102

Chiti, F., Fantacci, R., Giuli, D., Paganelli, F., & Rigazzi, G. (2017). *Communications protocol design for 5G vehicular networks. In 5G Mobile Communications* (pp. 625–649). Cham: Springer.

Dorri, A., Kanhere, S. S., Jurdak, R., & Gauravaram, P. (2017, March). Blockchain for IoT security and privacy: The case study of a smart home. In *Pervasive Computing and Communications Workshops (PerCom Workshops), 2017 IEEE International Conference on* (pp. 618-623). IEEE.

Eze, E. C., Zhang, S. J., Liu, E. J., & Eze, J. C. (2016). Advances in vehicular ad-hoc networks (VANETs): Challenges and road-map for future development. *International Journal of Automation and Computing, 13*(1), 1–18. doi:10.100711633-015-0913-y

Hasrouny, H., Bassil, C., Samhat, A. E., & Laouiti, A. (2017). Security risk analysis of a trust model for secure group leader-based communication in VANET. In *Vehicular Ad-Hoc Networks for Smart Cities* (pp. 71–83). Singapore: Springer. doi:10.1007/978-981-10-3503-6_6

Jan, M. A., Khan, F., Alam, M., & Usman, M. (2017). A payload-based mutual authentication scheme for Internet of Things. *Future Generation Computer Systems*.

Jimenez, J. M., Romero Martínez, J. O., Rego, A., Dilendra, A., & Lloret, J. (2015). Study of multimedia delivery over software defined networks. In Network Protocols and Algorithms (Vol. 7, No. 4, pp. 37-62). Macrothink Institute.

Katsaros, K., & Dianati, M. (2017). *A conceptual 5G vehicular networking architecture. In 5G Mobile Communications* (pp. 595–623). Cham: Springer.

Kim, J. H. (2017). A survey of IoT security: Risks, requirements, trends, and key technologies. *Journal of Industrial Integration and Management*, 2(02), 1750008. doi:10.1142/S2424862217500087

Kumar, S., & Dutta, K. (2016). Intrusion detection in mobile ad hoc networks: Techniques, systems, and future challenges. *Security and Communication Networks*, 9(14), 2484–2556. doi:10.1002ec.1484

Lei, A., Cruickshank, H., Cao, Y., Asuquo, P., Ogah, C. P. A., & Sun, Z. (2017). Blockchain-based dynamic key management for heterogeneous intelligent transportation systems. *IEEE Internet of Things Journal*, 4(6), 1832–1843. doi:10.1109/JIOT.2017.2740569

Liu, X., Cao, J., Yang, Y., & Jiang, S. (2018). CPS-Based Smart Warehouse for Industry 4.0: A Survey of the Underlying Technologies. *Computers*, 7(1), 13. doi:10.3390/computers7010013

Ma, H., Ding, H., Yang, Y., Mi, Z., Yang, J. Y., & Xiong, Z. (2016). Bayes-based ARP attack detection algorithm for cloud centers. *Tsinghua Science and Technology*, 21(1), 17–28. doi:10.1109/TST.2016.7399280

Menouar, H., Guvenc, I., Akkaya, K., Uluagac, A. S., Kadri, A., & Tuncer, A. (2017). UAV-enabled intelligent transportation systems for the smart city: Applications and challenges. *IEEE Communications Magazine*, 55(3), 22–28. doi:10.1109/MCOM.2017.1600238CM

Mozzaquatro, B. A., Jardim-Goncalves, R., & Agostinho, C. (2015, October). Towards a reference ontology for security in the internet of things. In *Measurements & Networking (M&N), 2015 IEEE International Workshop on* (pp. 1-6). IEEE. 10.1109/IWMN.2015.7322984

Namal, S., Ahmad, I., Saud, S., Jokinen, M., & Gurtov, A. (2016). Implementation of OpenFlow based cognitive radio network architecture: SDN&R. *Wireless Networks*, 22(2), 663–677. doi:10.100711276-015-0973-5

Nawa, K., Chandrasiri, N. P., Yanagihara, T., & Oguchi, K. (2014). Cyber physical system for vehicle application. *Transactions of the Institute of Measurement and Control*, 36(7), 898–905. doi:10.1177/0142331213510018

O'Donovan, P., Gallagher, C., Bruton, K., & O'Sullivan, D. T. (2018). A fog computing industrial cyber-physical system for embedded low-latency machine learning Industry 4.0 applications. *Manufacturing Letters*, 15, 139–142. doi:10.1016/j.mfglet.2018.01.005

Raut, S. B., & Malik, L. G. (2014, December). Survey on vehicle collision prediction in VANET. In *Computational Intelligence and Computing Research (ICCIC), 2014 IEEE International Conference on* (pp. 1-5). IEEE. 10.1109/ICCIC.2014.7238552

Samaila, M. G., Neto, M., Fernandes, D. A., Freire, M. M., & Inácio, P. R. (2017). Security challenges of the Internet of Things. In *Beyond the Internet of Things* (pp. 53–82). Cham: Springer. doi:10.1007/978-3-319-50758-3_3

Sfar, A. R., Natalizio, E., Challal, Y., & Chtourou, Z. (2018). A roadmap for security challenges in the Internet of Things. *Digital Communications and Networks*, 4(2), 118–137. doi:10.1016/j.dcan.2017.04.003

Song, H. M., Kim, H. R., & Kim, H. K. (2016, January). Intrusion detection system based on the analysis of time intervals of CAN messages for in-vehicle network. In *2016 international conference on information networking (ICOIN)* (pp. 63-68). IEEE.

Tewari, A., & Gupta, B. B. (2017). Cryptanalysis of a novel ultra-lightweight mutual authentication protocol for IoT devices using RFID tags. *The Journal of Supercomputing*, 73(3), 1085–1102. doi:10.100711227-016-1849-x

Thayananthan, V., Abdulkader, O., Jambi, K., & Bamahdi, A. M. (2017, June). Analysis of Cybersecurity based on Li-Fi in green data storage environments. In *Cyber Security and Cloud Computing (CSCloud), 2017 IEEE 4th International Conference on* (pp. 327-332). IEEE. 10.1109/CSCloud.2017.32

Thayananthan, V., & Albeshri, A. (2015). Big data security issues based on quantum cryptography and privacy with authentication for mobile data center. *Procedia Computer Science*, *50*, 149–156. doi:10.1016/j.procs.2015.04.077

Thayananthan, V., Alzahrani, A., & Qureshi, M. S. (2015). Efficient techniques of key management and quantum cryptography in RFID networks. *Security and Communication Networks*, *8*(4), 589–597. doi:10.1002ec.1005

Thayananthan, V., & Shaikh, R. A. (2016). Contextual Risk-based Decision Modeling for Vehicular Networks. *International Journal of Computer Network and Information Security*, *8*(9), 1–9. doi:10.5815/ijcnis.2016.09.01

Wang, L., Törngren, M., & Onori, M. (2015). Current status and advancement of cyber-physical systems in manufacturing. *Journal of Manufacturing Systems*, *37*, 517–527. doi:10.1016/j.jmsy.2015.04.008

Wu, J., Ota, K., Dong, M., & Li, C. (2016). A Hierarchical Security Framework for Defending Against Sophisticated Attacks on Wireless Sensor Networks in Smart Cities. *IEEE Access: Practical Innovations, Open Solutions*, *4*(4), 416–424. doi:10.1109/ACCESS.2016.2517321

Chapter 2
Smart Traffic Management and Secured Framework for Smart Cities

Mamata Rath
Birla Global University, India

ABSTRACT

Smart traffic administration has been a challenge for the engineers of smart applications in smart urban areas. Regardless of many smart applications such difficulties have not yet been completely comprehended because of an assortment of unpredicted traffic situations in various areas and also the necessity of progressive choice taking parameters. Because of street clog in traffic focuses, individuals in significant metropolitan urban communities are confronting issues amid heading out from one place to another. It makes unforeseen postponements amid travel, increased odds of mishaps, pointless fuel utilization, and unhygienic conditions because of contamination additionally corrupts the wellbeing of general individuals in a typical city situation. This chapter focuses on modern traffic systems and frameworks for smart cities and explains how various challenges are met with an improved mechanism.

DOI: 10.4018/978-1-5225-7189-6.ch002

INTRODUCTION

To maintain a strategic distance from such issues many smart urban communities are as of now executing traffic control frameworks that work on the standard of traffic computerization with aversion of the previously mentioned issues. The essential test lies in use of constant examination performed with on-line traffic data and effectively applying it to some the traffic stream. Presently a days, numerous managerial associations are receiving strategies to actualize the testing smart city observation in current township in social, specialized, monetary, and political areas by executing PC knowledge with huge information applications, Internet of Things and numerous other cutting edge innovations hence offering open doors for consistent refinement of the idea of smart city related applications. Smart urban communities use different advances to enhance the execution of wellbeing, transportation, vitality, training, water administrations and waste administration framework to enhance the expectations for everyday comforts of their natives. In transportation framework, the city government required better approaches to screen and oversee nearby traffic to give better transportation administrations to the general population. The computerized observing gadgets introduced in the city's traffic focuses catch pictures and video information ceaselessly. The expanding measure of traffic information currently presents challenges in the city's capacity to successfully oversee traffic. On the talked about setting, the present section displays a detail audit on various testing issues and applications amid outline and usage of smart traffic control framework and also use of huge information with uncommon spotlight on smart traffic control execution in heterogeneous network condition (Rath et. al, 2016), for example, distributed computing, (Rath et. al, 2018) remote network and smart matrix applications. Besides this part centers around an exceptionally basic issue in a smart traffic setting, for example, security (Rath et. al, 2018) and wellbeing of travelers amid traffic control with proposition of a novel traffic administration approach.

Essential attributes of planning an improved traffic control system incorporates associating traffic lights, signals and traffic control mechanism. It also focuses GIS empowered advanced guide of the smart city utilizing high computational energy of information examination as a key module. In such specific situation, the fundamental challenge lies in utilization of real time investigation on-line traffic data and accurately applying it to some essential traffic stream. In this specific situation, an enhanced traffic control and checking system has been proposed in the present article that performs insightful information investigation utilizing exceptionally adaptable versatile operator innovation . Under a VANET situation, the versatile operator executes a congestion control (Rath et. al, 2017) calculation to consistently mastermind the traffic by staying away from the congestion at the section of the smart traffic zone with the help of other interesting highlights, for example, counteractive action of mischances, wrongdoing, driver adaptability and security. Reproduction did utilizing MATLAB demonstrates empowering brings about terms of better execution to control and keep the congestion to a more noteworthy degree.

Information examination instruments takes information from the Traffic Management System and utilizing GIS mapping under real time bolster they give valuable data to the drivers in the vehicles and help diminishing the traffic congestion. Furthermore, fundamental vacationer data, for example, going by places, stopping zone and separation are additionally anticipated in real time premise on substantial advanced screens introduced at city focuses entrance focuses to control the drivers towards their goal. This spares fuel lastly to spare a considerable measure of time spent in looking different going to places. The smart living style in metro urban areas is additionally satisfied as nature progresses toward becoming contamination free and more sterile.

In the present innovation, Vehicular Ad-hoc Networks (VANETs) are ending up more well known applications in street traffic administration and control systems. The issue looked by smart urban areas as far as traffic congestion issues are better solvent by the utilization of VANETs as there is a network connectivity (Rath et. al, 2018) between the vehicle and the network framework. In this manner, unsurprising data with respect to street condition ahead and course data can be coordinated to the smart vehicles in travel and insightful choices can be taken adequate time before any issue happens. In different courses VANET in smart urban areas lessens the issues of congestion, mishaps, wrongdoing, stopping issues and populace overhead. Because of the general advancement of the remote innovation, their applications are colossal on vehicles and vehicles have been changed over to smart vehicles to be gotten to under smart traffic applications. The customary driving systems and drivers have likewise changed over to smart drivers(Rath et. al, 2014) with more specialized information of accepting smart signs from traffic controllers, understanding them and act as needs be. VANETs bolster adaptable correspondence amongst vehicles and traffic controlling systems in both framework based and in remote medium without settled foundation. The proposed traffic congestion arrangement in smart city utilizes an enhanced specialized arrangement with intense information investigation made by portable operator powerfully under VANET situation in a smart city.

In the advanced communication technology (Rath et. al,2018), Vehicular Ad-hoc Networks (VANETs) are becoming more popular applications in road traffic management and control systems. The problem faced by smart cities in terms of traffic congestion issues can be solved better by the use of VANETs as there is a network connectivity between the vehicle and the network infrastructure. Therefore, predictable information regarding road condition ahead and route information can be directed to the smart vehicles in transit and intelligent decisions can be taken sufficient time before any problem occurs. Figure1 represents a smart traffic control system under a VANET scenario showing vehicle to vehicle (V2V) communication (Rath et. al, 2018) and Vehicle to Road side unit (V2R) communication.

As the idea of smart urban areas is advancing, the requirement for computerization and successful conveyance of administrations is basic (Chahal et., al, 2017). Throughout the most recent couple of decades, requirement for safety and security of explorers out and about has been a noteworthy concern. Vehicular Ad Hoc Networks (VANETs) can assume a persuasive part in perceiving and executing such idea, by supporting safety, solace and infotainment administrations. In any case, the intricate and unbendable engineering of VANETs confronted an arrangement of difficulties, for example, high portability, irregular connectivity, heterogeneity of uses. In this specific circumstance, Software-defined networking (SDN) has risen as a programmable and adaptable network, which has as of late picked up consideration from inquire about groups, organizations, and businesses, in both wired network administration and heterogeneous remote correspondence. In an article survey has been carried out after analyzing and ordering various related SDN-construct inquire about works in light of remote networks uncommonly VANETs. Initially, a brief on the prerequisites of SDN over conventional networking is given, trailed by an elaboration on essential design and its layers. From there on, SDN applications in different remote network regions, for example, portable network and VANETs are depicted alongside an emphasis on breaking down and contrasting the ebb and flow

Figure 1. A traffic control system in VANET

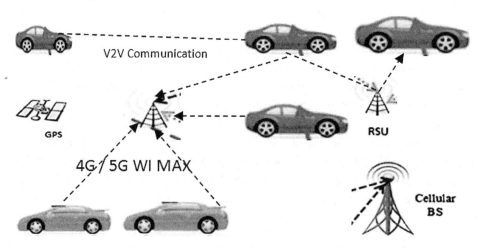

SDN-related research on various parameters. Moreover, it shows an audit of flow investigate activities to tackle difficulties of vehicular condition. The effect of SDN worldview alongside execution issues in vehicular correspondence and investigate likely utilize cases in view of SDN worldview.In other ways VANET in smart cities(Rath et.al, 2017) helps to reduce the problems of congestion, accidents, crime, parking problems and population overhead. Due to the overall development of the wireless technology, their applications are immense on vehicles and vehicles have been converted to smart vehicles to be accessed under smart traffic applications. The traditional driving systems and drivers have also converted to smart drivers with more technical knowledge of receiving smart signals from traffic controllers, understanding them and act accordingly. VANETs support flexible communication between vehicles and traffic controlling systems in both infrastructure based and in wireless medium without fixed infrastructure. The proposed traffic congestion solution in smart city uses an improved technical explanation to the problem with powerful data analytics made by mobile agent dynamically under VANET scenario in a smart city.

URBAN TRAFFIC CONTROL AND SMART GRID

Designing smart traffic control systems for smart cities is a challenging task for network developers. Use of control operators for coordinating parts in a network of collaborating dynamic frameworks is an emblematic approach. In this technique, every nearby control operator must guarantee that every single neighborhood detail are met, however in the meantime should guarantee that the distinctive segments help each other in accomplishing great worldwide conduct and great neighborhood conduct. This is shown by utilizing urban traffic control and smart electric power frameworks as illustrations. Brought together or various leveled control approaches are not powerful against disappointments in correspondence networks, and require doubtful presumptions on the information of every specialist about the general model. A totally decentralized approach, where every nearby control specialist egotistically attempts to accomplish its neighborhood particulars just, runs a high danger of worldwide associations that may destabilize the framework, making it difficult to accomplish the determinations. There are two ideal models for distributed input control (R. Boel et. al, 2013) that require next to no data trade and almost no worldwide model learning. The pioneer/supporter control worldview is delineated for urban traffic control: vigorously stacked pioneer operators send messages to their adherent neighbors asking for that these devotees give green just to companies of vehicles going towards the pioneer crossing point at those occasions when this will be ideal for the execution of the pioneer. Another coordination worldview is known as the organizing model prescient control (CMPC). A power transmission network might be viewed as that has been divided in collaborating areas, where CMPC is utilized keeping in mind the end goal to keep the spread of the unsettling influences following occurrences like line or machine disappointment. CMPC attempts to determine this by having every nearby control operator apply a model prescient controller, utilizing as on-line accessible data the nearby voltage and current estimations, as well as data on the arranged grouping of future control activities of neighboring specialists, conveyed to it now and again. This review talks about a portion of the minimum need for demonstrating, correspondence and control specialist set-up keeping in mind the end goal to vigorously accomplish determinations utilizing distributed control.

The information model (O. Boreiko et. al, 2017) which was created, is expected to portray information procedures and information streams, parameters, and properties of the control framework for traveler traffic enlistment in broad daylight transport of "smart" city that guarantees viable working of the created framework. The viable implementation of the information show includes utilizing the rundown information structures and XML.

Traffic over-burden is a squeezing problem for some urban areas these days and they are attempting to discover smart arrangements. In this manner urban communities are presenting traffic control frameworks that encourage a more unique travel through the city for travel traffic while advancing the control of individual intersections outfitted with light flagging. Assessment of traffic control parameters is normally in light of preparing of information from traffic overviews.

Urban city and congestion created with general smart society transport framework's creates disappointment, in taking care of the people's rising demand for a powerful transport methodology remain as significant reasons tempting the Sfax city [8] occupants to request for the private methods for transport instead of picking people in general transport benefit (80% versus 20%). It is really this reality which lies as the real hotspot for the astounding obstructed traffic and street clog, and additionally the constantly irritating pollution issues describing the city's condition. It is in this setting the proposed study can be set with a top to bottom research planned to examine the genuine condition of traffic on the light of hindrance focuses' assurance contemplate (red spots) as enduring on the city's street network. As a major aspect of the attempted exploratory investigation, a unique crusade has been actualized on leading body of the SORETRAS, the general population transport transporting organization's vehicles to gauge and assess the general overwhelming business speed. The measure (H. Elleuch et. al, 2017) has been directed through utilization of the ways' genuine term timing strategy, notwithstanding recognizable proof of the red spots and comparing stop span ID approach. The conceived structure includes concocting an extraordinary Smart control and urban administration by methods for a clever transportation framework. The last would serve to help in viably advancing the utilization of transport foundations through limiting street clog. It depends on the accompanying advances: obtaining the constant information of traffic status at each point as recognized from the reconnaissance cameras, gathering information, from that point, in an accumulation focus and as an extreme advance, investigating and settling on the correct choice with respect to the traffic circumstance. Such an arbitration, redirecting traffic, alongside establishment of police robots, watch dispatching and programmed direction.Notwithstanding many years of reformative enactment with respect to the security of drivers and enormous activities by developers to enhance vehicle security features, engine vehicle defect identified with speed killed around 3000 young people (matured 16-19) every year. A detailed report (M. Afify et. al, 2018) proposes an implanted smart control framework to constrain the vehicle speed to the most extreme admissible street speed. This can be accomplished by setting up a live correspondence between the voyage control framework and the locally available GPS module. Utilizing the Engine Control Unit (ECU) to control the electronic throttle and air valve, the most extreme rotational velocity of the motor can be restricted with the goal that the vehicle's speed does not outperform a street speed confine. By executing this installed framework to control the speed of the engine vehicle, this framework can possibly diminish the quantity of speeding-related mischances. The investigation assesses the practicability of the proposed inserted framework and talks about option controlling techniques. For this reason, a graphical UI (GUI) module is likewise created to make the framework more easy to use. The information used to test the framework was gathered through a GPS benefit from the boulevards around the grounds of the University of Central Oklahoma. Table 1 shows summary of Smart Traffic control systems with implementation details

Correspondence based prepare control (CBTC) frameworks utilize remote neighborhood for information transmission amongst trains and wayside gear. Since unavoidable bundle deferral and drop are presented in prepare wayside interchanges, information vulnerabilities in trains' states will prompt impromptu footing/braking requests, and also squander in electrical vitality (Sun et. al, 2017) . Also, with the presentation of regenerative braking innovation, control networks in CBTC frameworks are developing to smart lattices, and cost-mindful power administration ought to be utilized to decrease the aggregate monetary cost of devoured electrical vitality. In this paper, an intellectual control technique for CBTC frameworks with smart networks is displayed to upgrade both prepare activity execution and cost effectiveness. A psychological control framework show has been detailed for

Table 1. Summary of smart traffic control systems with implementation details

Sl No	Literature	Year	Application Type	Mechanism Used
1	L. A. Tawalbeh, *et al.*	2016	Improving Healthcare application using mobile cloud computing and big data analytics.	A cloudlet-based mobile cloud-computing infrastructure has been developed.
2	K. Siddique, *et al.*	2016	a benchmark evaluation of Hama's graph package and Apache Giraph using PageRank algorithm	comparative studies and empirical evaluations performed in big data applications.
3	Y. Wang, *et al.*	2016	a novel approach for clustering of electricity consumption behavior dynamics.	Markov model is applied to model the dynamic of electricity consumption and a clustering technique by fast search and find of density peaks (CFSFDP) is primarily carried out to obtain the typical dynamics of consumption behavior.
4	B. Lin, *et al.*	2016	a scheduling algorithm, which is called multiclouds partial critical paths with pretreatment (MCPCPP), for big data workflows in multiclouds is presented	This algorithm incorporates the concept of partial critical paths, and aims to minimize the execution cost of workflow while satisfying the defined deadline constraint
5	B. Tang, *et al.*	2017	a hierarchical distributed Fog Computing architecture to support the integration of massive number of infrastructure components and services in future smart cities	a smart pipeline monitoring system based on fiber optic sensors and sequential learning algorithms to detect events threatening pipeline safety
6	M. M. Islam, *et al.*	2017	an Ant Colony Optimization (ACO) based joint VM migration model for a heterogeneous, MCC based Smart Healthcare system in Smart City environment.	the user's mobility and provisioned VM resources in the cloud address the VM migration problem.
7	S. Shukla, *et al.*	2016	how big data analytics can be used to build a smart transportation system	Smartphones should be linked to smart traffic signals to achieve the objective of smart transportation system
8	P. Rizwan, *et al.*	2016	to provide better service by deploying traffic indicators to update the traffic details instantly	A mobile application is developed as user interface to explore the density of traffic at various places and provides an alternative way for managing the traffic.
9	D. Singh, *et al.*	2016	visual big data analytics which involves processing and analyzing large scale visual data such as images or videos to find semantic patterns that are useful for interpretation.	a framework for visual big data analytics for automatic detection of bike-riders without helmet in city traffic has been proposed.
10	R. A. Alshawish, *et al.*	2016	Reviews on the potentials where Big Data technology can drive a city to be smart.	visualizing the data in useful shape in order to improve any city's system application.
11	N. Ianuale, *et al.*	2016	The integration of lot of generated data from the heterogeneity and diversity of smart city data sources.	To describe smart city contexts through the various interlinked big data and networks.
12	Y. Sun, *et al.*	2016	Building Smart and connected communities SCC for a community to live in the present, plan for the future, and remember the past.	mobile crowdsensing and cyber-physical cloud computing as two most important IoT technologies in promoting smart and connected communities SCC .
13	W. Yuan *et al.*	2015	To develop a smart work performance measurement system.	New solutions and algorithms for indoor and outdoor location, GPS deviation improvement, and work performance measurement been used.
14	Z. Lv *et al.*	2016	A 3-D global browser is employed to load multiple types of demand data from the city, such as 3-D building model data, residents' information, and real-time and historical traffic data.	A 3-D analysis and visualization of the city's information are conducted on a platform. The GIS-based navigational scheme is used to access different available data sources.
15	M. V. Moreno *et al.*	2017	to provide profitable services of smart cities, such as the management of the energy consumption and comfort in smart buildings, and the detection of travel profiles in smart transport	Internet of Things based architecture is proposed to be applied to different smart cities applications.
16	J. M. Schleicher *et al.*	2016	To consider common problems of smart cities and how to address them.	Three representative types of smart city applications are outlined here, identifying key requirements and architectural guidelines for implementation.

CBTC frameworks (Sun et. al, 2017) . The information hole in subjective control is computed to break down how the prepare wayside interchanges influence the activity of trains. The Q-learning calculation is utilized in the proposed intellectual control strategy, and a joint target work made out of the information hole and the aggregate money related cost is a.pplied to create ideal approach. The medium-get to control layer retry-restrain adaption and footing procedure choice are embraced as psychological activities. Broad re-enactment results (Sun et. al, 2017) demonstrate that the cost productivity and prepare activity execution of CBTC frameworks are significantly enhanced utilizing our proposed subjective control strategy. Table 2 depicts Mechanism used in Urban Traffic Control System and Smart Grid.

Because of its extraordinary potential to enhance the general execution of information transmission with its dynamic and versatile range portion capacity in examination with numerous other networking innovations, intellectual radio (CR) networking innovation has been progressively utilized in networking and correspondence frameworks for smart matrices. Be that as it may, an optional client (SU) of a CR network must be pressed out from a channel when an essential client recovers the channel, which may happen in a randomized manner. The irregular interference of SU traffic may cause parcel misfortunes and postponements for SU information, and it will thus influence the strength of the checking and control of smart frameworks. The problem has been tended to (Liu et. al, 2015) the displaying and dependability examination of the programmed age control (AGC) of a smart lattice has been explored for which CR networks are utilized as the framework for the conglomeration and correspondence of both framework wide information and nearby estimation information. For this reason, a haphazardly exchanged power framework display is proposed for the AGC of the smart lattice. By demonstrating the CR network as an On-Off switch with visit times, the security of the AGC of the

Table 2. Mechanism used in urban traffic control system and smart grid

Literature	Smart Mechanism Used
R. Boel et. al,2013	Urban Traffic control and Smart Grid
O. Boreiko et. al, 2017	Information model of the control system for public transport
H.Elleuch et. al, 2017	Determination of obstruction points of road traffic in urban management
M.Afify et. al, 2018	Smart engine speed control system
W. Sun et . al, 2017	Cognitive control method for cost effective CBTC system
S. Liu et.al, 2015	Automatic generation control over CRN in smart grid

smart matrix is broke down. Specifically, the smart matrix has been broke down with two primary sorts of CR networks: 1) the waiting times are discretionary however limited and 2) the visit times take after a free and indistinguishable appropriation process. The adequate conditions are gotten for the solidness of the AGC of the smart lattice with these two CR networks, separately. Re-enactment results demonstrate the impacts of the CR networks on the dynamic execution of the AGC of the smart framework and represent the convenience of the created adequate conditions in the design of CR networks keeping in mind the end goal to guarantee the steadiness of the AGC of the smart matrix.

CONGESTION CONTROL SYSTEM IN SMART CITY

As the problem of urban traffic jamming problem increases, there is an urgent requirement of modern technique to enhance the traffic control system. The present techniques utilized, for example, based on clocks or human control are turned out to be sub-par compared to ease this emergency. A framework to control the traffic by estimating the ongoing vehicle quantity utilizing watchful border identification with advanced image processing techniques is arranged. Another approach (Tahmid et. al, 2017) offers noteworthy change accordingly time, vehicle administration, computerization, unwavering quality and generally speaking effectiveness over the current frameworks. Other than that, the total procedure from picture procurement to edge recognition lastly green flag allocation utilizing four example pictures of various traffic conditions is outlined with appropriate schematics and the last outcomes are checked by equipment execution.

The research results are displayed in the subject of estimation and investigation of the negative traffic impacts on nature in urban communities utilizing creative sensor networks with the yield to the traffic information and control frameworks and smart urban communities frameworks (Chikhardtova et. al, 2016) .Structure of the control framework for traveler traffic enrollment openly transport of "smart" city, which depends on a secluded guideline, was fabricated. Depiction and capacity of the principle parts in the auxiliary chart of the framework are exhibited. Well ordered calculation of the control framework for traveler traffic enrollment out in the open transport of "smart" city with particular clarification of the usefulness of the considerable number of segments of the framework structure at each phase of its activity is exhibited.

The line-up arrangement control of multi-wise traffic cones (ITCs) has been displayed (Lee et. al, 2017). ITCs are designed to be a portable robot that can be controlled to move to the coveted area by an administrator. The administrator summons ITCs to move to arrange through an application programming that demonstrates a guide and GUI on an individual smart telephone. The smart telephone is associated with a server through WiFi correspondence so the administrator sign in the server to control the developments of ITCs. The development of ITCs is controlled and shown on the guide since ITCs have GPS sensors so continuous application is plausible.

In current road scenario, congestion in traffic is a major problem in quicker way of life Kumar et. al, 2017) . One of the fundamental explanations behind the clog is extensive deferral or the time settled for the red light in the signal. The turnover time of relating light is as of now settled in the traffic framework and it did not depend on the quantity of vehicle on the specific bearing. The upgrade of traffic framework controller in a street intersection has been arranged utilizing microcontroller. This thought attempts to decrease the event of clog caused by traffic lights, to a degree The proposed thought depends on raspberry-pi. The proposed display contains IR transmitter and IR collector are settled at the conceivable bearing on the traffic signal streets. In light of the quantity of vehicles tally, the raspberry chooses and controls the traffic signal time span thus The vehicle tally delivered from raspberry information will be recorded. For revise classification, the record points of interest can be put away to the controller by illuminating raspberry-pi to the PC framework then it will send adjust deferral of signal into the LED lights. In future this model can be accustomed to offering information to voyagers about various zones and the traffic condition for the same.

TRAFFIC SURVEY AND IMPLEMENTATION OF NEW APPROACHES IN SMART CITY

Design, implementation and check of elective techniques (Ruzicka et. al, 2018) for traffic studies for the examination of traffic control frameworks for a situation investigation of Uherské Hradiste. In this city, a broad pilot venture occurred in summer and harvest time 2017 to test the new calculation for versatile traffic control at chosen light-controlled intersections of the city. The point of the designed traffic studies was to assess the parameters of the recently introduced traffic control frameworks and to contrast them and the control parameters of the first framework.

Lighting, both indoor and open air, expends a generous measure of vitality, making enhanced effectiveness a huge test. A promising methodology (Shahzad et. al, 2016) to address open air lighting is the smart control of open lighting. Smart lighting utilizing electronically controlled light-transmitting diode (LED) lights for versatile brightening and checking is being utilized to accomplish a vitality proficient framework. Be that as it may, the traffic designing coordinated with smart control for vitality streamlining has not been broadly utilized. In this paper, a novel idea of traffic-stream based smart (LED) road lighting for vitality improvement is proposed. The created smart lattice engineering based framework utilizes low power ZigBee work network to give most extreme vitality proficiency because of versatile traffic out and about. Besides, the adaptable remote network of smart LED lights offers enhanced dependability, lessened cost, and more client fulfillment. Keeping in mind the end goal to approve the execution, the proposed framework was actualized and tried in a genuine situation inside a college grounds. Trial results demonstrate that in correlation with the supplanted ordinary metal halide lighting, our framework is equipped for 68%-82% vitality funds relying upon the varieties in sunlight hours amongst summer and winter. A noteworthy decrease in ozone harming substances, enhanced in general framework unwavering quality, and diminished upkeep because of smart control recommends promising outcomes for future wide-territory arrangement.

Moving towards the advancement of smart city, different smart applications like smart home, smart human services, smart water system, Smart road lighting, smart stopping framework, Smart waste administration framework and so on are a piece of it (Hainaliker at. al, 2017). Out of these applications smart stopping framework is an imperative piece of purported smart city. Smart stopping framework permits holding the parking space ahead of time, which helps in decrease of time in looking through the parking space, decrease in traffic blockage, decrease in pollution, decrease in dissatisfaction of drivers and so on. A smart stopping framework (Hainalkar et. al, 2017) in view of web of things which not just enables the drivers to book a specific parking space yet in addition helps in programmed cashless charging, Hacking implication, post trip booking is proposed. The proposed framework likewise gives the updates of each stopping zones to traffic police which deals with the urban traffic problems. It framework gives every one of the highlights fundamental to overhauling the personal satisfaction of a person with respect to smart city.

Development in uncertainty, complex frameworks, and the knowledge sciences, especially smart city advances, have indicated incredible potential in helping to ease traffic blockage (Zhu et. al, 2016) . The general approach and the fundamental thoughts in building smart transportation for smart urban areas, especially ACP (counterfeit framework, computational investigation, and parallel execution)- based parallel transportation administration and control frameworks (PTMS), are exhibited. PTMS can be extended to the new age of insightful transportation frameworks. The fundamental parts of the proposed engineering in incorporate social flag and social traffic, ITS mists and administrations, agent-based traffic control, and transportation information mechanization. Some specialized points of interest of these segments are talked about. At long last, one contextual analysis is presented, and the viability is broke down.

With the fast advancement of Chinese economy and car industry, urban traffic clog has turned out to be progressively genuine. Along these lines, how to adequately mitigate the traffic blockage and enhance the effectiveness of vehicles has turned into the fundamental concern. Traffic signal control is one of the successful approaches to unravel urban traffic blockage. A traffic signal control strategy in light of Action-Dependent Heuristic Dynamic Programming (ADHDP) is researched (Cao et. al, 2016) . The control calculation is reenacted on two crossing points, both of which have two stages with four passageway approaches. The PC reproduction results demonstrate that the control technique has the better capacity of on-line learning contrasted and customary Fix-Time Control, and can successfully enhance the normal speed of vehicles, and diminish travel time and mitigate the traffic weight.

As a contextual analysis, a blockage control framework has been considered (Son et. al, 2017) . As urbanization builds, traffic clog and street security problems end up genuine social problems. Korea has 1.6 times higher urbanization rate than the world normal because of quick industrialization. Consequently, it is basic to enhance the nature of traffic to finish Smart City with traffic blockage cost of 2.62% of GDP, national coordinations cost of 12.5% of GDP, and traffic mishap cost of 1.53% of GDP in 2008. In this paper, we propose a protected LTE - based specialized strategy for smart - convergence development for open and ideal task of traffic signal information. Throughout the previous 25 years, Korea has been overseeing signal working conditions and blame conditions with a low-speed wired correspondence foundation between traffic lights and traffic control focuses. So as to share signal information between traffic light and traffic control place for remote signal task (Awadella et. al, 2017) and ideal signal activity, we are making an achievement in rapid remote correspondence framework. Traffic signaling framework design is required by security. The real field test results by building up the innovation to settle the security vulnerabilities that are concerned when utilizing existing LTE as correspondence foundation between neighborhood controllers and the remote controller. It is normal that the outcomes are critical and can altogether lessen shipping costs in future urban clog circumstances.

INTELLIGENT TRAFFIC CONTROL METHODS IN SMART CITY

A traffic control framework which makes utilization of video observation is depicted (Goudar et. al, 2017) whose application will investigate the information of picture/ video film which can find people who are disturbing the movement rules. Since, customary PC vision methodology can't investigate gigantic measure of information created constantly, there is a necessity for enormous information representation like pictures or recordings and so forth. A structure has been anticipated called shrewd traffic administration framework for the route preoccupation in either path when there is a traffic clog, and to locate the all inclusive community who are harming the traffic action rules like riding the bicycle without cap, and who are crossing the traffic signals and so forth, these can be identified consequently with the assistance of cameras and sensors since it will assist us with identifying the individual effectively despite the fact that when there is a nonappearance of traffic police in the city, and the points of interest ought to be extricated from the database of the RTO which will make the traffic framework as smarter and will influence the city as smarter city and the smooth stream of traffic to will be normal.

Figure 2. An intelligent traffic management system

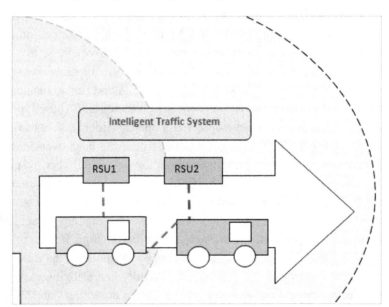

As shown in Figure 2, in an in intelligent traffic management system, there are RSU (Road side units) deployed to convey wireless messages from the intelligent vehicles / smart vehicles. They have V2V (vehicle to vehicle) communication model between them as per smart traffic system protocol. Design in (Zhijie et al, 2016) depends on single chip microcomputer keen traffic control framework, that capacities totally on two primary viewpoints, typical traffic stream, individually control the north-south and east-west bearing of red, green and yellow lights out, When the north-south crossing point traffic is huge, can expand the north-south convergence of the green light time, when things convergence traffic is huge, can build the convergence of the green light time, back to ordinary after the end. The design is utilized to finish the implementation of canny traffic control framework, meet the smart framework prerequisite, unwavering quality and continuous needs.

Over the previous decades, expanding traffic volume presents numerous difficulties, dynamic traffic light administration is one of them. Not at all like outside nations, street traffic states of India are heterogeneous because of which identification, including and characterization of vehicles ongoing has progressed toward becoming difficult (Dubey et. al, 2017). Because of settled and predefined nature of traffic light exchanging traffic blockage is frequently. In nutshell, an enhanced and improved traffic control framework is required. Presenting Beagle Bone Black/ Rasp Pi to the traffic light framework gives various customizations to transform a conventional traffic light into a smart one. To control traffic at street crossing points, framework containing microcontroller is set up on traffic light. Picture handling calculations, for example, Haar Cascade and Background Subtraction are utilized to control clock. The traffic control framework display has been produced for the fundamental crossing points of street. The continuous traffic picture, traffic thickness and different insights will be sent to server. The information can be communicated from server whenever on request through advanced arrangements. The gadget gives an anticipated course to client by reaching distinctive traffic lights in transit of client. The adage of this research is to deal with shifted traffic circumstances proficiently and to spare time and cash of client.

Versatile Traffic Monitoring and Controlling System (ATMC) is a shabby, programmed and down to earth traffic observing and controlling framework which requires no human supervision at all (Mishra et. al, 2009). Driving permit would be supplanted by the smart card, which would function as e-scratch required for beginning the vehicle. On the off chance that a vehicle breaks the traffic flag, this framework gets the information about the vehicle utilizing RF correspondence and detaches the beginning circuit of vehicle from the battery. This is a versatile framework which detects the traffic and afterward modifies the traffic light spans as needs be.

For traffic Supervisory Control and Data Acquisition System, present day detecting innovation has been received into the dynamic hypothesis for the mix of human, vehicle, and roadway needs. A novel analyst gadget for an insightful traffic framework (ITS) is created (Chih-Ju et al, 2015) to productive vehicle administration. The examining ranges and misconstruing blunders comparing to the deliberate stage point from the smart detecting framework was considered. The novel gadget, comprised of one arrangement of Doppler radar modules, can supplant the push base with the sign capacity for street walkers when they are traverse the road in the wellbeing and fast way. Also, the vehicle on the principle pathway is running easily and does not stop by the traffic sign controlled by seconds because of little transportation on the branch street. It is a decent method to maintain a strategic distance from the auto collision, which is a useful development to diminish traffic stick and air pollution and additionally improving the productivity of oil because of higher passing rate of green lights (Chih-Ju et. al, 2015). From the trial results, the smart time-sharing gadget has shown that the benefit of sign control for vehicle stream on the crossing point crosswise over street and furthermore compute the ideal seconds in light of the normal measure of transportation administration.

TRAFFIC LIGHT SYSTEM WITH NOVEL APPROACHES

Road lighting framework is one of the primary divisions of vitality buyers in the territory of the electrical vitality utilization in the nation. Since reasonable lighting framework assumes a critical part regarding individual and standardized savings in the city, giving an ideal administration plan in accordance with the point of sparing utilization of electrical vitality while giving suitable lighting levels, can be an awesome help to the field of vitality utilization. Creators (Helali et. al, 2014) have endeavored to give a complete and by and large arrangement for controlling and wise dealing with the road lighting framework in view of sections traffic utilizing holonic multi agent frameworks. The approach controls the lighting framework by astute agents in a holonic association mounted on the traffic foundation. This control is done at various levels in unified and distributed shape. There are two levels in the holonic multi agent framework. In the abnormal state of the framework, there are geographical holons and their heads with vital errands. In the low level of the structure, there are learning agents with strategic errands in geographical holons. They are fundamental controllers introduced in the lanes. They frame high utilization and low utilization holons powerfully, as per their choice for controlling the road lighting. The head of geographical holon additionally administer the learning agents and arrangement of the utilization holons. We demonstrate that our approach can control the road lighting progressively and proficiently in various conditions. The consequences of executing this approach on a reenacted network of boulevards and their traffic, demonstrates that this strategy can prompt huge abatements in electrical vitality utilization.

Traffic congestion and traffic related accidents gives rise to difficult problems particularly in swarmed urban communities which are very challenging. Traffic light is essential component in control traffic move through determine pausing and going time, settled traffic light time frameworks is repulsive control path, since number of autos isn't consistency with each traffic light, along these lines prompt irregularity system (Zaid et. al, 2017). Perceptive transportation framework including smart approach to control traffic light time in view of number of autos in each traffic light,a programmed computation has been used to control traffic light time in light of counterfeit keen strategies and picture for autos on traffic lights, this estimation is approved by comparing its outcomes and manual results. Applying the proposed computation in transportation framework has directed the traffic stream and decreased congestion and holding up time wasted in streets.

Blockage in traffic is a difficult problem these days. In spite of the fact that it appears to swarm all over the place, uber urban areas are the ones most influenced by it. Furthermore, it's consistently expanding nature makes it basic to know the street traffic thickness progressively for better signal control and powerful traffic administration (Kanungo et. al, 2014) . There can be diverse reasons for blockage in traffic like inadequate capacity, over the top request, extensive Red Light deferrals and so on. While inadequate capacity and unreasonable request are some place interrelated, the deferral of particular light is hard coded and not subject to traffic. Hence the requirement for reproducing and streamlining traffic control to more readily suit this expanding request emerges. Lately, video checking and reconnaissance frameworks have been generally utilized in traffic administration for voyager's information, slope metering and refreshes progressively. The traffic thickness estimation and vehicle classification can likewise be accomplished utilizing video checking frameworks. A strategy to utilize live video feed from the cameras at traffic intersections for ongoing traffic thickness computation utilizing video and picture handling has been introduced that spotlights on the calculation for exchanging the traffic lights as indicated by vehicle thickness on street, along these lines going for decreasing the traffic clog on streets which will help bring down the quantity of mishaps. Thus it will give safe travel to individuals and decrease fuel utilization and holding up time. It will likewise give critical information which will help in future street arranging and examination. In additionally arranges different traffic lights can be synchronized with each other with a point of even less traffic blockage and free stream of traffic.

CONCLUSION

Traffic blockage is a challenging problem that causes crowdedness, time squander, and at times unsafe mishaps. Many smart traffic information framework have been proposed those screens the traffic environment and get the best goal course, the slightest swarmed course, to spare time and diminish the quantity of mischance. Most of them are image processing based to get the status of the diverse courses. Video streams are prepared, examined, and after that the decided best course are routinely put away on a server that can be remotely gotten to. These frameworks have an electronic application that offices its administrations and gives helpful information about the streets to the approved individual. Many such models are presented in the above article ad their challenges are explained in a methodical approach.

REFERENCES

Afify, M., Abuabed, A. S. A., & Alsbou, N. (2018). Smart engine speed control system with ECU system interface. *IEEE International Instrumentation and Measurement Technology Conference (I2MTC)*, 1-6. 10.1109/I2MTC.2018.8409871

Alshawish, R. A., Alfagih, S. A. M., & Musbah, M. S. (2016). Big data applications in smart cities. *International Conference on Engineering & MIS (ICEMIS)*, 1-7.

Awadalla, M. H. A. (2017). Design of a smart traffic information system. *International Conference on Intelligent Computing, Instrumentation and Control Technologies (ICICICT)*, 757-762.

Boel, R., Marinica, N., Moradzadeh, M., & Sutarto, H. (2013). Some paradigms for coordinating feedback control with applications to urban traffic control and smart grids. *2013 3rd International Conference on Instrumentation Control and Automation (ICA)*, 1-6. 10.1109/ICA.2013.6734036

Boreiko, O., & Teslyuk, V. (2017). Information model of the control system for passenger traffic registration of public transport in the "smart" city. *12th International Scientific and Technical Conference on Computer Sciences and Information Technologies (CSIT)*, 113-116. 10.1109/STC-CSIT.2017.8098749

Cao, L. (2016). Two intersections traffic signal control method based on ADHDP. *IEEE International Conference on Vehicular Electronics and Safety (ICVES)*, 1-5. 10.1109/ICVES.2016.7548166

Chahal, Harit, Mishra, Sangaiah, & Zheng. (2017). A Survey on software-defined networking in vehicular ad hoc networks: Challenges, applications and use cases. *Sustainable Cities and Society, 35*, 830-840. doi:10.1016/j.scs.2017.07.007

Chih-Ju, C., Sheng-Hao, S., Kuo-Hsiung, T., & To-Cheng, L. (2015). A novel SCADA system design and application for intelligent traffic control. *The 27th Chinese Control and Decision Conference*, 726-730. 10.1109/CCDC.2015.7162015

Cikhardtová, K., Bělinová, Z., Tichý, T., & Růžička, J. (2016). Evaluation of traffic control impact on smart cities environment. *2016 Smart Cities Symposium Prague (SCSP)*, 1-4. 10.1109/SCSP.2016.7501011

Dubey, Akshdeep, & Rane. (2017). Implementation of an intelligent traffic control system and real time traffic statistics broadcasting. *International conference of Electronics, Communication and Aerospace Technology (ICECA)*, 33-37. 10.1109/ICECA.2017.8212827

Elleuch, H., & Rouis, J. (2017). Devising a smart control and urban management network based on determination of the obstruction points of road traffic in Sfax. *International Conference on Smart, Monitored and Controlled Cities (SM2C)*, 111-116. 10.1109/SM2C.2017.8071830

Goudar, R. H., & Megha, H. N. (2017). Next generation intelligent traffic management system and analysis for smart cities. *International Conference On Smart Technologies For Smart Nation (SmartTechCon)*, 999-1003. 10.1109/SmartTechCon.2017.8358521

Hainalkar, G. N., & Vanjale, M. S. (2017). Smart parking system with pre & post reservation, billing and traffic app. *International Conference on Intelligent Computing and Control Systems*, 500-505.

Ianuale, N., Schiavon, D., & Capobianco, E. (2016). Smart Cities, Big Data, and Communities: Reasoning From the Viewpoint of Attractors. IEEE Access, 4, 41-47. doi:10.1109/ACCESS.2015.2500733

Islam, Razzaque, Hassan, Nagy, & Song. (2017). Mobile Cloud-Based Big Healthcare Data Processing in Smart Cities. *IEEE Access*.

Kanungo, A., Sharma, A., & Singla, C. (2014). *Smart traffic lights switching and traffic density calculation using video processing*. Chandigarh: Recent Advances in Engineering and Computational Sciences. doi:10.1109/RAECS.2014.6799542

Kumar, K. K., Durai, S., Vadivel, M. T., & Kumar, K. A. (2017). Smart traffic system using raspberry pi by applying dynamic color changer algorithm. *IEEE International Conference on Smart Technologies and Management for Computing, Communication, Controls, Energy and Materials (ICSTM)*, 146-150. 10.1109/ICSTM.2017.8089141

Lee, H., & Jung, S. (2017). Line-up formation control of intelligent traffic cones. *17th International Conference on Control, Automation and Systems (ICCAS)*, 616-618. 10.23919/ICCAS.2017.8204303

Lin, B., Guo, W., Xiong, N., Chen, G., Vasilakos, A. V., & Zhang, H. (2016). A Pretreatment Workflow Scheduling Approach for Big Data Applications in Multicloud Environments. *IEEE eTransactions on Network and Service Management, 13*(3), 581–594. doi:10.1109/TNSM.2016.2554143

Liu, S., Liu, P. X., & Saddik, A. E. (2015). Modeling and Stability Analysis of Automatic Generation Control Over Cognitive Radio Networks in Smart Grids. *IEEE Transactions on Systems, Man, and Cybernetics. Systems, 45*(2), 223–234. doi:10.1109/TSMC.2014.2351372

Lv, Z. (2016). Managing Big City Information Based on WebVRGIS. IEEE Access, 4, 407-415.

Mamata, R. B. P. (2018). Communication Improvement and Traffic Control Based on V2I in Smart City Framework. *International Journal of Vehicular Telematics and Infotainment Systems, 2*(1).

Mamata, R. (2018). A Methodical Analysis of Application of Emerging Ubiquitous Computing Technology With Fog Computing and IoT in Diversified Fields and Challenges of Cloud Computing. *International Journal of Information Communication Technologies and Human Development, 10*(2).

Mishra, S., & Lohani, V. (2009). Adaptive Traffic Monitoring and Controlling System (ATMC). *International Conference on Advanced Computer Control,* 74-78. 10.1109/ICACC.2009.125

Moghadam, M. H., & Mozayani, N. (2011). A street lighting control system based on holonic structures and traffic system. *3rd International Conference on Computer Research and Development,* 92-96.

Moreno, M. V., Terroso-Saenz, F., Gonzalez-Vidal, A., Valdes-Vela, M., Skarmeta, A. F., Zamora, M. A., & Chang, V. (2017). Applicability of Big Data Techniques to Smart Cities Deployments. *IEEE Transactions on Industrial Informatics, 13*(2), 800–809. doi:10.1109/TII.2016.2605581

Rath & Oreku. (2018). Security Issues in Mobile Devices and Mobile Adhoc Networks. In Mobile Technologies and Socio-Economic Development in Emerging Nations. IGI Global. doi:10.4018/978-1-5225-4029-8.ch009

Rath. (2018). Effective Routing in Mobile Ad-hoc Networks With Power and End-to-End Delay Optimization: Well Matched With Modern Digital IoT Technology Attacks and Control in MANET. In *Advances in Data Communications and Networking for Digital Business Transformation*. IGI Global. Doi:10.4018/978-1-5225-5323-6.ch007

Rath, M. (2017). Resource provision and QoS support with added security for client side applications in cloud computing. *International Journal of Information Technology*, *9*(3), 1–8.

Rath, M. (2018). An Exhaustive Study and Analysis of Assorted Application and Challenges in Fog Computing and Emerging Ubiquitous Computing Technology. *International Journal of Applied Evolutionary Computation*, *9*(2), 17-32. Retrieved from www.igi-global.com/ijaec

Rath, M., & Panda, M. R. (2017). MAQ system development in mobile ad-hoc networks using mobile agents. *IEEE 2nd International Conference on Contemporary Computing and Informatics (IC3I)*, 794-798.

Rath, M., & Pati, B. (2017). *Load balanced routing scheme for MANETs with power and delay optimization. International Journal of Communication Network and Distributed Systems,* 19.

Rath, M., Pati, B., & Pattanayak, B. K. (2016). Inter-Layer Communication Based QoS Platform for Real Time Multimedia Applications in MANET. Wireless Communications, Signal Processing and Networking (IEEE WiSPNET), 613-617. doi:10.1109/WiSPNET.2016.7566203

Rath, M., Pati, B., & Pattanayak, B. K. (2017). Cross layer based QoS platform for multimedia transmission in MANET. *11th International Conference on Intelligent Systems and Control (ISCO)*, 402-407. 10.1109/ISCO.2017.7856026

Rath, M., & Pattanayak, B. (2017). MAQ: A Mobile Agent Based QoS Platform for MANETs. *International Journal of Business Data Communications and Networking, IGI Global*, *13*(1), 1–8. doi:10.4018/IJBDCN.2017010101

Rath, M., & Pattanayak, B. K. (2014). A methodical survey on real time applications in MANETS: Focussing On Key Issues. *International Conference on, High Performance Computing and Applications (IEEE ICHPCA)*, 1-5. 10.1109/ICHPCA.2014.7045301

Rath, M., & Pattanayak, B. K. (2018). Monitoring of QoS in MANET Based Real Time Applications. In Information and Communication Technology for Intelligent Systems (vol. 2). Springer. doi:10.1007/978-3-319-63645-0_64

Rath, M., & Pattanayak, B. K. (2018). SCICS: A Soft Computing Based Intelligent Communication System in VANET. Smart Secure Systems – IoT and Analytics Perspective. *Communications in Computer and Information Science, 808*, 255–261. doi:10.1007/978-981-10-7635-0_19

Rath, M., Pattanayak, B. K., & Pati, B. (2017). *Energetic Routing Protocol Design for Real-time Transmission in Mobile Ad hoc Network. In Computing and Network Sustainability, Lecture Notes in Networks and Systems* (Vol. 12). Singapore: Springer.

Rath, M., Swain, J., Pati, B., & Pattanayak, B. K. (2018). *Attacks and Control in MANET. In Handbook of Research on Network Forensics and Analysis Techniques* (pp. 19–37). IGI Global. doi:10.4018/978-1-5225-4100-4.ch002

Rizwan, P., Suresh, K., & Babu, M. R. (2016). Real-time smart traffic management system for smart cities by using Internet of Things and big data. *International Conference on Emerging Technological Trends (ICETT)*, 1-7. 10.1109/ICETT.2016.7873660

Růžička, J., Šilar, J., Bělinová, Z., & Langr, M. (2018). Methods of traffic surveys in cities for comparison of traffic control systems — A case study. *Smart City Symposium Prague (SCSP)*, 1-6. 10.1109/SCSP.2018.8402666

Schleicher, J. M., Vögler, M., Dustdar, S., & Inzinger, C. (2016). Application Architecture for the Internet of Cities: Blueprints for Future Smart City Applications. *IEEE Internet Computing, 20*(6), 68–75. doi:10.1109/MIC.2016.130

Shahzad, G., Yang, H., Ahmad, A. W., & Lee, C. (2016). Energy-Efficient Intelligent Street Lighting System Using Traffic-Adaptive Control. *IEEE Sensors Journal, 16*(13), 5397–5405. doi:10.1109/JSEN.2016.2557345

Shukla, S., Balachandran, K., & Sumitha, V. S. (2016). A framework for smart transportation using Big Data. *2016 International Conference on ICT in Business Industry & Government (ICTBIG)*, 1-3. 10.1109/ICTBIG.2016.7892720

Siddique, K., Akhtar, Z., Yoon, E. J., Jeong, Y. S., Dasgupta, D., & Kim, Y. (2016). Apache Hama: An Emerging Bulk Synchronous Parallel Computing Framework for Big Data Applications. IEEE Access, 4, 8879-8887. doi:10.1109/ACCESS.2016.2631549

Singh, D., Vishnu, C., & Mohan, C. K. (2016). Visual Big Data Analytics for Traffic Monitoring in Smart City. *15th IEEE International Conference on Machine Learning and Applications (ICMLA)*, 886-891. 10.1109/ICMLA.2016.0159

Son, M., & Jung, H. (2017). Development and Construction of Security-Enhanced LTE traffic signal system in Korea. *2nd International Conference on Computer and Communication Systems (ICCCS)*, 91-95. 10.1109/CCOMS.2017.8075274

Sun, W., Yu, F. R., Tang, T., & You, S. (2017). A Cognitive Control Method for Cost-Efficient CBTC Systems With Smart Grids. *IEEE Transactions on Intelligent Transportation Systems*, *18*(3), 568–582. doi:10.1109/TITS.2016.2586938

Sun, Y., Song, H., Jara, A. J., & Bie, R. (2016). Internet of Things and Big Data Analytics for Smart and Connected Communities. IEEE Access, 4, 766-773. doi:10.1109/ACCESS.2016.2529723

Tahmid, T., & Hossain, E. (2017). Density based smart traffic control system using canny edge detection algorithm for congregating traffic information. *3rd International Conference on Electrical Information and Communication Technology (EICT)*, 1-5. 10.1109/EICT.2017.8275131

Tang, Chen, Hefferman, Pei, Tao, He, & Yang. (2017).Incorporating Intelligence in Fog Computing for Big Data Analysis in Smart Cities. *IEEE Transactions on Industrial Informatics*.

Tawalbeh, L. A., Mehmood, R., Benkhlifa, E., & Song, H. (2016). Mobile Cloud Computing Model and Big Data Analysis for Healthcare Applications. IEEE Access, 4, 6171-6180. doi:10.1109/ACCESS.2016.2613278

Wang, Y., Chen, Q., Kang, C., & Xia, Q. (2016). Clustering of Electricity Consumption Behavior Dynamics Toward Big Data Applications. *IEEE Transactions on Smart Grid*, *7*(5), 2437–2447. doi:10.1109/TSG.2016.2548565

Yuan, W. (2015). A Smart Work Performance Measurement System for Police Officers. IEEE Access, 3, 1755-1764. doi:10.1109/ACCESS.2015.2481927

Zaid, A. A., Suhweil, Y., & Yaman, M. A. (2017). Smart controlling for traffic light time. *IEEE Jordan Conference on Applied Electrical Engineering and Computing Technologies (AEECT)*, 1-5.

ZhiJie & RuiBing. (2016). Intelligent Traffic Control System Based Single Chip Microcomputer. *International Conference on Intelligent Transportation, Big Data & Smart City (ICITBS)*, 577-579.

Zhu, F., Li, Z., Chen, S., & Xiong, G. (2016). Parallel Transportation Management and Control System and Its Applications in Building Smart Cities. *IEEE Transactions on Intelligent Transportation Systems, 17*(6), 1576–1585. doi:10.1109/TITS.2015.2506156

Chapter 3
Survey on Risk–Based Decision–Making Models for Trust Management in VANETs

Junaid Mohammad Qurashi
King Abdulaziz University, Saudi Arabia

ABSTRACT

Ubiquitous use of wireless technology and ad-hoc networks have paved the way for intelligent transportation systems also known as vehicular ad-hoc networks (VANETs). Several trust-based frameworks have been proposed to counter the challenges posed by such fast mobile networks. However, the dynamic nature of VANETs make it difficult to maintain security and reliability solely based on trust within peers. Decision-making upon collaborative communications is critical to functioning of VANETs in safe, secured, and reliable manner. Decision taken over malicious or wrong information could lead to serious consequences. Hence, risk management within paradigm of trust becomes an important factor to be considered. In this chapter, a survey of the existing works having incorporated risk factor in their trust models has been explored to give an overview of approaches utilized. The parameters chosen in these models are analyzed and categorized based on the approaches modeled. Finally, future research directions will be presented.

DOI: 10.4018/978-1-5225-7189-6.ch003

INTRODUCTION

With ever growing population and significant increase in the number of vehicles on the road, traffic efficiency and road-safety have become important issues to be addressed. Several studies have been conducted on the number of accidents, giving harrowing statistics to the number of lives lost due to road accidents each year ("Wikipedia on road traffic safety, n.d.). In Saudi Arabia alone, there is an accident every minute, claiming 20 deaths every day (Gazette, 2018).

Recent advancement in technology and better connectivity over wireless connection have given rise to set of technologies to tackle such problems. More vehicles connected with Wi-Fi devices and equipped with GPS technology has enabled vehicle to vehicle (V2V) or vehicle to road side units (RSUs) communication, thus forming peer-to-peer network known as vehicular ad-hoc network (VANET). Connected vehicles can communicate and share information about the condition of the road or any relevant information to avoid accidents and ensure efficient traffic as well road safety. Several applications have been developed to aid the driver through all the information collected besides providing entertainment (Al-Sultan, Al-Doori, Al-Bayatti, & Zedan, 2013). These applications could be classified as traffic related, safety related, and entertainment related (Shaikh, 2016). Thus, VANET has garnered significant interest and hence an interesting area of research in tackling issues pertaining to road safety, traffic management as well as comfort (Zeadally, Hunt, Chen, Irwin, & Hassan, 2010). Several projects have been established in industrial and academic arenas e.g. TRIG Project (London, 2015), GST, PreVent and Car-to-Car Consortium (Car 2 Car Consortium, n.d.) to solve the problems pertaining to VANET and establish safer and efficient mobility. Car manufacturers have already started to equip their vehicle with devices that help them to establish vehicular connectivity and also propose algorithms that issue a warning if there is highly likely possibility of a crash or an accident (Zhang J., A Survey on Trust Management for VANETs, 2011).

Lot of effort is being put to ensure secure and reliable delivery of the information shared through peers in VANET. However, evaluating the quality of the message shared, for instance, message forwarded by a malicious peer that leads to false warning could bring traffic to halt or even cause a fatal accident, the notion of trust has been incorporated in various models (Wei & Chen, 2014; Cohen, Zhang, Finnson, Tran, & Minhas, 2014; Mui, Mohtashemi, & Halberstadt, 2002) to seek out the peers that are dishonest or exhibiting malicious behavior. Several works have been established in the discipline and quite a number of comprehensive surveys have also been published related to the work.

As argued by authors like Shaikh (2016) and Thayananthan & Shaikh (2016) risk is an important factor that needs to be incorporated in trust-based models as the decision logic in most of the proposed models are straight forward, with message having highest trust value being accepted. There are few limitations to the existing trust-based approach, for example, there is no way of determining if the trust values have been influenced by each of the peers involved or even if the values have been from a compromised peer. Therefore, it is evident to incorporate factors that would consider parameters like weather, condition of the road, type of the road or driver behavior etc. (Fitzgerald & Landfeldt, 2015) to determine the risk values before taking a decision to change a lane or any other safety measures that could prevent unwanted circumstances like collision, rather than having just to depend on the trust value of the peers. Another scenario would be that of receiving contradicting messages from peers with same trust values with no other reliable option to take decision in real time scenario, such cases could lead to fatal consequences. Moreover, as stated in Shaikh (2016) most of the researches do not clearly state how to determine the minimum trust threshold that is used for decision making.

Several works have been established that incorporate risk into their decision modelling for VANETS. This paper will present a survey about the decision-making models based on approaches taken and parameters defined upon which risk is calculated to aid in decision-making process. The contribution of the paper are as follows:

1. A review on risk-based decision methods in VANETs.
2. A discussion on taxonomy of approaches applied and parameters taken into consideration.

METHODOLOGY

Online databases like IEEE Explorer, Springer, ACM, were used to search the paper using keywords like, "risk-based", "decision method", "VANETS", "trust management", "vehicle decision-making", "autonomous vehicles" and few more. Use of Google search engine including Google Scholar was extensively used to search for the related articles. Thorough literature review was done mainly focused on approaches applied and parameters chosen, comparison of which will be provided in the Appendix (Table 1). Papers were analyzed keeping in view of the limitation of each approach to enrichen the literature review. Tools like Zotero were used to organize the papers. MS-Excel was used to maintain the literature review and MS-Word was used for documentation as well as for referencing and citations.

TRUST MANAGEMENT IN VANETS

The concept of "Trust" once native to social sciences, has been studied in the field for many years. Particular to the field, notion of trust is restricted to human-to-human relationship. However, with evolving time, proliferation of information technology and with advent of artificial intelligence, the notion of trust is no longer restricted to social sciences. Notion of "trust" is becoming highly relevant to human-to-machine or machine-to-machine relationships and hence the precedence of trust into the field of information systems has led to significant research in recent past few years. As mentioned by (Marsh & Dibben, 2005), trust in information system could be attributed to security, privacy, efficiency, productivity and even comfort. Further in the context of information science and technology, authors (Marsh & Dibben, 2005) define "Trust" as a concept that *"concerns a positive expectation regarding the behavior of somebody or something in a situation that entails risk to the trusting party"*. According to (Cho, Swami, & Chen, 2011) the term, *"Trust Management" was first introduced by* (Blaze, Feigenbaum, & Lacy, 1996) *and clarifies that, "Trust management provides a unified approach for specifying and interpreting security policies, credentials, and relationships"*. Thus, Trust Management in VANETS could be defined as the establishment of trust among different peers and with the objective of relying on them in taking safety and efficiency related decisions while at the same time avoiding or mitigating any risk associated with malicious data being propagated by dishonest or compromised peers by identifying them so as to maintain the reliability of VANETS.

The concept of trust becomes eminent in VANET owing to inherent security threats any network is vulnerable to and also in addition to the characteristic that peers are interdependent on each other for maintaining safety and reliability of overall system. These identified threats (Tangade, Sun, & Manvi, 2013) could lead to compromising any of the peers which in turn could cause severe damage including loss of life. However, establishing trust in VANETS has its own set of challenges owing to the inherent nature of VANETS being highly dynamic in terms of mobility, decentralized and rapid increase or decrease in density of vehicles. Furthermore, rapidly changing road conditions or surrounding in VANETS could easily render peer as untrustworthy when information at the time of peer forwarding it would have hold true but cannot be held true for other peers few moments later. Moreover, interaction between two peers might happen only once and may not recur with the same peer in future. Interaction more than once is likely approach in establishing trust which isn't the case with VANETS (Zhang J., 2012). Most of the trust-based models in VANETS are based on voting and reputation scheme assuming that the peers interact more than once. While some other trust-models are based on gathering the information about the surrounding on the assumption that all the peers have complete information of the surroundings, which may not be the case at all times. Some trust-based models are based on centralized authority to establish trust on peers which again may not be the best approach in VANETS owing to their dynamic and distributed nature. Thorough research on the subject and latest available in terms of quality survey has been presented in (Zhang J., 2011) and could be referred for further details on trust management within VANETs.

LIMITATIONS OF TRUST MANAGEMENT APPROACHES

Limitations to the trust management systems in VANETs arise due to its ephemeral and dynamic nature. Trusted Management schemes are mostly based on reputation or voting schemes. Owing to the short-lived span of the interaction in VANETS, establishing trust becomes challenging among the peers. Moreover, millions of peer-vehicles plying over the roads having interacted once might not get to interact with the same peer again. For the details on the limitations of Trust Management schemes, the following works may be referred (Huang, Ruj, Cavenaghi, & Nayak, Limitations of Trust Management Schemes in VANET and Countermeasures, 2011). The challenges have been summarized as of the following:

1.   In VANET, peers are in contact for a brief amount of time, and the interaction with another peer may occur only once with very less probability of occurring it again with the same peer.
2.   Another behavior that could be attributed to peers is that they may be selfish but not malicious, hence it becomes difficult to gauge the malicious behavior.
3.   Most of the trust-based models deploy voting mechanisms to determine trust. Problem arises when there are not enough number of peers to achieve minimum threshold number of votes owing to dynamic nature of VANETs.

Countering inherent problems related to trustworthiness in trust-based models in peer-to-peer (P2P) has been an interesting research area. For example, in Wu, He, Zhang, & Xu (2009), the authors devised a trust-based model that incorporated risk-factor. Risk is being calculated by degree of interaction with the peers on the mobile P2P networks based on cost and benefit analysis. In Yan, (2007), the author proposed a reputation-based trust supporting framework for assessing the trustworthiness of peers based on a transaction-based feedback system, and a decentralized implementation of such a model over a structured P2P network.

OVERVIEW OF RISK ANALYSIS IN INFORMATION TECHNOLOGY

Oxford dictionary defines risk as the, "possibility of loss, injury or other adverse or unwelcome circumstance, a chance or situation for such a possibility" (Wikipedia, 2018). National Institute of Standards and Technology defines risk as the, "net negative impact of the exercise of a vulnerability, considering both the impact of occurrence". Furthermore, in Kaplan & Garrick (1981), risk has been defined as the notion of uncertainty and damage.

Risk = uncertainty + damage

Whereas, Risk Management has been defined as the procedure to identify risk, estimating the risk and taking steps to mitigate risk or reduce the risk to an acceptable level (Stoneburner, Goguen, & Feringa, 2002). TVRA methodology declares risk as the product of likelihood and impact. Likelihood can be explained as the possibility of occurrence of an identified threat whereas as impact as the effect a threat has on the system or subject (deMeer & Rennoch, 2011).

Risk = likelihood * impact

In context of VANETS, risk analysis can be defined as the threshold up to which the level of risk can be tolerated. Threshold values determined through probability of various parameters taken into considerations become the deciding factor to accept or avoid certain or high probable losses. It also depends on willingness of the peer to take risks (Wu, et al., 2009).

PARAMETERS CHOSEN ACROSS THE MODELS

For Intelligent Transport Systems to be completely reliable, that allow safety and are free from prone to error or malicious behaviour; the systems and applications installed on the vehicle in VANETS have to be able to take account of most of the situational aspects or parameters involved that would deem the system to be reliable and safe. To model all the aspects that would determine the system to be intelligent and robust enough is a challenging task, especially, when situations to be assessed are dynamic and real time decision-making is required. However extensive work is being done in both academic and industrial arenas' to counter the challenges posed by such systems; several projects have been initiated that model different contexts to take account of situations involving a vehicle travelling on the road. As stated in McCall & Trivedi (2007), while the vehicle is on the road, risk could be identified as a factor arising from following three different contexts:

- **Environment Context:** It includes different climatic conditions and road infrastructure.
- **Vehicle Context:** It broadly includes the parameters involving the vehicle itself, for example, the make and model of the vehicle, speed of the vehicle, type of the vehicle based on size, time to collision, etc. Also includes the type of applications and gadgets vehicles are fitted with. Though, some article, have included road scenarios in vehicle context as well.
- **Driver Attitude:** Driver attitude is the most important yet complex factor to be modelled. Few works have attributed driver context but not limited to: drivers age, experience and gender.

The papers surveyed have chosen different parameters depending upon the problems they are trying to address. A comparison of the parameters chosen in different works has been provided in the Appendix.

Basic definition of some of the common parameters chosen have been defined below. These definitions have been adapted from Thayananthan & Shaikh (2016) and also as inferred from surveying other research papers.

Road

Roads are the path-ways where the vehicles are restricted to have their trajectories limited to. Roads could be bi-directional. Characteristics like, dry, icy wet, road border could be attributed to roads depending on environmental condition and also whether road is congested depending upon the traffic condition.

Lanes

Lane could be inferred as the division of the roads into different speed limits and have attributes that follow for the roads as well viz, curved, downhill, uphill and intersection.

Weather

Weather determines different climatic conditions that could affect the severity of risk and following variances could be attributed to it: foggy, raining, snowing or windy.

Driver Behavior

Driver behavior is most complex to model, owing to the number of factors that could be used to determine human behavior attributed to a driver. The common attributes considered are age and experience. However, few works have taken account more attributes including emotions.

Time

Time corresponds to the time of the day, viz. evening, night or day. Different time of the day may have different levels of risks involved, for example, if a vehicle is heading towards west on a straight road, sunlight over the wind-shield could increase the severity of risk.

Time-to-Collision (TTC)

In deterministic approaches, Time-to-Collision has been defined as the time it takes two vehicles to collide if they continue to travel on the same trajectory with the same speed.

Time-to-React (TTR)

In deterministic approaches, Time-to-React has been defined as the time it takes to react to any threat and take risk-mitigating actions.

Time-to-Brake (TTB)

Time-to-Brake is defined as the time period during which brake can be applied and collision can be avoided.

Time-to-Leave (TTL)

Time-to-Leave corresponds to the amount of time taken to leave the accident-prone area.

TAXONOMY OF RISK-BASED METHODS

Different approaches have been utilized in calculating risk-assessment which employ different methods and models. In this paper for the sake of simplicity, these approaches are broadly classified under dynamic and static approaches. Static because they don't take into account the probability or randomness that is attributed to VANETS for large. Perhaps the primary reason that most of the risk-based methods are dynamic, meaning that these research works have take randomness into account, hence are probabilistic or hybrid. Static approaches are further classified into deterministic and framework-based approaches. Under deterministic approaches, models established on risk-mitigating factors involving vehicle maneuvers like TTC, TTB, TTR etc are proposed. Deterministic models do not explicitly model uncertainties and are mostly based on some initial input to calculate risk related to collision or accident (Glaser, Vanholme, Mammar, Gruyer, & Nouveliere, 2010; Noh & An, 2017). The possibility to collision is calculated as a binary prediction, that is, either having a value of true or false. Under, Framework based, authors have just proposed architectures that could be implemented using mathematical model, however, in such works no such mathematical approach is implied.

Dynamic approaches form all those models that take randomness into account. Dynamic Approaches are further classified under Fuzzy based approaches, Probabilistic and Hybrid approaches. Fuzzy based approaches are based on fuzzy logic. Probabilistic approaches have been implemented using mathematical models based on Makarov process, Bayesian-networks and Probability Distribution methods. These models involve the probability or uncertainties.

Henceforth, this article covers the review on papers based on the approaches taken and parameters considered for risk-assessment, therefore, classification formed is as depicted in the Figure 1. This survey also takes the works on decision-making methods into consideration which fall under intelligent transport systems. The following section covers the works falling under corresponding approaches.

Figure 1. Taxonomy for risk-based approaches

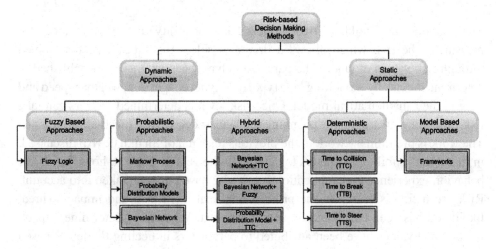

DYNAMIC APPROACHES

Fuzzy Based Approaches

Shaikh (2016) proposed a fuzzy risk-based solution to determine a minimum threshold, calculated through risk assessment based on sensitivity of the application, environmental conditions, and driver's attitude. The types of the application discussed in the work have been given predetermined sensitively levels (low, medium and high) based on the type of the application which are classified as infotainment application, traffic application and safety applications; with infotainment having the lowest and safety application highest level of sensitivity. Similar attributes have been assigned to the driver's attitude based on experience and age. Risk assessment is calculated as a product of threat likelihood and impact, where likelihood is accumulative of vehicle context (road conditions, time of the day, speed of the vehicle and type of lane) and impact, which is obtained through predetermined values of application sensitivity low and high ranging from values 0 to 1. The values have been modeled using fuzzy logic and different scenarios created to find out the impact of various parameters on cumulative risk. Unlike, few other models, gender and speed limit have not been considered.

Naranjo, González, García, & Pedro (2008) proposed a fuzzy based decision-making system that allows automatic lane change on straight path in overtaking a vehicle. The system depends on GPS for navigation and information gathering from other vehicles taking part in overtaking. Three types of driving maneuvers modelled based on fuzzy logic have been applied viz. bend, straight road, or lane change.

Probabilistic Approaches

Thayananthan & Shaikh (2016) proposed a probability distribution model for measuring the risk which allows a driver to select the safest lane determined through risk values of lanes. The parameters have been extensively established to determine the risk by modelling lane type, road type, congestion, weather speed and time in their mathematical model. Once risk level is determined, drivers can take risk-mitigating measures like changing the lane or decreasing speed of the vehicle. The work is preliminary work to the previous research of Shaikh (2016) discussed, in that it uses similar parameters to determine the vehicle context, however, driver behavior, experience and application sensitivity have not been taken into account. Risk estimation is based on the product of threat likelihood and impact. Threat likelihood has been calculated by vehicle context. Whereas to determine impact, preassigned values have been attributed to parameters indicating the levels based on severity of risk. For example, a dry road has the severity impact level of low (i), whereas an icy road has severity level of high (iii). Moreover, the research contributes in determining the recommended secure lane gaps for types of vehicles, viz small, medium and large.

Chen, Irissappane, & Zhang (2017) proposed a model to counter the challenges posed by fake or untrustworthy messages sent in VANETS by rogue or malicious peers. Often when a peer is encountered by an untrustworthy message, to make a risk-mitigation maneuver e.g. to decide to change the lane or to stay in the same lane, requires exchanging more message to increase the belief about an event resulting higher consumption of resources and longer decision-making process and hence delayed response. To counter such uncertainties, authors propose a Partially Observable Markov Decision Process (POMDB) based approach. The model maintains probability distribution about whether a traffic event has occurred. It updates the confidence on the event by processing the messages about the event. The model also assumes that the neighboring vehicles can exhibit different behaviors, i.e., truthful, malicious etc., in providing event information. Based on the event belief, the model then decides to re-route or stay on the current route.

Fitzgerald & Landfeldt (2015) claim to have introduce the concept of *risk limits* to be able maintain the low probability of crash. Authors state that risk limits are directly proportional to the probability of occurrence of an accident. Furthermore, it is stated that the risk in traffic management is related to probability of exposure of the vehicle to other vehicles, probability of crash, probability of injury and injury outcome. However, the work is limited to base the calculation of risk on probability of crash and likelihood of injury; authors justify the selection as most amenable factors to be calculated for risk mitigation. Based on qualitative analysis on data from CrashLink database which collects the data on traffic accidents, following parameters were selected to calculate the risk values: driver age, driver gender, vehicle make, number of occupants of the vehicle, vehicle speed at the time of the crash, vehicle type and year of the manufacture of the vehicle. To calculate the risk values, authors determined the occurrence of each of the factors established against the number of vehicles in accidents within the data of CrashLink database. To establish risk limit, these values were modeled into the simulation environment to achieve a minimum threshold to avoid crash by simulating risk mitigating factors, decreasing speed, increasing headway or changing lane. The work established different risk values for different vehicle types. However, the work does not take environmental condition into account which in real world scenario is a highly occurrence and a deciding factor for accidents.

Fitzgerald & Landfeldt (2012) extend their work (Fitzgerald & Landfeldt, 2015), by coupling the risk-limit values obtained from other vehicles in VANETS. To counter the possibility of having received corrupt risk-limits from compromised or malicious peers, authors propose of converging the risk-limits obtained from a number of vehicles in the network to that with individual set risk-limits obtained from various factors like driver attitude, make of the vehicle, lane type, speed of the vehicle as described in previous work (Fitzgerald & Landfeldt, 2015). The proposed model works by computing a weighted average of risk estimates from surrounding vehicles to that of independent risk values calculated from its sensors.

Fitzgerald & Landfeldt (2013) proposed an algorithm to calculate the probability of an accident (accident risk) related to a link. Risk values are assigned to each link based on the attributes, such as type, condition and geometry, and the surrounding environment. Accident risk is calculated as an accumulative of vehicle risk and link risk. The work is an extension done to the previous discussed work [19]. Upon approaching intersection information vehicle calculates the maximum travel speed for the link, based on position and speeds of the vehicles already on the link. While maintaining the risk limits, effective speed for the link is calculated. The link with highest allowable speed is chosen to mitigate the risk.

Gindele, Brechtel, & Dillmann (2010) propose a Bayesian model to predict the next trajectory for the vehicle based on situation context and behavior of the driver. Research takes interaction of the vehicles into account. Situation context describe the properties of the vehicle surroundings, for example, the safe distance between the vehicles. Behavior of the driver is modeled against the most common maneuvers in highway scenarios and consist of "free-ride", "following", "acceleration-phase", "sheer out", "overtake", and "sheer in", as depicted below in Figure 2.

Figure 2. Behaviors with trajectories
(Gindele, Brechtel, & Dillmann, 2010).
Note: A: free-ride; B: following; C: acceleration phase; D: sheer-out; E: overtake; F: sheer-in

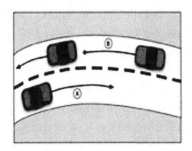 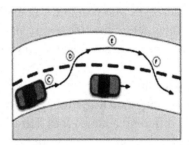

Niehaus & Stengel (1994) propose probabilistic mathematical model to handle uncertainties and provide an automated decision making. The framework has been applied to previous work of Niehaus & Stengel (1991) based on deterministic approach. The approach takes account of surrounding traffic, current road condition (snowy, icy) and driver attitude based on cruising and desired levels of safety and aggressiveness. Model proposes modules to assess the situation of the traffic (relative position of the vehicle in front) which determines the situation to be safe or unsafe. The model determines the desired risk-mitigating factors like change lane or continue straight and adequate acceleration in safe or unsafe situation.

McCall & Trivedi (2007) proposed a system to prevent rear collision by modeling the behavior of the driver in Bayesian network based on parameters steering wheel position and pedal actions to determine driver behavior. Other parameters that have been evaluated to model the surrounding situation are: steering angle, wheel speed, longitudinal acceleration, brake pedal pressure and accelerator pedal position. Parameters have been modelled in Bayesian framework to determine the probability of braking action required and probability of intention of the driver for applying brakes.

Hybrid Approaches

Noh & An (2017) proposed a risk assessment algorithm for automatic lane change maneuvers to reliably establish potential collision risks of observed vehicles and recommend a lane change maneuver. Risk of each lane is determined by the behavior of the dangerous vehicle present in front or rear of the subject vehicle. Authors have included deterministic risk-metrics like time-to-collision (TTC) and minimal safety margin to take account of high variations in vehicle speeds and minimal safety within the vicinity of the subject vehicle respectively. To account for the uncertainties and likelihood for each risk metric, risk assessment is modelled using Bayesian networks based on three risk levels assigned to each vehicle: dangerous, attentive and safe. However, the work doesn't take account of risk associated with different types of road, congestion and environmental conditions.

Schneider, Wilde, & Naab (2008) proposed a Bayesian network model and fuzzy method to handle driver based and sensor-based uncertainties. The parameters to identify the uncertainties and necessary mitigation required is defined by: fast approach, braking reaction, lane change. However, authors didn't explicitly mention the parameters chosen to calculate the probabilities for driver situation.

Noh, An, & Han (2015) propose a probabilistic model to determine the risk associated with traffic on the road by assessing the situation assessment. Situations assessment can be inferred as the probability distribution of each region surrounding the vehicle and the probability distribution of each lane. Based on the situation assessment, recommendation for driving behavior can be done. The project is modelled on the information gathered from the sensors attached to the vehicle in the following positions: current forward, current backward, left forward, left backward, right forward, and right backward, thus estimating threat levels of each assigned region as the probability of the region state by using two deterministic factors as a threat measure: time-to-brake and minimal safety margin. The combination of probabilities reported by each of these sensors determine the probability distribution of the state of each lane.

Berthelot, Tamke, Dang, & Breuel (2012) proposed probability distribution model to obtain threshold values for TTC under different circumstances. The limitation of the work is that it takes account the values based only one object which may not be the case in real world scenario.

STATIC APPROACHES

Deterministic Approaches

Glaser, et al. (2010) have taken a deterministic approach. The model is applied to do decision-making for driving change maneuvers based on collision risk and lane change risk. To determine the possibility of the collision, authors have taken two parameters inter-vehicular-time, time-to-collision and reaction distance. Risk involving each maneuver is calculated based on the average of risk associated with speed range and lane. Driver behavior is modelled after the commands of the driver like (pedal position, pedal speed) and state of the vehicle (acceleration, speed). The proposed model helps to determine the optimal trajectory path for the vehicles considering the road environment and vehicles. The ranking of the maneuvers is based on a fast risk evaluation on every lane.

Hillenbrand, Spieker, & Kroschel (2006) proposed a model that aids in decision making for forward collisions based on the deterministic metric time-to-react (TTR). The derivation of the TTR in the paper is dependent on other metrics like TTC (Time-to-collision), which in turn is derived from TTE (Time-to-enter) and TTD (Time-to-Disappear). TTR is defined as the time it takes the driver to take risk mitigating maneuver to avoid colliding with an object, based on objects future motion and physical constraints due to vehicle dynamics (speed, type etc.).

Framework Approaches

Reichardt (2008), author has proposed an architecture that integrates the emotional status of a driver and models its influence on the driving style. Main modules of the architecture have been divided are: situation assessment, emotional assessment and risk assessment. For the situations assessment following parameters have been considered: traffic jams(congestion), accidents blocking the lane(congestion), lane endings or beginnings and upcoming entry/exit lanes (intersection). Aspects like "desirability", "praiseworthiness", "appealigness", "sense of reality", "proximity" and "unexpectedness" have been modeled to measure the emotions of the driver. All these attributes have been derived and defined OCC model (Damásio, 1994). The make of the vehicle (vehicle type) whether it is a passenger vehicle or goods vehicle, or a motorcycle has also been taken into account. And, finally for the static elements for traffic jam itself, parameters like, lane width, number of lanes, road border, downhill grade, and curvature are considered.

FUTURE RESEARCH AND CHALLENGES

Several approaches and methods have been employed to strengthen the existing trust-based models which pose challenges in establishing the safety of the road traffic owing to the ephemeral and dynamic nature of VANETs. The future direction of the research would be in establishing and modelling the parameters using methods/ models which could yield more precise risk-limit values. The approaches could be a mix of one or more methods including probabilistic or deterministic methods.

The challenge would be in identifying an apt combination of different approaches that yield better results mapping them in the real-world scenario. Further, the more the number of parameters taken into consideration, the closer we are in accurately predicting the risk values or gathering awareness about the surroundings in VANETs. However, few challenges are posed in this area: first identifying the parameters that have direct impact in decision making and second modelling these risks associated with the parameters. Even though several parameters have been identified, but not all have been successfully modeled, emotional status of the driver being one such example. Furthermore, another aspect to notice is that all the calculations based on the mathematical models proposed are to be processed on low-cost hardware that are installed on automobiles, hence, any models proposed shouldn't limit the computational capacity of control units. Very few works have been published concerning the incorporation of risk-based models in decision-making for VANETS. The area of work is fairly new and offers lot opportunities to explore the field.

CONCLUSION

In conclusion, this is the first survey based on risk-based decision methods applied in VANETS under paradigm of trust-based models to the best of author's knowledge. The decision methods have broadly been classified under approaches being used. Most of the approaches are probabilistic to account for the uncertainties in VANETS that exist in real world scenarios. Paper has also given a brief overview on understanding of the risk assessment in information technology. To increase the credibility of the trust-based model, incorporation of the risk-based decision methods is a must. Currently, not much of the risk-based decision methods take all of the contexts (environment context, driver context and vehicle context) into consideration owing to the complexities it offers. Few works have been published that may be improvised to give a comprehensive solution to better decision-making in VANETS.

REFERENCES

Al-Sultan, S., Al-Doori, M. M., Al-Bayatti, A. H., & Zedan, H. (2013). A comprehensive survey on vehicular Ad Hoc network. *Journal of Network and Computer Applications*, *37*(1), 380–392. doi:10.1016/j.jnca.2013.02.036

Berthelot, A., Tamke, A., Dang, T., & Breuel, G. (2012). A novel approach for the probabilistic computation of Time-To-Collision. *Intelligent Vehicles Symposium*. 10.1109/IVS.2012.6232221

Blaze, M., Feigenbaum, J., & Lacy, J. (1996). Decentralized trust management. In *Proceedings 1996 IEEE Symposium on Security and Privacy*. IEEE. 10.1109/SECPRI.1996.502679

Car 2 Car Consortium. (n.d.). Retrieved from https://www.car-2-car.org

Chen, S., Irissappane, A. A., & Zhang, J. (2017). POMDP-Based Decision Making for Fast Event Handling in VANETs. *The Thirty-Second AAAI Conference on Artificial Intelligence*.

Cho, J.-H., Swami, A., & Chen, I.-R. (2011). A Survey on Trust Management for Mobile Ad Hoc Networks. *IEEE Communications Surveys and Tutorials*, *13*(4), 562–583. doi:10.1109/SURV.2011.092110.00088

Cohen, R., Zhang, J., Finnson, J., Tran, T., & Minhas, U. F. (2014). A trust-based framework for vehicular travel with non-binary reports and its validation via an extensive simulation testbed. Journal of Trust Management. doi:10.118640493-014-0010-0

Damásio, A. (1994). *Descartes' Error: emotion, reason, and the human brain*. New York: G. P. Putnam's Sons.

Dashtinezhad, S., Nadeem, T., Dorohonceanu, B., Borcea, C., Kang, P., & Iftode, L. (2004). TrafficView: a driver assistant device for traffic monitoring based on car-to-car communication. *IEEE 59th Vehicular Technology Conference*. 10.1109/VETECS.2004.1391464

deMeer, J., & Rennoch, A. (2011). The ETSI TVRA Security-Measurement Methodology by means of TTCN-3 Notation. *10th TTCN-3 User Conference*.

Fitzgerald, E., & Landfeldt, B. (2012). A System for Coupled Road Traffic Utility Maximisation and Risk Management Using VANET. *15th International IEEE Conference on Intelligent Transportation Systems*. 10.1109/ITSC.2012.6338630

Fitzgerald, E., & Landfeldt, B. (2013). On Road Network Utility Based on Risk-Aware Link Choice. *International IEEE Conference on Intelligent Transportation Systems*. 10.1109/ITSC.2013.6728361

Fitzgerald, E., & Landfeldt, B. (2015). Increasing Road Traffic Throughput through Dynamic Traffic Accident Risk Mitigation. *Journal of Transportation Technologies*, 5(5), 223–239. doi:10.4236/jtts.2015.54021

Fitzgerald, E. S., & Landfeldt, B. (2015). Increasing Road Traffic Throughput through Dynamic Traffic Accident Risk Mitigation. *Journal of Transportation Technologies*, 05(04), 223–239. doi:10.4236/jtts.2015.54021

GazetteS. (2018). Retrieved 2018, from http://saudigazette.com.sa/article/524118/SAUDI-ARABIA/One-accident-every-minute-20-deaths-daily-on-Saudi-roads

Gindele, T., Brechtel, S., & Dillmann, R. (2010). A Probabilistic Model for Estimating Driver Behaviors and Vehicle Trajectories in Traffic Environments. *Annual Conference on Intelligent Transportation Systems*. 10.1109/ITSC.2010.5625262

Glaser, S., Vanholme, B., Mammar, S., Gruyer, D., & Nouveliere, L. (2010). Maneuver-Based Trajectory Planning for Highly Autonomous Vehicles on Real Road with Traffic and Driver Interaction. *IEEE Transactions on Intelligent Transportation Systems*, 11(3), 589–606. doi:10.1109/TITS.2010.2046037

Hillenbrand, J., Spieker, A. M., & Kroschel, K. (2006). A Multilevel Collision Mitigation Approach—Its Situation Assessment, Decision Making, and Performance Tradeoffs. *IEEE Transactions on Intelligent Transportation Systems*, 7(4), 528–540. doi:10.1109/TITS.2006.883115

Huang, Z., Ruj, S., Cavenaghi, M., & Nayak, A. (2011). Limitations of Trust Management Schemes in VANET and Countermeasures. IEEE 22nd International Symposium on Personal, Indoor and Mobile Radio Communications. doi:10.1109/PIMRC.2011.6139695

Kaplan, S., & Garrick, B. J. (1981). *On The Quantitative Definition of Risk*. Academic Press.

LondonM. U. (2015). *TRIG Project*. Retrieved from http://www.vanet.mdx.ac.uk/research/trig-project/

Marsh, S., & Dibben, M. R. (2005). The Role of Trust in Information Science and Technology. *Annual Review of Information Science & Technology.* doi:10.1002/aris.1440370111

McCall, J. C., & Trivedi, M. M. (2007). Driver Behavior and Situation Aware Brake Assistance for Intelligent Vehicles. *Proceedings of the IEEE.* 10.1109/JPROC.2006.888388

Mui, L., Mohtashemi, M., & Halberstadt, A. (2002). A computational model of trust and reputation. *Proceedings of the 35th Hawaii International Conference on System Science (HICSS).* 10.1109/HICSS.2002.994181

Naranjo, J. E., Gonzalez, C., Garcia, R., & Pedro, T. (2008). Lane-Change Fuzzy Control in Autonomous Vehicles for the Overtaking Maneuver. *IEEE Transactions on Intelligent Transportation Systems, 9*(3), 438–450. doi:10.1109/TITS.2008.922880

Niehaus, A., & Stengel, R. (1991). An expert system for automated highway driving. *IEEE Control Systems Magazine, 11*(3), 53–61. doi:10.1109/37.75579

Niehaus, A., & Stengel, R. F. (1994). Probability-Based Decision Making for Automated Highway Driving. *IEEE Transactions on Vehicular Technology, 43*(3), 626–634. doi:10.1109/25.312814

Noh, S., & An, K. (2017). Risk Assessment for Automatic Lane Change Maneuvers on Highways. *International Conference on Robotics and Automation (ICRA).* 10.1109/ICRA.2017.7989031

Noh, S., An, K., & Han, W. (2015). High-Level Data Fusion based Probabilistic Situation Assessment for Highly Automated Driving. *18th International Conference on Intelligent Transportation Systems.* 10.1109/ITSC.2015.259

Reichardt, D. M. (2008). Approaching Driver Models Which Integrate Models Of Emotion And Risk. *IEEE Intelligent Vehicles Symposium.* 10.1109/IVS.2008.4621284

Road traffic safety. (n.d.). In *Wikipedia.* Retrieved from https://en.wikipedia.org/wiki/Road_traffic_safety

Schneider, J., Wilde, A., & Naab, K. (2008). Probabilistic Approach for Modeling and Identifying Driving Situations. *IEEE Intelligent Vehicles Symposium*. 10.1109/IVS.2008.4621145

Shaikh, R. A. (2016). Fuzzy Risk-based Decision Method for Vehicular Ad Hoc Networks. *International Journal of Advanced Computer Science and Applications*, 7(9), 54–62. doi:10.14569/IJACSA.2016.070908

Stoneburner, G., Goguen, A. Y., & Feringa, A. (2002). *Risk Management Guide for Information Technology Systems*. National Institute of Standards and Technology. doi:10.6028/NIST.SP.800-30

Tangade, Sun, & Manvi. (2013). A Survey on Attacks, Security and Trust Management Solutions in VANETs. In *2013 Fourth International Conference on Computing, Communications and Networking Technologies (ICCCNT)*. IEEE. 10.1109/ICCCNT.2013.6726668

Thayananthan, V., & Shaikh, R. A. (2016). Contextual Risk-based Decision Modeling for Vehicular Networks. *International Journal of Computer Network and Information Security*, 8(9), 1–9. doi:10.5815/ijcnis.2016.09.01

Vlacic, A. F. (n.d.). Multiple Criteria-Based Real-Time Decision Making by Autonomous City Vehicles. Institute of Integrated and Intelligent Systems. *Griffith University*.

Wei, Y.-C., & Chen, Y.-M. (2014). Adaptive decision making for improving trust establishment in VANET. *The 16th Asia-Pacific Network Operations and Management Symposium*. 10.1109/APNOMS.2014.6996523

Wikipedia. (2018). Retrieved November Saturday, 2018, from https://en.wikipedia.org/wiki/Risk

Wu, X., He, J., Zhang, X., & Xu, F. (2009). A Distributed Decision-Making Mechanism for Wireless P2P Networks. *Journal of Communications and Networks (Seoul)*, *11*(4), 359–367. doi:10.1109/JCN.2009.6391349

Yan, Z. (2007). *Trust Management for Mobile Computing Platforms*. Academic Press.

Zeadally, S., Hunt, R., Chen, Y.-S., Irwin, A., & Hassan, A. (2010). *Vehicular ad hoc networks (VANETS): status, results, and challenges*. Springer Science.

Zhang, J. (2011). A Survey on Trust Management for VANETs. *International Conference on Advanced Information Networking and Applications*. 10.1109/AINA.2011.86

Zhang, J. (2012). *Trust Management for VANETs: Challenges, Desired Properties and Future Directions*. Singapore: IGI Global. doi:10.4018/jdst.2012010104

APPENDIX

Table 1. Comparison of Approaches and Parameters used in the chapter

Title	Road			Lane				Vehicle						Weather	Driver				Application		APPROACH
	Congestion	Dry	Icy	Type	Curve	Straight	Intersection	Model	Proximity	Speed	Year of Manufacture	Length	Type		Age	Gender	Emotions	Experience	Type	Sensitivity Level	
(Fitzgerald & Landfeldt, 2013)	√	-		√	-	-	√	-	-	√	-		-	-	-	-	-	-	-	-	PDM
(Thayananthan & Shaikh, 2016)	√	√	√	√	-	-	-	√	√	√	-	√	√	√	-	-	-	-	-	-	PDM
(Naranjo et.al., 2008)	√	-	-	-	√	-	-	-	-	-	-	-	-	-	-	-	-	-	-	-	FUZZY
(Shaikh, 2016)	√	√	-	√	√	√	√	-	√	√	-	√	-	√	√	-	-	√	√	√	FUZZY
(Fitzgerald & Landfeldt, 2012)	√	-	-	√	-	-	-	√	-	√	√	-	√	-	√	√	-	√	-	-	PDM
(C, McCall, & Trivedi, 2007)	√	-	-	-	-	√	-	√	√	-	-	-	-	-	-	-	-	-	-	-	PDM
(Chen, Irissappane, & Zhang, 2017)	√	-	-	-	-	-	-	-	-	-	-	-	-	-	-	-	-	-	-	-	PDM
(Gindele & Dillmann, 2010)	-	-	-	-	-	-	√	-	√	-	-	-	-	-	-	-	-	-	-	-	BN
(Stengel, Niehaus, & Robert, 1994)	√	√	√	√	-	-	-	-	√	√	-	-	-	-	-	-	√	-	-	-	PDM
(Glaser et.al., 2010)	√	-	-	-	√	√	√	-	√	-	-	-	-	-	-	-	-	-	-	-	TTC
(Noh, An, & Han, 2015),	√	-	-	-	-	√	-	-	√	√	-	-	-	-	-	-	-	-	-	-	PDM+TTB

NOTE:

- PDM : Probability Distribution Model
- BN : Bayesian Network
- TTC : Time-to-Collide
- TTB : Time-to-Break
- D : Deterministic
- FW : Framework

- Fuzzy based approaches
- Probabilistic approaches
- Deterministic approaches
- Hybrid approaches
- Framework based

Chapter 4
Detecting Intrusions in Cyber–Physical Systems of Smart Cities:
Challenges and Directions

Ismail Butun

iD https://orcid.org/0000-0002-1723-5741
Mid Sweden University, Sweden

Patrik Österberg
Mid Sweden University, Sweden

ABSTRACT

Interfacing the smart cities with cyber-physical systems (CPSs) improves cyber infrastructures while introducing security vulnerabilities that may lead to severe problems such as system failure, privacy violation, and/or issues related to data integrity if security and privacy are not addressed properly. In order for the CPSs of smart cities to be designed with proactive intelligence against such vulnerabilities, anomaly detection approaches need to be employed. This chapter will provide a brief overview of the security vulnerabilities in CPSs of smart cities. Following a thorough discussion on the applicability of conventional anomaly detection schemes in CPSs of smart cities, possible adoption of distributed anomaly detection systems by CPSs of smart cities will be discussed along with a comprehensive survey of the state of the art. The chapter will discuss challenges in tailoring appropriate anomaly detection schemes for CPSs of smart cities and provide insights into future directions for the researchers working in this field.

DOI: 10.4018/978-1-5225-7189-6.ch004

INTRODUCTION

Today, the pace of technology is incredibly high and brings new terms and notions every often. For instance; cyber-cities, cyber-infrastructures, cyber-facilities, Internet of Things (IoT), Industrial IoT (IIoT), Web of Things (WoT), Internet of Everything (IoE) are namely a few which are related to this current topic of smart cities.

Cyber is the critical term mentioning that the thing it refers to is related to computing technology and emphasizes some artificial smartness. Therefore, cyber-cities, cyber-infrastructures, and cyber-facilities eventually are the counterparts of the terms they emphasize (i.e. cities, infrastructures, and facilities, respectively) and relate to the smarter, automated and technologically improved versions of them (Kim, 2012). Cyber-Physical System (CPS) is one of the main pillars of all cyber-related notions such as cyber-cities, cyber-infrastructures, and cyber-facilities (Poovendran, 2010).

According to UN reports, current cities of the developed and under-developed countries are on the edge of livable limits in terms of scalability, environment, and security, owing to the fast population growth among the world (Khatoun and Zeadally, 2017). Although the main objective of a smart city is to improve the quality of lives of its habitants, it may help our world to relieve from the over-population stress by providing efficiently managed cities along with sustainable resources (energy, water, etc.). That is being said; the security and privacy of the people constitute one of the biggest concerns and challenges to be faced in the rapid development of the smart cities.

IoE is bringing together people, processes, data (raw or processed), and things (cameras, sensors, actuators, etc.) to make network connections more relevant and valuable than ever before, turning information into actions that create new capabilities, richer experiences, and unprecedented economic opportunity for businesses, individuals, and countries. Whereas, the IoT is the network of physical objects accessed through the Internet. These objects contain embedded technology to interact with internal states or the external environment. In other words, when objects can sense and communicate, it changes how and where decisions are made, and who makes them. For example, Learning Thermostat by Nest Inc. (2018) learns what temperature you like and builds a regulating (to heat up or cool down the house) schedule around yours as acting like a personal assistant. In this way, not only the comfort level in the house is improved but also the overall (reduces the heating up or cooling down the house during your away time) energy consumption is drastically decreased.

Similarly to what the application-layer web is to the network-layer Internet, the WoT provides an application layer that simplifies the creation of IoT applications and therefore is a term that needs to be considered under the IoT umbrella. When IoT gets into the industry domain with higher communication and security standards, it is called Industrial IoT (IIoT). In addition, IoE further advances the power of the Internet to improve business and industry outcomes, and ultimately make people's lives better by adding to the progress of IoT. Nevertheless, as devices are getting more connected and collect more data, privacy and security concerns will increase too. How companies and entities decide to balance customer privacy with this wealth of IoE data will be critical (Banafa, 2016).

CPS, in general, is interrelated to IoT and in specific cases to IIoT, in which CPS utilizes the IoT or IIoT to command and control several tasks related to the automation of the real world duties such as in the process control of sewage-drainer systems, water treatment facilities, etc. Therefore, in this chapter, CPS has been thought of as an upper umbrella to represent both IoT and IIoT whereas IoE is somewhat out of the CPS's scope. CPS is an upper class of several sub-networks such as IoT, Wireless Sensor Network (WSN), smart grid, etc., and possesses all the security vulnerabilities they might have. Therefore, in securing CPS, all sub-component networks also need to be considered (Cardenas, 2009). The smart grid system is a very promising candidate to replace legacy power grid, however, it possesses its own risks due to the cyber nature (Mo, 2012). For instance, an attack on smart grid of a CPS would not only reveal private user consumption profiles but also may alter or change some of the users updated electricity consumption data and eventually can harm the provider economically (Sridhar, 2006). Somehow, advanced monitoring procedures need to be devised to detect and identify the malfunctioning of the network components or corruptions at measurements caused by intruders (Pasqualetti, 2011). Therefore, robust and resilient power systems, smart grid systems[1], eventually CPSs need to be developed (Zhu, 2011). Security functions, especially the Intrusion Detection System (IDS) is one of the most powerful tools to provide these specifications to CPSs of smart cities. Equipped with robust, resilient, and secure CPSs, smart cities will eventually be recognized by the public without hesitation.

Cyber-attacks sometimes intend to reveal secret information but can also be aimed at damaging the communication abilities of the target network. An example is Vampire Attacks (Vasserman, 2013) in the category of Denial-of-Service (DoS) attacks. In this type of attack, the attacker intends to drain batteries of the target WSN nodes, ultimately causing DoS in the overall WSN. In CPSs of smart cities, if possible, this kind of attacks need to be prevented first; otherwise, they need to be detected and mitigated, in order to be able to provide a stable working condition.

No matter how complicated and advanced security measures a network may have, there is a possibility that hackers will infiltrate that system sooner or later. Hence, IDSs are needed at that moment. They tell whether there exists any unauthorized access to the system from within or outside of the network premises. However, IDSs are not a magical tool to secure your system right away. They spot, identify and report all kind of intrusions towards a network, which is very valuable information for network administrators in order to close the security breaches that the network may possess.

In contrast to common knowledge, most hackers are computer programs rather than humans. Therefore, most of the IDSs end up recognizing these computer programs rather than catching the humans behind them, missing the real deal. Human hackers are an imminent threat to every existing computer network and system as they can execute attacks that are more sophisticated and hide their existence at the same time (Albanese, 2004).

In history, very important events of intrusions have happened. An example is Stuxnet (Langner, 2011) incidence in which the Iranian nuclear enrichment facilities are targeted. There, specifically, the Stuxnet virus was devised to be active and operable at Siemens' Supervisory Control And Data Acquisition (SCADA) systems only. The impact of a cyber-attacks on SCADA[2] systems is very critical and reported in Ten et al.'s work (2008), two years earlier than the Stuxnet incidence. According to authors finding, compliance with the requirements to meet the standards had become increasingly challenging, as the SCADA system became more dispersed in wide areas. Hence, Stuxnet infiltration happened through Microsoft Windows installed computers that had the internet connection, which somehow further infected the computers located at intranet of the nuclear facilities. These SCADA systems were responsible for controlling the turn speed of the centrifuges which are used in uranium enrichment. Stuxnet virus caused these centrifuges to get out of control rapidly and eventually resulted in explosions with loss of property and possibly human lives. The summary of this incidence is; lack of cybersecurity measures (such as firewalls and IDSs) in critical infrastructures may lead to very harmful attacks, which might have serious consequences such as loss of property and human lives as well as industrial espionage.

As shown in *Figure 1*, the Provisioning for Information Security Services consists of three basic steps: prevention, detection, and mitigation (Butun, 2013). In contrast to common thought, Intrusion Prevention Systems (IPSs) such as firewalls are not fully trustable and not efficient to prevent all kinds of intrusions towards computing systems. As discussed above, history has witnessed many severe incidences of hacking towards important targets such as nuclear enrichment facilities, banking systems, online payment systems, etc. Therefore, IDS comes into the picture to offer further remedies to the problem. IDS helps systems security administrators to detect the security breaches in their systems, if any, in a timely manner so that they can mitigate the risks by counteracting (disabling some ports, limiting the access to important system files, etc.) against those attackers. Hence, for cybersecurity, IDS is as important as the IPS.

CPS of smart cities is a futuristic notion comprised of several smart and connected computing devices that need to be secured. Therefore, in this chapter, the importance of the IDS for the CPS of smart cities is stressed.

In this chapter, vulnerabilities in the CPS of smart cities will be provided and the importance of the IDS for the CPS will be stressed. Then, the system diagnosis methodologies for the IDS will be presented. This is followed by a detailed description of the various IDS classifications. First, the classification related to the source of audit data will take place. Second, the classification related to the detection methodologies will be presented. Afterward, Honeypot systems as an IDS agent will be discussed and explained. In addition, the chapter will be concluded by discussing the applicability of the presented systems to the CPS of the smart cities. Finally, the chapter will be concluded and future research directions will be projected for the readers.

Figure 1. The Provisioning for Information Security Services

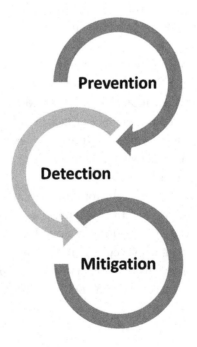

SECURITY VULNERABILITIES IN CYBER-PHYSICAL SYSTEMS OF SMART CITIES

CPS of smart cities may include critical infrastructures such as water treatment facilities, electric power turbines, public transportation systems, smart buildings, etc. Cybersecurity of such infrastructures has prime importance, since the failure of these systems may threaten property or more importantly, human lives.

As shown in *Figure 2*, with the rise of industry 4.0 and adoption of IIoT by industrial facilities and smart factories will constitute a significant component of smart cities. Especially, critical infrastructures of smart cities (nuclear power plant, water treatment facility, etc.) will be a great target for cyber attackers, thieves, blackmailers, ransom hunters and terrorists. These ill-mannered attackers are generally seeking social justice, populism, ransom money, or act of hatred.

Owing to their physical limitations or technological constraints, data among sensors and actuators in CPS of smart cities can be transmitted over networks that were not designed to handle security from the beginning design step (Ding, 2018). Besides, the interconnection of several varieties of devices, nodes, and networks of

Figure 2. An illustration regarding the implementation of IIoT and Industry 4.0 to a smart factory application

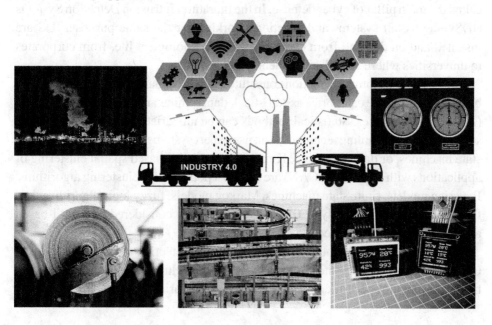

CPSs may cause not only interoperability problems but more importantly security and privacy problems. All of these vulnerabilities may introduce severe security problems and challenges that need to be considered for the CPS of smart cities.

Additionally, as CPS is one of the main soft components of smart cities, it might constitute a single point of failure attracting cyber attackers. Intrusions can always happen in computer systems and CPSs are no different (Mitchell, 2013). Therefore, if an intrusion happens in a CPS, it needs to be detected in a timely manner to prevent further serious damages and losses. Especially in CPS installed critical infrastructures, rapid detection of the intrusions and related counter-action has prime importance, as it may affect millions of lives (Cardenas, 2008). Consider a CPS installed water treatment facility in a smart city. What happens if cyber hackers intrude in those systems and cause the release of an excess amount of chlorine or just simply cause sewage water to be released in an untreated way, which would risk the lives of human and animals using that water (Stamp, 2003). This and many other historical incidences have shown that security of critical infrastructures, especially CPS of smart cities, needs to be taken very seriously.

Intrusion Detection and Its Importance for the Cyber-Physical Systems

Intrusion detection is a very important part of the cyber security systems and can be called the main pillar of cyber defense. In the literature, Intrusion Detection Systems (IDSs) denote all systems and functions working for the same purpose. IDS are installed and on demand from every aspect of technological life, from corporates to universities where the IT department exists.

Conventional anomaly detection constitutes one of the main classes of IDS and can be divided into three categories as statistical, data mining and artificial intelligence (AI)-based techniques. Statistical methods can be univariate, multivariate or time-series based. Data mining techniques can use expert systems, description languages, state machines, or data clustering. K-means and density-based spatial clustering of applications with noise (DBSCAN) are the most popular data clustering algorithms. Bayesian models, finite state machines, Markov models, fuzzy logic algorithms and Principle Component Analysis (PCA) algorithms are considered as the AI-based approaches in anomaly detection.

It is not feasible to adopt conventional anomaly detection techniques for use in CPSs. Due to that, they create big data sets with high volume (due to communication between large numbers of connected objects), high variety (due to collaboratively working heterogeneous sensing environments) and high velocity (due to continuous data collection for monitoring and actuation as needed). Prediction models on large data sets require distributed approaches. Rule-based techniques are fast approaches that scale well in distributed anomaly detection. However, derivation of rules may require additional effort. There are several distributed rule-based anomaly detection approaches that also take advantage of machine learning. Distributed and collaborative anomaly detection techniques via agent-based intelligence provide rapid response to detect and intercept anomalies. However, they require high processing power and memory usage and eventually high-end hardware installations.

A number of IDS categorizations have been proposed by Butun et al.'s work (2014) and in other works such as Stallings (2015) and Pathan (2014). The seven categories of IDS presented by Butun et al. (2014) are as follows: Intruder type, intrusion type, detection methodology, the source of audit data, computing location of the collected anomaly data, infrastructure and finally, usage frequency. In this chapter, especially two of these categories will be concentrated on, which are important and related to the current context; source of audit data and detection methodology.

The reason behind this selection is, that the source of audit data classification provides a very good categorization in terms of network architecture where IDS is planned to be installed. Whereas, the detection methodology classification provides intensive details of detection methodologies, which may be considered as the core of the IDS.

SYSTEM DIAGNOSIS METHODOLOGIES FOR INTRUSION DETECTION

It has been thought that it would be obvious to tell whether intruders have visited a system. Unfortunately, this is not true for most of the time. Intruders who intend to have access to your system continuously, or during a long duration of time, want to hide their activities as much as they can. Agnostic to its type of IDS categorization, there are several system diagnosis methodologies to reveal intruders that are adopted by almost all of the IDSs. The four mostly utilized system diagnosis methodologies for IDSs are as follows:

- **File Integrity Checking:** This is one of the strongest tools of IDSs that can successfully detect unauthorized modification (tampered with) of critical system files (in Windows systems mostly *.dll* and *.bat* files) as well as the data files. In order to do this kind of comparison, the integrity checker needs to store all stable-known versions of the critical system files into a secure folder. If a secure folder does not exist, Read Only Memory (ROM) devices such as CD-ROM or DVD-ROM can be used. Alternatively, an even better practice is to hash all of those necessary files and store the hash values in a protected folder. An alarm is triggered when a mismatch occurs in the process of comparisons.

- **Network Scanning:** These are the programs, which examine critical network systems and services for configuration errors and vulnerabilities. It is just like as a weapon, benefits people when at good hands, whereas harmful when used by the intruders. Network administrators utilize it to reveal the vulnerabilities of their network and to develop the defense based on the result, whereas the hackers use network scanning to recognize the security gaps of a target network. Nmap, Nessus, SAINT, SANTA, and SATAN are some well-recognized network scanners on the market.
- **Network Sniffing:** These tools capture network traffic for the purpose of analysis and intrusion detection. Network sniffing devices leave hackers in a dilemma. As in the case of network scanners, network sniffers are great hacking tools, at the same time they are also awesome intrusion detection tools. If a network sniffer exists on a network, then the network administrator has a better chance at detecting the hacking.
- **Log Analysis:** It is the activity of collecting and analyzing diagnostic status information from the network devices and software. Logging is the most important concept for IDSs and for their recovery process. Without the help of logging, the only way to learn what the problem is to see it while happening or to observe its consequences afterward. There are two analysis methods used for logging: manual and automated.

CLASSIFICATION OF IDS ACCORDING TO THE SOURCE OF AUDIT DATA

Audit data is the data of evidence of network intrusion, such as log records, access records, network routing information, etc. IDSs can be classified into three categories according to their source of audit data; host-based IDS, network-based IDS, and hybrid IDS. For this classification, Stallings and Brown (2015) provided a very detailed explanation and adopted by this text. However, before this, data resources of detection sensors need to be described first:

Data Resources for Detection Sensors

The heart of any IDS consists of a detection sensor which traces and counts unusual events happening on the pre-determined set of triggers. Here, most commonly used data sources for the detection sensors are summarized as follows:

- **Access to the Registry:** Creation of the logs for accessing the registry records in an operating system, Microsoft Windows.
- **Integrity Check of the Files:** Critical files in a system are detected and a check-sum of all these is created. Then, after a long run, deviation from this check-sum indicates an active intrusion into the system.
- **Log File Records:** Logs of the user account activities are recorded in typical operating systems and these can be used for the detection of abnormal behavior to indicate intrusions.
- **System call Traces:** In Unix based operating systems, these call traces are highly informative about the user activities in a system.

Host-Based Intrusion Detection

Basically, host-based intrusion detection methodology works on the hosting computer or system by collecting the evidence right at the source and providing the alerts in case of a triggering event. Pieces of evidence are generally related to that host, such as processes and service identifiers related to applications, along with their specific system and function calls.

Network-Based Intrusion Detection

Network-based intrusion detection methodology works on a special partition of a network that is generally divided by a router or switch. Here, it monitors and analyses the content of the packets flowing through the network and upper layers of the OSI protocol stack[3] in order to detect the suspicious and unusual activities. Network-based intrusion detection can identify more events and activities because of its strategical positioning on the network, therefore should be considered as more complex and high-level methodology compared to the host-based one.

Hybrid Intrusion Detection

Hybrid intrusion detection methodology merges both host-based and network-based IDS methodologies in a unified and centralized manner, to further enhance overall detection probability. For this centralized detection mechanism, three main detection-triggering events exist:

- **Distributed Detection and Inference Events:** In an event of outsider attack (intrusion), this mechanism is triggered in a pre-selected set by the network traffic to notify the security administrators.
- **Policy Enforcement Point Events:** This mechanism is installed on trusted systems such as protected terminals or gateways, to collect and evaluate the pieces of evidence of neighboring device activities. They can be thought of as the watchdogs of the neighborhood, working at an upper level compared to the host-based IDSs.
- **Summary Events:** In enterprise networks, there exist several dedicated machines (such as routers, special IDS nodes) aimed at collecting events from various types of resources to feed the central IDS policy.

CLASSIFICATION OF IDS ACCORDING TO THE DETECTION METHODOLOGY

IDS can be classified into three categories according to their detection methodology; anomaly detection-based, misuse detection-based, and specification detection-based. Sobh (2006) identified the main distinction between the anomaly detection-based and misuse detection-based intrusion detection as follows: "anomaly detection systems try to detect the effect of bad behavior but misuse detection systems try to recognize known bad behavior".

Anomaly Detection-Based Intrusion Detection

Anomaly detection-based intrusion detection helps with both computer and network intrusions along with misuse. This is achieved by monitoring the system activities continuously and sifting them as either normal or abnormal. The sifting (classification) depends on the rules, rather than signatures (or patterns). These systems try to detect any type of behavior that is out of normal system operation. This is in contrast to the signature-based intrusion detection, which can only detect the attacks for which a known signature has been created previously.

Conventional anomaly detection can be divided into three categories: Statistical, data mining and artificial intelligence (AI)-based techniques. *Figure 3* summarizes all anomaly detection methodologies available in the literature, most of which are employed by the well-known IDS systems on the market.

Statistical Detection

Statistical detection techniques are classified into three sub-categories: univariate, multivariate, time-series:

- **Univariate Statistical Detection:** In this detection model, all parameters related to detection are modeled as the independent Gaussian random variables. The detection is sought through the examination of the distribution of the cases on only one parameter at a time.
- **Multivariate Statistical Detection:** In this detection model, two or more metrics are correlated and evaluated to obtain a deeper conclusion regarding the intrusions. The detection is sought through the examination of the distribution of the cases on two or more parameters simultaneously. This provides deeper and enhanced observation of the events, compared to the univariate detection; however, it requires more enhanced data processing ability.
- **Time-Series Statistical Detection:** Time series analysis includes methods for analyzing time series data for the sake of extracting understandable statistics and other metrics of the data presented. In IDS, this detection model is utilized as follows: Event counters record all the incidences of inter-arrival times and happening times of the pre-defined sets of events and these event-related records are further analyzed later on. The result of the analysis includes correlation of the intra- and inter-dependence of the multiple types of events, which reveals significant evidence of the intrusions provided in chronological order.

Detection With Data Mining

Data mining (knowledge) based detection techniques depend on the availability of the prior data of the normal network operation conditions as well as the previously recorded behavior under certain attack vectors. These detection techniques are classified into four sub-categories:

- **Expert Systems:** In general, expert systems are devised to manage complex problems by reasoning among the variety of knowledge, represented mostly as if-then rules rather than through the legacy procedural code. When used as a detection model, audit data is classified by the expert system according to the predefined if-then rules to be solved by the reasoning approach.

Figure 3. Taxonomy of the conventional anomaly detection techniques

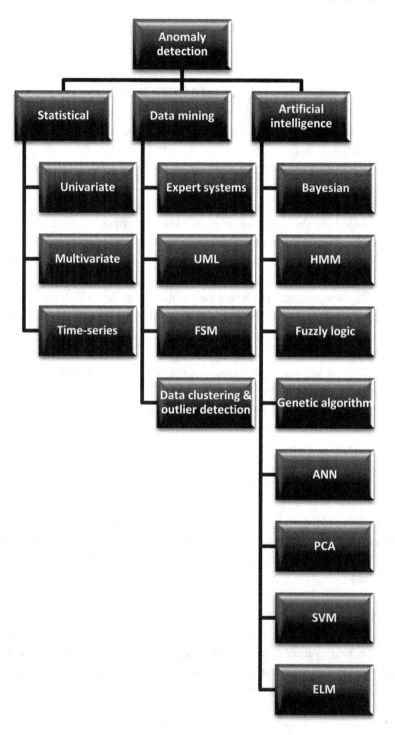

- **Description Languages:** Description languages such as Unified Modeling Language (UML) diagrams can be generated based on some data specifications to help with the detection of intrusions. Data mining over a versioned UML repository would detect change patterns among modeled elements at different abstraction levels.

- **Finite State Machine (FSM):** An abstract machine that can be only in one of a finite number of states at any given time. The FSM can switch from one state to another according to some external stimulants; the change from one state to another is referred to as 'transition'. An FSM is declared by a list of its states, its initial state, and the terms for each transition. When used as a detection model, based on available data sets, states and transitions related to the FSM are generated. Some of the states and state transitions are expected to help with catching the intrusion events.

- **Data Clustering and Outlier Detection:** In order to help with further grouping and classification of attacks, this machine-learning model can be leveraged for intrusion detection. Data clustering is applied to the observed data according to the pre-defined similarities or distance measures. Any event point that is not covered by any cluster is announced as the outlier (hence the intruder).

Detection With Artificial Intelligence

In artificial intelligence based detection techniques, an implicit or explicit model of the analyzed data patterns is generated. In order to improve the intrusion detection performance on the basis of the previous results, these models are periodically updated. Artificial intelligence detection techniques are classified into eight sub-categories:

- **Bayesian Networks:** In *Probability Theory*, *Bayes' Theorem* is described as the probability of an event, based on the prior knowledge of conditions, which would be related to that event. Hence, a Bayesian Network is a probabilistic statistical model to represent a set of variables and their conditional dependencies. When used as an IDS model, probabilistic relationships among the events are determined and modeled as the Bayesian network variables along with their dependencies.

- **Hidden Markov Models (HMM):** When used as an IDS model, *stochastic Markov Theory* is employed to create states that are inter-related with some transition probabilities. The topology of the network, as well as the capabilities of the overall IDS, can be modeled and observed by using the HMM.

- **Fuzzy Logic:** Boolean logic has the truth-values of variables, which are represented by two integers: 0 or 1. In contrast to Boolean, *Fuzzy logic* is a kind of multi-valued logic in which the truth-values of variables can be any real number between 0 and 1. It is employed to handle the concept of partial truth, where the truth-value may range between completely true and completely false. When used as an IDS model, uncertainty and approximation methodologies are used to evaluate the event conditions whether they are intrusions or not.

- **Genetic Algorithms:** A *Genetic algorithm* is a metaheuristic inspired by the natural selection that belongs to the larger class of evolutionary algorithms. Genetic algorithms are commonly employed to obtain high-quality solutions for search and optimization problems while dependent on bio-inspired operators such as the mutation, crossover, and selection. When used as an IDS model, the mentioned biologically inspired evolutionary theory is employed. Chromosomes and their mutation spread over multiple lifetimes are used to model past and current intrusion events to predict the future ones.

- **Artificial Neural Networks (ANN):** Learning abilities of the neural networks in the human brain influenced the mathematicians to use them in solving complex problems. In ANN, some specific data sets are used to construct and train the neural network nodes on single or multiple layers in order to solve a specific task. When used as an IDS model, the mentioned specific task is the detection of intrusion events.

- **Principal Component Analysis (PCA):** In statistics, PCA is a procedure that converts a set of observations of correlated variables into a set of values of linearly uncorrelated variables called the *principal components*, by employing an orthogonal transformation. When used as an IDS model, a dimensionality reduction technique is used to catch the intrusions.

- **Support Vector Machines (SVM):** When used as an IDS model, supervised learning models called *SVM* with associated learning algorithms that analyze network traffic data are used for the classification and regression analysis of the intrusions. For instance, Zhang (2004) has shown that SVM can be effectively used for detecting intrusions. Authors have used SVM for feature ranking and intrusion detection rules generation. Their proposed scheme yielded a high detection rate for some attacks, and also their feature ranking algorithm which has used SVM for IDS was reported to be simple and fast.

- **Extreme Learning Machines (ELM):** When used as an IDS model, feedforward neural networks are used for the classification, regression, clustering, sparse approximation, compression and feature learning of the attack vectors towards a system or a network, with a single layer or multiple layers of hidden nodes, where the parameters of the hidden nodes need not be tuned.

Decision Making on Anomaly Alerts of Anomaly-Based Intrusion Detection

There are four situations according to the decisions made on anomaly alerts (Pathan, 2014):

1. **True Positive:** These are the intrusive activities truly declared as abnormal. In intrusion cases, IDS should catch these hackers by triggering an alert. This is the ideal condition, which the IDS should work at.
2. **True Negative:** These are the normal user activities truly declared as normal. In normal operation conditions, IDS should not trigger an alert.
3. **False Positive:** These are the normal user activities falsely declared as abnormal. It is a bad performance indicator for the IDS.
4. **False Negative:** These are the intrusive activities falsely declared as normal. It is a dangerous situation since the intruders stay undetected.

In ideal conditions, it is desirable to have zero instances of false positives and false negatives. However, in real operation conditions, all of the four decisions occur with nonzero-probabilities as shown in *Figure 4*. Therefore, there is always a trade-off in the anomaly alert settings, related to the anomaly detection algorithm and the parameters it uses. In most IDS, the significant threshold point of the anomaly detection algorithm needs to be determined by using a heuristic approach.

Misuse Detection-Based Intrusion Detection

The profiles of the historically known attacks are transformed into the signatures and used as a matching point to detect future attacks. For brute forcing attack to crack a password, an example signature can be a number of failed login attempts within a short period of time.

Figure 4. Possible outputs of the IDS: Four types of anomaly decisions defined as the probability distributions of the normal and abnormal events

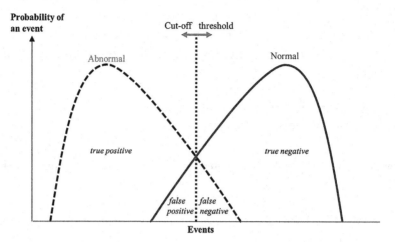

Misuse detection type is advantageous in a sense that it can effectively detect the previously known attacks; therefore, overall average detection has a low false-positive rate. However, this type of detection is un-effective against the new attack types. As stressed in Butun et al.'s work (2014), misuse detection type is similar to the anti-virus systems which are efficient against the pre-known attack types but almost useless against the brand-new attacks.

The most used attack signatures by the misuse detection algorithms are:

- **Interval Rule:** Inter-arrival times of the consecutive messages must be below some certain threshold.
- **Retransmission Rule:** Nodes should cooperatively work to transmit the messages on the route.
- **Integrity Rule:** The integrity of the original message should be verifiable at the receiver.
- **Delay Rule:** If a packet (in general, messages consists of several packets) is delayed a certain amount of time, then it needs to be re-transmitted by the source.
- **Repetition Rule:** Retransmissions are allowed up to some certain number.
- **Radio Transmission Range:** Messages from the outsiders should be detected by comparing the Received Signal Strength Indicator[4] (RSSI) values of the received signals.

- **Jamming Rule:** There should be a maximum threshold to define the normal packet collision rate.

Specification Detection-Based Intrusion Detection

In this intrusion detection methodology, a special blend of constraint and specification set is defined related to an application, program, procedure or protocol. Then later on, in normal operation conditions, deviations from the pre-defined set of specifications are observed and alerts are created in case of a deviating event. As discussed in Butun et al.'s work (2014), this intrusion detection methodology exhibits low false-positive rate while being able to detect the newly introduced attacks.

Specification detection-based intrusion detection combines the advantages of both anomaly and misuse detection-based intrusion detection methodologies, by employing manually devised constraints and specifications to obtain legal system behavior. Specification detection-based intrusion detection is similar to the anomaly detection-based intrusion detection in the sense that both detect attacks through the deviations from normal user behavior. However, as specification detection-based intrusion detection uses manually devised constraints and specifications, it has a lower false-alarm rate. Ultimately, the stressed low false-alarm rate comes with the cost of the time-consuming task of developing the constraints and specifications.

HONEYPOT SYSTEMS

Honeypot systems, or simply honeypots, are decoy systems designed to lure the intruders that they are attacking the real system. However, in reality, they are attacking the decoy system and their behavior is being monitored. Honeypots are also considered as IDS, as they are used for detection of the intrusions towards the computing systems.

In the event of a honeypot being compromised, it will have no major consequences and honeypots can typically be easily restored. It is a common practice to employ virtual machines as honeypots. Here are a few reasons: Several virtual machines can be hosted on one physical system, the virtual systems have short responding time, and finally less amount of code is required for virtual machine implementations, reducing the complexity of the system security.

Honeypots are generally categorized into two categories:

1. **Low-Interaction Honeypots:** They simulate only the services frequently requested by the attackers. Hence, low-interaction honeypots have a low demand for resources. Virtual systems can be efficiently employed for this purpose.

2.   **High-Interaction Honeypots:** They are imitating the functionalities of regular systems that host a bunch of services. Attackers spend and waste their valuable time in exploring these services. As mentioned earlier, multiple honeypots can be hosted on a single physical machine by employing virtual machines. Most frequently, high-interaction honeypots provide more security compared to low-interaction ones by possessing difficulty to be detected. Nevertheless, these benefits come at a price.

DISCUSSION REGARDING THE USAGE OF IDS FOR CPS OF SMART CITIES

Hybrid IDS is the best option for intrusion detection in the CPS of smart cities. The reason is, high priority facilities need extra protection and customized host-based IDS can be used for them. The rest of the systems, i.e. medium or low priority, can be protected through network-based IDS that will be installed at the common gateways.

In a cloud-supported CPS application of a smart city, artificial intelligence based anomaly-detecting IDS methodologies would be efficiently employed to protect especially the critical infrastructures, as these systems quickly adapt and learn new attack patterns on the go. This would protect smart cities from zero-day attacks, which are not known to anybody in the world but devised just at that moment to take down the CPS of smart cities.

Rapid responding is important for critical infrastructures in the CPS of smart cities. Therefore, Zimmer et al.'s proposal (2010) of an enhanced security measure in deeply embedded real-time systems would be a practical solution for the IDS of CPS. The proposed scheme provides elevated security assurance through two levels of instrumentation that enables detection of anomalies, such as timing dilations exceeding *Worst-Case Execution Time* bounds. More formally, the proposed scheme detects the execution of unauthorized instructions in real-time CPS environments.

Intrusions do not always originate from the outside; sometimes it is sourced from the inside. This kind of intrusions is called insider attacks and constitutes one of the most dangerous class. As stressed in Henningsen et al.'s work (2018), insider attacks are very harmful and threatening for the IIoT systems. Hence, the same is valid for the CPS of smart cities. Misbehavior detection comes here as a remedy. Robust misbehavior detection algorithms need to be devised, by using smart computing methodologies, such as the ones presented in this chapter. Artificial intelligence based IDS algorithms might be a good path in developing defenses against the insider attacks in the CPS of smart cities.

Recently, Deng et al. (2017) proposed a distributed intrusion detection algorithm, which utilized the cloud environment by running genetic algorithms to catch intrusions towards CPSs. The inclusion of the genetic algorithm for IDS solutions for CPS of smart cities is promising because Artificial Intelligence is expected to be one of the dominating technologies over the next decade.

Haller and Genge (2017) argued that IDS for Industrial CPSs should take advantage of *Sensitivity Analyses* and *Cross-Association Optimization* methods. By this way, authors claim that the total cost and complexity of IDS for Industrial CPSs can be reduced. As smart cities constitute a giant operational environment in terms of collected data size and number of components to be used from servers to the end elements along with users, any kind of cost change in a tiny element would have a high impact in the bigger picture. Since all these kind of big installations depend on budget, any cost reducing attempt like the one mentioned above would be appreciated by the municipals and the funders of the smart cities.

Graph-based anomaly detection techniques mentioned in Sudrich et al.'s work (2017) can be used as IDS in the CPS of smart cities. However, this requires an analysis of several homogenous and heterogeneous graph models for various use cases in the context of smart cities. Besides, determined abnormalities should be further matched with the ongoing intrusions in order to obtain a solid IDS database for future comparisons. Considering the data sizes to be handled, this is a challenging and non-trivial task.

As CPS of smart cities is mostly considered as critical infrastructure, they need to be protected by all means. Therefore, as a precaution, deployment of *high-interaction honeypots* (please refer to the previous sub-section for further details) would be a brilliant option for IDS. These honeypots can lure intruders and attract the intruding traffic towards them thereby buying some time for network security administrators to identify and mitigate the ongoing risks. An example implementation of high-interaction honeypots for the CPS of smart cities is depicted in *Figure-5*.

CONCLUSION

Intrusions are nightmares to the managers of cyber facilities, network operators, cyber warehouses, and thereby will be the same for the governors of the cyber cities. Hackers will be giving a tough time for the information security crew, the ones that are working for the critical infrastructures, especially for CPS of smart cities. IDS helps security administrators in that manner, detecting the intruders in a timely fashion. However, IDSs are not magical systems that can mitigate the threats

Figure 5. Implementation of honeypots to the cps of smart cities

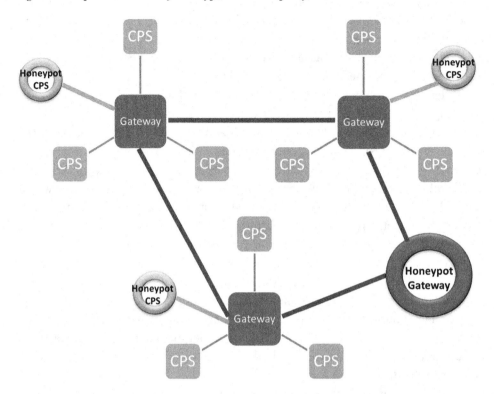

by themselves. They operate like watchdogs, letting the security personnel know if something bad or strange is happening in the system rapidly in the real-time or close to the real-time. Following the detection event of intrusions, it is security administrators' responsibility to mitigate those detected threats by using some other specific tools of mitigation, not the IDS's. Those specific mitigation tools might be; planning of an effective cyber-incident response, applying the patches and updates through patch-management systems in a timely manner (Al, 2012), attesting the application white/black-listing (Field, 2013), and minimizing the administrative privileges, etc.

FUTURE RESEARCH DIRECTIONS

CPS of smart cities is a hot topic for municipalities and governments, as well as researchers and developers. As applications of CPS widely spread, more discussions and topics related to CPS will be revealed. Especially, research topics such as the one presented in this chapter (IDS) will provide a great opportunity for the researchers to delve into. There will be many different applications of the IDS for CPS of smart cities and this will most likely stay a topic of research interest for a while in the near future. By then, smart cities will be hosting possibly several millions of lives and goods worth billions of dollars including infrastructure, property, equipment, etc. Hence, those cities will need to be protected against the cyber-attacks in a timely manner. As a solution to this chaotic problem, robust and rapid IDS algorithms need to be developed and those algorithms must be applied through the agile software development techniques for the CPS of smart cities.

REFERENCES

Al F., Dalloro, L., Ludwig, H., Claus, J., Frohlich, R., & Butun I. (2012). *Networking elements as a patch distribution platform for distributed automation and control domains*. Patent App. PCT/US2012/043,084.

Albanese, J., & Sonnenreich, W. (2004). *Network Security Illustrated*. McGraw-Hill.

Banafa, A. (2016). *The Internet of Everything (IoE)*. Retrieved from https://www.bbvaopenmind.com/en/the-internet-of-everything-ioe/

Butun, I. (2013). *Prevention and Detection of Intrusions in Wireless Sensor Networks* (Ph.D. Dissertation). University of South Florida.

Butun, I., Morgera, S. D., & Sankar, R. (2014). A survey of intrusion detection systems in wireless sensor networks. *IEEE Communications Surveys and Tutorials*, *16*(1), 266–282. doi:10.1109/SURV.2013.050113.00191

Cardenas, A., Amin, S., Sinopoli, B., Giani, A., Perrig, A., & Sastry, S. (2009, July). Challenges for securing cyber physical systems. In *Workshop on future directions in cyber-physical systems security* (*Vol. 5*). Academic Press.

Cardenas, A. A., Amin, S., & Sastry, S. (2008, July). Research Challenges for the Security of Control Systems. HotSec.

Deng, S., Zhou, A. H., Yue, D., Hu, B., & Zhu, L. P. (2017). Distributed intrusion detection based on hybrid gene expression programming and cloud computing in a cyber physical power system. *IET Control Theory & Applications*, *11*(11), 1822–1829. doi:10.1049/iet-cta.2016.1401

Ding, D., Han, Q. L., Xiang, Y., Ge, X., & Zhang, X. M. (2018). A survey on security control and attack detection for industrial cyber-physical systems. *Neurocomputing*, *275*, 1674–1683. doi:10.1016/j.neucom.2017.10.009

Field, S. A. (2013). *Tagging obtained content for white and black listing*. U.S. Patent No. 8,544,086. Washington, DC: U.S. Patent and Trademark Office.

Haller, P., & Genge, B. (2017). Using sensitivity analysis and cross-association for the design of intrusion detection systems in industrial cyber-physical systems. *IEEE Access: Practical Innovations, Open Solutions*, *5*, 9336–9347. doi:10.1109/ACCESS.2017.2703906

Henningsen, S., Dietzel, S., & Scheuermann, B. (2018). Misbehavior Detection in Industrial Wireless Networks: Challenges and Directions. *Mobile Networks and Applications*, 1–7.

Khatoun, R., & Zeadally, S. (2017). Cybersecurity and privacy solutions in smart cities. *IEEE Communications Magazine, 55*(3), 51–59. doi:10.1109/MCOM.2017.1600297CM

Kim, K. D., & Kumar, P. R. (2012). Cyber–physical systems: A perspective at the centennial. *Proceedings of the IEEE, 100*, 1287-1308.

Langner, R. (2011). Stuxnet: Dissecting a cyberwarfare weapon. *IEEE Security and Privacy, 9*(3), 49–51. doi:10.1109/MSP.2011.67

Mitchell, R., & Chen, R. (2013). Effect of intrusion detection and response on reliability of cyber physical systems. *IEEE Transactions on Reliability, 62*(1), 199–210. doi:10.1109/TR.2013.2240891

Mo, Y., Kim, T. H. J., Brancik, K., Dickinson, D., Lee, H., Perrig, A., & Sinopoli, B. (2012). Cyber–physical security of a smart grid infrastructure. *Proceedings of the IEEE, 100*(1), 195–209. doi:10.1109/JPROC.2011.2161428

Nest Inc. (2018). *Nest Learning Thermostat, programs itself, and then pays for itself.* Retrieved from https://nest.com/thermostats/nest-learning-thermostat/overview/

Pasqualetti, F., Dörfler, F., & Bullo, F. (2011, December). Cyber-physical attacks in power networks: Models, fundamental limitations and monitor design. In *Decision and Control and European Control Conference (CDC-ECC), 2011 50th IEEE Conference on* (pp. 2195-2201). IEEE. 10.1109/CDC.2011.6160641

Pathan, A. S. K. (Ed.). (2014). *The state of the art in intrusion prevention and detection.* CRC Press. doi:10.1201/b16390

Poovendran, R. A. D. H. A. (2010). Cyber–physical systems: Close encounters between two parallel worlds. *Proceedings of the IEEE, 98*(8), 1363–1366. doi:10.1109/JPROC.2010.2050377

Sobh, T. S. (2006). Wired and wireless intrusion detection system: Classifications, good characteristics and state-of-the-art. *Elsevier J. Computer Standards & Interfaces, 28*(6), 670–694. doi:10.1016/j.csi.2005.07.002

Sridhar, S., Hahn, A., & Govindarasu, M. (2012). Cyber–physical system security for the electric power grid. *Proceedings of the IEEE, 100*(1), 210–224. doi:10.1109/JPROC.2011.2165269

Stallings, W., & Brown, L. (2015). *Computer security: principles and practice* (4th ed.). Pearson Education.

Stamp, J., Dillinger, J., Young, W., & DePoy, J. (2003). *Common vulnerabilities in critical infrastructure control systems. SAND2003-1772C.* Sandia National Laboratories.

Sudrich, S., Borges, J., & Beigl, M. (2017, August). Graph-based anomaly detection for smart cities: A survey. In *2017 IEEE SmartWorld, Ubiquitous Intelligence & Computing, Advanced & Trusted Computed, Scalable Computing & Communications, Cloud & Big Data Computing, Internet of People and Smart City Innovation (SmartWorld/SCALCOM/UIC/ATC/CBDCom/IOP/SCI).* IEEE.

Ten, C. W., Liu, C. C., & Manimaran, G. (2008). Vulnerability assessment of cybersecurity for SCADA systems. *IEEE Transactions on Power Systems, 23*(4), 1836–1846. doi:10.1109/TPWRS.2008.2002298

Vasserman, E. Y., & Hopper, N. (2013). Vampire attacks: Draining life from wireless ad hoc sensor networks. *IEEE Transactions on Mobile Computing, 12*(2), 318–332. doi:10.1109/TMC.2011.274

Zhang, L. H., Zhang, G. H., Yu, L., Zhang, J., & Bai, Y. C. (2004). Intrusion detection using rough set classification. *Journal of Zhejiang University. Science A, 5*(9), 1076–1086. doi:10.1631/jzus.2004.1076 PMID:15323002

Zhu, Q., & Başar, T. (2011, December). Robust and resilient control design for cyber-physical systems with an application to power systems. In *Decision and Control and European Control Conference (CDC-ECC), 2011 50th IEEE Conference on* (pp. 4066-4071). IEEE. 10.1109/CDC.2011.6161031

Zimmer, C., Bhat, B., Mueller, F., & Mohan, S. (2010, April). Time-based intrusion detection in cyber-physical systems. In *Proceedings of the 1st ACM/IEEE International Conference on Cyber-Physical Systems* (pp. 109-118). ACM. 10.1145/1795194.1795210

ADDITIONAL READING

Butun, I., Morgera, S. D., & Sankar, R. (2014). A survey of intrusion detection systems in wireless sensor networks. *IEEE Communications Surveys and Tutorials, 16*(1), 266–282. doi:10.1109/SURV.2013.050113.00191

Endorf, C., Schultz, E., & Mellander, J. (2004). *Intrusion detection & prevention* (pp. 1–247). Emeryville, CA: McGraw-Hill/Osborne.

Ghorbani, A. A., Lu, W., & Tavallaee, M. (2009). *Network intrusion detection and prevention: concepts and techniques* (Vol. 47). Springer Science & Business Media.

Hu, F. (2013). *Cyber-physical systems: integrated computing and engineering design.* CRC Press. doi:10.1201/b15552

Kumar, V., Srivastava, J., & Lazarevic, A. (Eds.). (2006). *Managing cyber threats: issues, approaches, and challenges* (Vol. 5). Springer Science & Business Media.

Lyon, G. F. (2009). *Nmap network scanning: The official Nmap project guide to network discovery and security scanning.* Insecure.

Pathan, A. S. K. (Ed.). (2014). *The state of the art in intrusion prevention and detection.* CRC press. doi:10.1201/b16390

Scarfone, K., & Mell, P. (2007). Guide to intrusion detection and prevention systems (idps). *NIST special publication, 800*(2007), 94.

Şen, S., & Clark, J. A. (2009). Intrusion detection in mobile ad hoc networks. In *Guide to wireless ad hoc networks* (pp. 427–454). London: Springer. doi:10.1007/978-1-84800-328-6_17

Traoré, I., Awad, A., & Woungang, I. (Eds.). (2017). *Information Security Practices: Emerging Threats and Perspectives.* Springer. doi:10.1007/978-3-319-48947-6

KEY TERMS AND DEFINITIONS

Agent: An agent is a program installed at the host to do the following tasks; event filtering, event aggregation, normalization of the aggregated events, and sending the aggregated results regarding the events to the data analyzer of a centralized intrusion detection program for the further inspection and decision making.

Anomaly: An abnormal behavior of a system or user that is deviating from the normal set of usage. An anomaly can be an indication of a malfunction, an error or more importantly an intrusion.

Denial of Service (DoS): A class of attack in which targeted network or system disconnects and quits from the intended mode of operation. It is one of the most dangerous attacks capable of taking down any type of network that is of any size.

Event: An expected or unexpected happening related to the systems that are in operation. An example of events maybe; arrival of connection commands, the request for the permission to some certain files, the request for escalation in the permissions, etc.

Firewall: A software or hardware that is designed to block the unwanted or unauthorized network traffic between computer networks or hosts. Firewalls are mostly considered as a part of IPS that constitutes the first line of defense in the provisioning of the information security services.

Intrusion: An event where unauthorized users, generally referred to as hackers, gather information or access rights that he/she is normally not allowed to.

Log File: A file that keeps records of events that happen in an operating system or in other software. Logging is the function of keeping a log in a specific place. Following an intrusion event, log files help information security officers to reveal what went wrong and which damages happened during the intruding activity.

Security Risk Assessment: Identifying the vulnerabilities of a system along with the possible worst-case scenarios as well as their probabilities and the evaluation of total property losses in case of such events. This activity generally performed during the establishment of security services for a network or computer system as a part of the provisioning of information security services.

Sensors: These are responsible for collecting evidence regarding the events. It is one of the most critical components of the IDS. The input for a sensor may be any component of a network that could generate useful information regarding intrusions. Some kind of input for a sensor may be composed of the log files, network packets, traces of the system calls, etc. Sensors also send the intrusion-related data to the data analyzer of the IDS[5].

Snort: It is a lightweight and dedicated host-based IDS, which is composed of the four components: packet decoder, the detection engine, logger, alert generator. Snort employs an easy and flexible rule definition language that creates rules used by the intrusion detection engine.

ENDNOTES

1   Security of *smart grid systems* is another topic itself and out the scope of this chapter. More interested readers in the subject may refer to the references provided.

2   *SCADA systems* are one of the advanced Programmable Logic Controller (PLC) systems that are been employed widely by the automation industry.

3   *OSI protocol stack* is a joint ISO and ITU-T standard for computer networks and communications protocols that represents a layered description for them. OSI protocol stack consists of seven conceptual layers.

4   In Telecommunications theory, *Received Signal Strength Indicator (RSSI)* is a measurement of the power existing in a received radio signal.

5   The *sensors* referred to here should not be confused with the ones in Wireless Sensor Networks or with the other physical sensor types. Here, what is meant with *sensor* is a piece of software or program that is performing 'intrusion sensing' activity, that is why the name is given.

Chapter 5
Secure Routing Protocols Using Trust–Based Mechanisms in the Internet of Things for Smart City Environment Challenges and Future Trends

Aminu Bello Usman
York St. John University, UK

Jairo A. Gutierrez
Auckland University of Technology, New Zealand

Abdullahi Baffa Bichi
Bayero University, Nigeria

ABSTRACT

The internet of things (IoT) is expected to influence both architecture and infrastructure of current and future smart cities vision. Thus, the requirement and effectiveness of making cities smarter demands suitable provision of secure and efficient communication networks between IoT networking devices. Trust-based routing protocols play an important role in IoT for secure information exchange and communications between IoT networking elements. Thus, this chapter presents the foundation of trust-based protocols from social science to IoT for secure smart city environments. The chapter outlines and discusses the key ideas, notions, and theories that may help the reader to understand the current status and the possible future trends of trust-based protocols in IoT networks for smart cities. The chapter also discusses the implications, requirements, and future research challenges of trust-based protocols in IoT for smart cities.

DOI: 10.4018/978-1-5225-7189-6.ch005

INTRODUCTION

Smart City is an innovative new technological paradigm that uses Information and Communication Technologies (ICTs), and different types of electronic data collection sensors to supply information, which can be used to manage assets and resources efficiently, provide digital connectivity and improve quality of lives. Recently, significant technological, economic and ecological changes have generated interest in smart cities, including financial restructuring, climate change, environmental monitoring, smart social housing, and many more. Subsequently, in the intricate architecture of the smart city, the application of the Internet of Things is of interest, as it responds to the efficiency, security, and the provision of low-risk technology choices in building smart city environment and solutions, thus realising the so-called Smart environment. For example, some building managers across the world are more frequently looking to incorporate IoT devices and solutions into their infrastructures to reduce costs and improve the quality of their buildings.

The Internet of things (IoT) is the network of physical devices, vehicles, home appliances and other items embedded with electronics, software, sensors, actuators, and connectivity which enables these objects to connect and exchange data (Gubbi, Buyya, Marusic, & Palaniswami, 2013). The Internet of Things (IoT) system connects the physical world into cyberspace via sensors, Radio Frequency Identification (RFID) tags, to provide environmental information, body conditions, etc., in smart cities. Subsequently, the potential applications of IoT in building smart cities is limitless, and the growth of IoT technologies will profoundly accelerate the adoption and the efficient design of smart cities in the coming years.

There are several application areas of IoT for smart cities in the areas of commercial, military and domestic purposes: for example, Smart Appliances, Smart Cars, Wearable Devices, Connected Cars, Sensor Actuator Networks (SANETs), etc. A high-level illustration of an IoT-based smart city can be seen Fig. 1 (Mehmood et al., 2017). From the figure, the applications of IoT in smart cities were grouped into personal and home utilities, smart transportation, etc,. The personal and home applications of IoT include e-healthcare pervasive computing devices and services, which help doctors monitor patients remotely; utilities applications such as smart grid, smart metering/monitoring, water network monitoring, and video-based surveillance. Given the different application domains, efficient routing and secure data forwarding between devices in IoT networks are essential. (Bello, Liu, Bai, & Narayanan, 2015b). The IoT devices and systems are not complicated but designing and building communication between them can be a complex task. Subsequently, building secure and effective communication among the IoT wide variety of devices is essential in designing a smart city environment. Furthermore, the generalizability of the communication mode of the Internet of Things (IoT) follows the Device-to-Device network paradigm: a communication mode that allows two or more devices to communicate among themselves directly instead of through a central wireless access point. Thus, the deployment of IoT systems tend to give rise to spatially heterogeneous device distributions and employs a distributed, multi-hop network architecture in which devices are equally privileged participants in the networking and routing processes. Next section present the brief about the communication technologies and protocols for IoT in smart cities environments.

Communication Technologies and Protocols for IoT in Smart Cities Environments

As described earlier, the realization of IoT-based smart city environment is based on reliable communication protocols to transport data across the network. The communication between IoT devices can operate either by short-range or wide-range communication mode. Some of the short-range communications technologies

Figure 1. An illustration of an IoT-based smart city

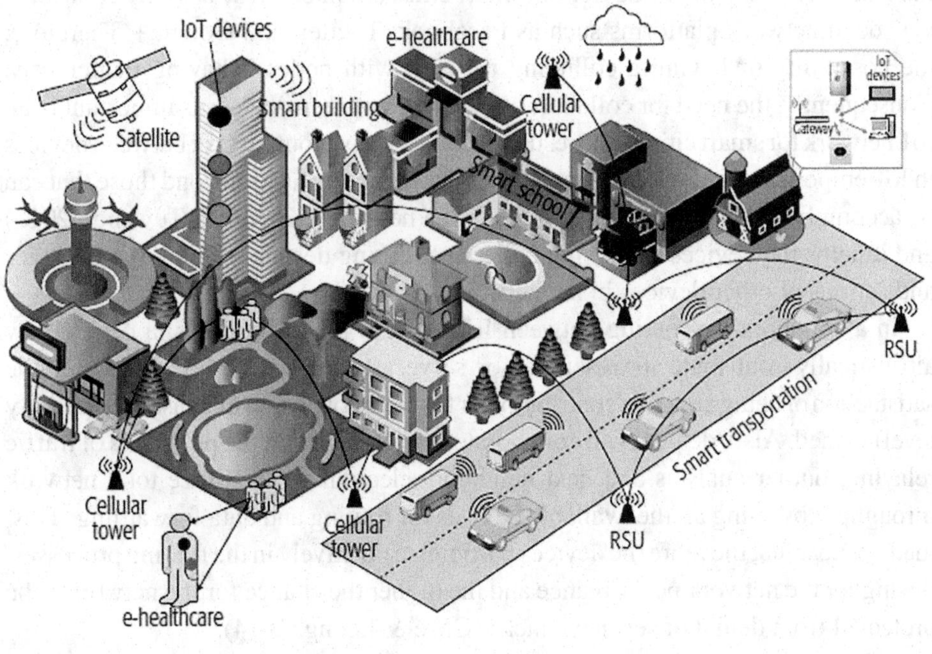

and protocols used for smart metering networks, e-health-care, and vehicular communication includes WiMAX, and IEEE 802.11p, ZigBee, Blue-tooth, Wi-Fi, etc. Other Wide-range communication technologies include Global System for Mobile Communication (GSM) and general packet radio service (GPRS), Long Term Evolution (LTE). Other protocols such as the LoRaWAN protocol, SIGFOX and the Internet Protocol (IPv6) and Low-power Wireless Personal Area Networks (LoWPAN) allows the smallest devices with limited processing ability to transmit information wirelessly using an internet across IoT network environments. However, there have been several discussions in the domain of secure and reliable data forwarding strategies between IoT devices using the above-mentioned technologies and protocols, and what is not yet clear is how the communication strategies between wireless devices can accommodate the rapid advancement and dynamic changes of wireless mobile architecture, protocols and applications. How can a device in the network understand the status of its communication partners (genuine or malicious) (Zhou, Zou, Song, Wang, & Yu, 2016)? Can a device trust any device in the network for data handling? How can the devices in the IoT network perform self-organization processes to provide a secure network paradigm with a high-quality service? Moreover, some of the challenging aspects of protocol design in IoT include the dynamic and distributed nature of devices, secure packet forwarding, network traffic control, selecting a good relaying device for data forwarding (Usman, 2018). Furthermore, users of IoT devices in smart cities are likely to be socially connected via social network platforms such as Facebook, Twitter, and Google+ Therefore, the possibility of having a colluding network with nodes behaving maliciously. Consequently, the need for collaborative routing between devices is important in an IoT network for smart cities. The resultant collaborative routing task between devices in IoT empowers the devices to engage in greater routing tasks beyond those that can be accomplished by individual devices in the network (Kraijak & Tuwanut, 2015) and it helps the devices in making collective routing decisions about the behavior and actions of other devices in the network.

In a collaborative routing between IoT devices for smart cities, a device may altruistically contribute its resources or serve as a suitable relay device for the satisfaction of being an active contributor, or for the recognition (increase in popularity level) gained. Also, devices can collaborate and cooperate in the processes of traffic relaying, outlier analysis and next neighbor selection to maximize total network throughput by using all the available devices for routing and data forwarding. This, made it clear that the more the devices participate positively in the routing processes, the higher the network performance and the higher the chance for the network to be protected from denial of service attacks (Chze & Leong, 2014).

On the other hand, the collaborative routing mechanism, along with its advantages brings some challenges in IoT networks such as information errors and losses caused by the component failure of devices in the network, external interference, wireless transmission error and excessive packet drops which can adversely affect the delivery performance of data communication in the IoT network. Therefore, the success of collaborative routing mechanisms between wireless IoT devices mainly depends on the extent to which the devices can make an efficient routing decision through identifying good relaying devices, non-selfish devices and reliable devices in the network(Yan, Zhang, & Vasilakos, 2014).

To this end, several pieces of literature have proposed that the dynamic, autonomous nature of devices in IoT networks for smart cities makes it difficult for the devices to have a predefined basis for the self-routing decision which can result in different non-uniform request distributions and may lead to poor routing and unbalanced load distribution in the IoT networks. In addition, due to certain inherent IoT devices' dynamic attributes (energy, mobility, connectivity, etc.), a device can succeed in promoting different selfish behavior in the network which can contribute to degrading the quality of communication between the devices (Hui, Sherratt, & Sánchez, 2017).

Therefore, a distributed alternative mechanism is needed to achieve a high level of quality of service while minimizing the resources used in IoT routing process.

In this line, the premise for self-cooperative, altruist behavior monitoring is the trustworthiness of a device, which needs to be established, observed and monitored for the devices' routing behavior control and surveillance . The motivation of the trust system in IoT for smart cities can be perceived as a soft-security solution; a security mechanism that can mitigate misbehaving attacks, discriminatory attacks against the reputation of IoT devices in the networks. Trust systems can increase the probability of network performance and quality of service and provide devices with the best strategy for choosing a best-relaying device for routing collaboration (Lin et al., 2017).

Trust is a relationship between trustee and trustor which can easily be interpreted from the actions of the devices involved (Usman & Gutierrez, 2016). However, the nature and the process of trust and reputation management in collaborative routing networks requires a series of message transfers between the devices in the network; thus, the management of trust and reputation systems needs to take into consideration the requirements of the application, the relationships between the devices (devices' connectivity and location of devices in the network), the devices' attributes, and the devices' specifications for efficient trust-based routing management.

Trust and Reputation Foundation in IoT for Smart Cities

Although trust is an underdeveloped concept in computer science and information technology, one can argue that the concept of trust is as old as the existence of human beings. There are also promising theoretical formulations and empirical studies in different related studies such as (Mayer, Davis, & Schoorman, 1995) that can support the formulation and modelling of trust and reputation in IoT network environments. In this regard, the concept of trust and reputation seems to be becoming an important consideration in the design of autonomous agents in the network.

In the first place, trust has to do with the belief, uncertainty, intention and willingness to cooperate or not to cooperate (Mui, Mohtashemi, & Halberstadt, 2002). These attributes are mainly behavioral characteristics of human beings which cannot be accurately predicted with a high degree of accuracy. Further, the study of network structure, topology and the relationship between devices has been explored in different fields of study such as communication networks robustness of networks (Bello et al., 2015b), and optimal routing decision (Moraru, Leone, Nikoletseas, & Rolim, 2007), and information flow and dissemination in social networks to mention but few.

Every trust and reputation model in IoT has its own specific characteristics and requirements; nonetheless, their pattern and abstract scheme can be generalized as presented in Figure 2. The figure shows the series of processes of a trust model in a distributed IoT system. The first method is responsible for gathering information from other devices in the network for trust evaluation. This includes gathering local information (direct experience) and global information through indirect experience or recommendations. Based on the collected information in the first method, the score and ranking methods are used to score each device along the path to a service-providing device (A. B. Usman & J. Gutierrez, 2018). The result of ranking the trust level of the devices is that this step will return the most trustworthy service provider or the most trust-worthy device in terms of routing handling. The transactions and evaluation method can then be applied for the devices to get the required service from the selected most trustworthy devices (devices with a higher-ranking score). After the transactions, the receiving devices can then evaluate the service received from the service provider and score it accordingly (higher, lower) depending on the trust evaluation method. Finally, the punishment and reward methods perform the function of rewarding good serving devices and punishing the device that is providing a lower quality service. It is however, worth emphasizing that some trust algorithms do not apply the reward and punishment phases in trust evaluation.

Figure 2. Generic trust and reputation model

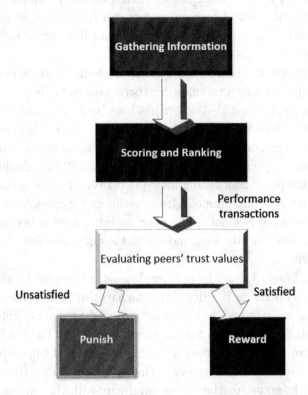

Trust From Social Sciences to IoT for Smart Cities

Like many disciplines, trust-based protocols in IoT have developed their own lexicon, partially inherited from the social sciences fields. For a complete introduction to trust, reputation and cooperation from social trust to digital trust we refer the reader to (Yan & Holtmanns, 2008). However, some trust-theoretic notions are required for our discussion about the original context of trust between agents from the social science perspective.

One possible justification for adopting trust in routing process between devices in modern computing paradigm may originate from the non-trivial dynamic interaction between devices in the network. Indeed, different networks of wireless devices such as IoT, Mobile Adhoc Networks (MANETs), Vehicle Adhoc Networks (VANETs), Delay Tolerant Networks (DTNs) and Robot Networks, are not necessarily static networks. The networks can grow or decay, and the networks' topologies evolve with the possibility of hubs emerging or declining, and communities or clique's formation. It has, therefore, become apparent that it is important to consider the collective dynamic behavior of network devices for trust-based self-organization routing processes.

Worthy of notice, the idea of trust between agents in the network has to be understood regarding the agent relationships between devices. In most social science research, the concept of social structure has long been agreed to be the source of social capital which plays a vital role in sustaining trust relationships between individuals and organizations to create values, trust and transfer of knowledge, and enhance creativity (Tavakolifard & Almeroth, 2012). In this line, there have been several attempts to address the trade-off between closed structure and open structure in which structure promotes trust, norms and cooperation between social elements (Handfield & Bechtel, 2002). The analysis of these two social network group terms encompasses theories, models and assumptions that are based on the relational concept.

For example, Figure 3, presents an example of open structure. From the figure, an actor q occupies a brokerage position and it can have three distinct benefits: Control Benefits (having access to control other devices in the network), Information and Referral Benefits (it can serve as a good recommender) and Uniting Benefit which are all significant factors that promote inter-organizational trust. There are some theories from the social science perspective including the work of Burt (Handfield & Bechtel, 2002) who argued that having a central actor with a higher communicability index in the network can facilitate the transfer of information and provide reachability benefits in the network of devices. On this line, there are some non-distributed trust routing algorithms that follow this idea of centralised reputation systems such as those in (Thenmozhi & Somasundaram, 2016) and (Tajeddine, Kayssi, Chehab, Elhajj, & Itani, 2015). In these types of trust model, the central device (usually the most important device in the network) will evaluate the devices trust behaviour and perform all the tasks of trust aggregation and trust recommendation in the network. Although, the concept of a centralised trust management works well in a simple IoT network where a single central device is needed for trust and reputation evaluation, it is less suitable in a distributed IoT network and it inherits number of problems such as a single point failure (when a central device is compromised or fails). In the next sub-section, we study the closed structure and its related properties for efficient trust evaluation.

The concept of closed structure was first introduced by (Coleman, 2000) based on the observation that in a closed structure, there is a high degree for every device in the network knowing every other device directly or indirectly, and, as such, the devices are more likely to trust each other and predict other devices' behaviour than in an open structure. Extending the theoretical work of (Coleman, 2000) is the finding of Simmel (Simmel, 2011) who argued that the triad is the fundamental unit of social analysis, and hypothesized that the strength or quality of relationships and trust are not based on the content of the relationship, but rather on the cohesive structure of the relationship. Simmel articulated several reasons and features of triad relationships and concluded that while a triad is the smallest form of group, increasing group size does not significantly alter its critical features.

Figure 4 illustrates a closed structure, a form of distributed trust management. The study in (Engle, 2001) presented Simmel's work regarding Similien ties (closed structure) and structural holes (a form of open structure) and explored their roles in terms of task interdependence and effectiveness, the study showing that Simmelian ties promote development of group norms such as norms of cooperation and reciprocity, and also promote trust, cohesion and transitivity which are all positively correlated to interdependency and redundancy. Therefore, to lay the foundation of our concept of integrating device routing attributes as trust evaluation factors, we seek to discuss the two important properties of closed structures: transitivity and reciprocity.

Figure 3. Simple illustration of centralised trust management in IoT

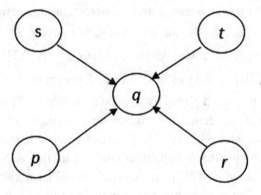

Figure 4. Simple illustration of distributed trust management in IoT

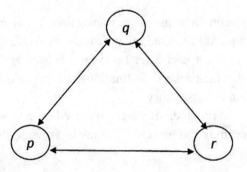

Trust-Based Routing Protocols in IoT for Smart Cities

Before the implementation of trust-based routing protocols in IoT, the job of the IoT devices in routing process was to forward packets from the sources to the destinations. Trust-based routing protocol encourages that, in addition to the forwarding the packets from the sources to the destination, intelligent forwarding decisions can be incorporated to increase routing performance and security considerations. A trust-based routing protocol in IoT is, therefore, a routing scheme in which the IoT devices in the network can integrate their opinions about the behaviour, reliability and the trustworthiness of their neighbours or any device in the network. The devices' opinion is aggregated and quantified as a trust metric or aggregated trust value.

The trust-based forwarding strategy between wireless devices in IoT can be either end-to-end or hop-by-hop. In end-to-end trust-based routing forwarding, the set of forwarding devices can be determined on the fly on a packet basis using certain criteria (Airehrour, Gutierrez, & Ray, 2017). With hop-by-hop forwarding strategy, each forwarding device (including the packet sources) determine its own forwarder set based on the trust level of the devices. A selected forwarding device will again do this process until the messages reach the destination.

For example, when two devices p and q interact, they can establish a local trust between themselves. Let $t_{p,q}$ be the trust value that device p places in device q based on its prior experience with device q, where $t_{p,q} \in [0,1]$: $p \neq q$. The local trust value $t_{p,q}$ of different devices who once had an experience with device q can then be aggregated to build a global trust value of device q. The global trust value of a device is a devices' credentials that it can use to interact with other devices who have never interacted with it before. The devices can also use the global trust value of a device to rate and predict its behaviour (Bello, Liu, Bai, & Narayanan, 2015a).

As shown in (Usman, 2018), to compute the local trust values, the peers can access the success of their previous transactions from the details in their subjects' delivery vectors. When two peers p and q interact, they can establish a local trust between themselves based on the following three conditions:

1. If the status of the previous messages transactions between trustor and trustee is 0 (meaning a packet received by a trustee is no longer in trustees' buffer), and there was a recent encounter between a trustee and other peers in the network (probably the packets' destination); we can assume that the outcome of the transaction is satisfactory.
2. If the status of the previous messages received by a trustee shows it is still in its buffer (meaning the messages' status is 1) while the trustee is not the

destination of the message and there was no recent encounter between trustee and other peers in the network, then the outcome of the transaction is considered as a neutral or undecided.

3.   If the status of the previous messages received by a trustee shows it is still in its buffer, while there is an encounter history of a trustee meeting the messages' destination the outcome of the transaction is not satisfactory.

Subsequently, each time peer p interact with a peer q, the satisfactory level of interactions between the devices can increases or decreases based on the routing behaviours of devices, i.e. $X_{p,q}$ if satisfactory and $X'_{p,q}$ otherwise. Therefore, the satisfactory level of interactions between device p and q can be computed as $S_{p,q}$ $= \left| X_{p,q} - X'_{p,q} \right|$. Subsequently, the normalized local trust value between p, q can be computed using the following equation (1).

$$t_{p,q} = \frac{\max\left(S_{p,q}, 0\right)}{\sum_q \max\left(S_{p,q}, 0\right)}, \left|\vec{t}_p\right| : \sum_{q=1}^{N} t_{p,q} = 1 \tag{1}$$

Subsequently, the global trust between the devices can be aggregated based on the notion of transitive trust as shown in equation (2).

$$T_{p,q} = \sum_q t_{p,q} * t_{q,r} \tag{2}$$

Therefore, each peer will maintain the local trust observation vectors of its subjects' trust values as follows:

$$\vec{t}_p = \left(t_{p,q}, \ldots, t_{p,N}\right)^T, 0 \le t_{p,q} \le 1 \tag{3}$$

Note: the local trust value ($t_{p,q}$) in equation (1) represents the normalised local trust value device p has about device q and other devices in the network. $T_{p,q}$ in equation (2) is a global (transitive trust) of q computed by p based on trust that p has about q. Therefore, every peer can use his global observation vector's elements ($\vec{t}_p$) presented in equation (3) to compute the global trust values of devices in the network.

In related development, the work in (A. B. Usman & J. J. A. o. O. R. Gutierrez, 2018) suggests that, in a trust-based routing protocol for IoT, it will be additional trust evaluation reliability if peers can include the routing attributes of their subjects. The routing attributes can be devices energy, buffer etc.

Given a set of peers' attributes A, the function a: V $\rightarrow$ $R_+^{\{|A|\}}$ assigns to each peer a list of its attributes values A = {a1 = v1, a2 = v2, . . ., ai = vi}. Each attribute ai $\in$ A, has a value vi: vi $\in$ R_+. The tuple of the device q attributes can be represented as (a_q) Where a $\in$ A represent the attribute value of peer q. If the maximum possible attribute value of device q is represented as $\tau \max_{(a_q)}$ and the most recent attributes' value as τcurr (a_q). We can use the following equation (4) to compute the value of devices attributes for trust evaluation.

$$a_q = \frac{\tau \max_{(a_q)}}{\tau \max_{(a_q)} - \tau curr\left(a_q\right)} \tag{4}$$

From the resultant global trust algorithm presented in Eq. (2) and the attributes' model Eq. (4), the global-reliable trust value can be computed using equation (5).

$$T_{p,q} = a_q * \sum_q t_{p,q} * t_{q,r} \tag{5}$$

where a_q represents the attributes of peer q computed by peer p and $T_{p,q}$ represents the reliability transitive trust of peer q computed by peer p.

Different trust-based routing protocols which address quality of service degradation, selfish and non-cooperative routing behaviors, and end-to-end delay aspects have been proposed in the literature. Some of the serious discussions and analysis about trust-based routing protocols are based on game theory (Yan & Holtmanns, 2008), fuzzy logic (Yan & Holtmanns, 2008) or Bayesian networks (Daniel, Zapata-Rivera, & McCalla, 2007). Some of the possible reasons why the above-mentioned techniques are commonly used in trust-based modelling include: (i) both game theory and fuzzy logic can be used for modelling cooperative behavior in complex networks (ii) they can be used for learning the dynamic pattern of devices' states and interactions for proper predictions (Usman & Gutierrez, 2018b).

For example, (Dai, Jia, & Qin, 2009) employed the uses of fuzzy inference rules to improve the routing protocols with fuzzy dynamic programming in MANETs. Though the concept was highly referenced in the literature, the model did not consider the dynamic nature and behavior of D2D due to the problems of multipara meter optimization limitations and the computational complexity of fuzzy systems.

Additionally, there have been some longitudinal studies that proposed trust-based routing using different techniques. For example, the study in (Zhan, Shi, & Deng, 2012) proposed a Trust-Aware Routing Framework for WSNs (TARF). TARF provides a trustworthy and energy-efficient route between a sender and a receiving node. Also, the work of (Sarkar & Datta, 2012) proposed a trust-based protocol for energy-efficient routing decision; the proposed protocol reduced delay and routing overheads; however, the routing decision in the model only considers the selection of the shortest link to avoid the depletion of energy.

Furthermore, in the proposed trust model of (Zahariadis, Leligou, Trakadas, & Voliotis, 2010), when a misbehaving device is detected, the neighboring devices can isolate the misbehaving device for data forwarding or any other cooperative function. Nevertheless, the model only considers counting the systematic failure of a device as a method of learning the capability of the device, not the attributes. In this type of model, the learning parameters can be faulty due to the dynamic change of device behavior (good or bad). The work of (Sen, 2010) proposed a reputation system-based solution for trust aware routing, which implements a new monitoring strategy called an efficient monitoring procedure. The model considers the reputation value as a factor in the routing decision that may not explicitly reveal the devices capacity for handling the routing decision. Recently, due to the rapid development of complex and autonomous D2D networks, the concepts of trust, reputation and cooperation continue to receive an enormous attention in the field of D2D networking. To sum it all up, several trust-aware models based on energy awareness, location awareness, and delay tolerance will continue to emerge. Considerably more work is needed for further in-depth analysis to advance this research area (trust-based routing protocols).

Trust-Based Protocols Based on Direct Neighbors' Experience

Many of the ad-hoc network trust models are naively based on a trust-your neighbor relation. In this type of trust model, the entire trust management system (origination, managing and expiration) usually has a short lifespan, and the devices may lack a comprehensive knowledge of the overall neighbors' trust level. As a result, most of the direct trust models only work in an environment where all the nodes are self-organized and mobile (e.g., military and law enforcement applications) which limits their functionalities to some specific areas.

Consider the diagram in Figure 5, a direct trust relationship between device p and s can be derived since there exist a direct contact between the devices. Likewise, the direct trust between r and q , q and s or between u and t is obtainable through a direct trust aggregation method. In such a protocol, the trust management system (from origination to expiration) usually has a short lifespan, and the devices may lack a comprehensive knowledge of the overall devices' trust level in the network.

Recently, several attempts were made by many authors to propose a different improvement in the various aspects of direct trust and reputation algorithms. For example, the study in (Ho, Wright, & Das, 2012) introduced a zone-based trust management agreement scheme in wireless sensor networks. The scheme was designed to detect and revoke groups of compromised nodes within the trust formation and forwarding phase. Each node directly interacts with the neighboring nodes for the trust report event and stores the report in a knowledge cache. The proposed protocol comprises of a zone discovery phase, trust formation and forwarding phases. Before making a final judgement, a trustor will always compute the difference between the probability distribution function of the neighborhood trust and the probability distribution function of the information received from its neighbors at every slot of time $\left(say, T\right)$. The total trust factor can be determined based on the deviation between the reports of the observation using the information theoretic metric KL-divergence.

Figure 5. Direct and indirect trust management in IoT

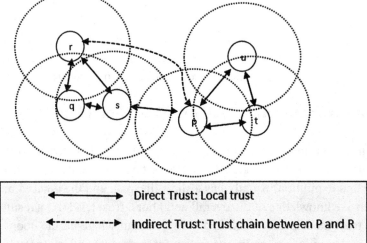

Also, the work in (Srinivasan, Li, & Wu, 2008) proposed a novel, Connected Dominating Set (CDS) based reputation monitoring system which employs a CDS-based monitoring backbone to securely aggregate the reputation of sensors without subjecting them to energy depletion or reputation pollution attacks. In addition, apart from constraints that are application-specific, the concept of direct trust suffers from the following setbacks that may limit its application in a distributed and autonomous wireless network: a) Notion of prediction: device p can either trust device q or distrust device q since it has no other means of trusting device q ; b) device p can only compute device q 's trust value under the condition that device p trusts device q ; C) Energy depletion problem: the amount of energy needed for a wireless node to accomplish trust management processes (trust aggregation and trust evaluation) with all other neighboring devices in a distributed network will be high, since the trust between devices can only be based on their direct contacts and the energy needed for the node to communicate with other devices is proportional to its distances to other devices in the network.

Referral in Trust-Based Protocols in IoT for Smart Cities

As mentioned in the above section, trust evaluation in IoT for smart cities based on direct experience between IoT devices is not often obtainable to the assessors, and sometimes having a direct experience between all IoT devices in large-scale Smart Cities networks is non-trivial. Therefore, aggregating trust opinions of other devices on a particular agent remains a fundamental building block of successful trust and reputation systems. One common issue in trust aggregation based on recommendations is the reliance on device interactions to evolve the global trust value of an agent. Inherently, this creates an undesirable dependency, where the device interaction and connectivity plays a significant role in the accuracy of the resultant trust score(Kamvar, Schlosser, & Garcia-Molina, 2003).

Further, because the process of reputation and recommendation aggregation involves a series of message transfers across the chain of devices, the management of the recommendation may prompt many challenges. Also, in the recommendation processes, knowing who made the recommendation and the time at which recommenders make a recommendation is essential for determining the accuracy of the recommendations' value and to minimize the problem of trust decay respectively. Thus, modelling the recommendation or referral between devices in the network might not be completed without visiting some related concepts of device interactions and connectivity for proffer trust evaluations. In the next subsections, we present some related surveys that specifically focus on trust and reputation models that are of great interest to this chapter.

Summary of Related Surveys in Trust-Based Protocols

There are various comprehensive survey articles on trust and reputation management that focus on D2D networks. Here, we highlight some of the surveys that are related to our work. For example, the survey in (A. B. Usman & J. Gutierrez, 2018) provide a comprehensive augments the trust concept and definition from various field of studies and proposed models in the literature and provide general conceptual phases and methods of trust management toward-trust-based protocols, in the context of Pervasive and Mobile Computing. The survey addresses a broad range of techniques, methods, models, applications and desired futures of trust-based protocols.

The survey in (Wang & Vassileva, 2007) reported a systematic review of various trust and reputation systems for D2D-based web services and the survey classified different trust and reputation systems. The survey concluded that collaborative filtering would be better for the implementation of reputation in recommender systems technology.

Also, the work of (Hoffman, Zage, & Nita-Rotaru, 2009) provides a survey on the different frameworks for general decomposition of existing reputation systems. The survey classifies different attacks by identifying which reputation system components and design choices are vulnerable to attacks. The survey also explored different defense mechanisms against those attacks for the future reputation system.

Subsequently, the survey in (Yu, Shen, Miao, Leung, & Niyato, 2010) focuses on computational trust models for wireless communication networks (WSNs) and cognitive radio networks (CRNs). The survey reported two major categories of trust and reputation: individual-level trust and system-level trust. Based on the survey, the majority of TRM models focus mostly on the detailed aspects of individual-level trust due to the following reasons: i) the individual trust mechanism facilitates first-hand interaction and experiences between the devices, ii) aggregates testimonies from witness nodes about potential interaction partners and evaluates trustworthiness of potential interaction partners based on the available past experience, while some trust models consider system interaction protocols to enforce cooperation among the nodes in a network.

(Cho, Swami, & Chen, 2011) presented a survey of trust management for mobile ad hoc networks in distributed environments. The survey reported different potential attacks and performance metrics of trust and reputation models based on composite cognitive, social information and communication taking into account severe resource constraints of mobile ad hoc networks such as computing power, bandwidth, quality of service, node mobility, topology changes and propagation channel conditions. Furthermore, the author concluded that exploring the social relationship in evaluating trust and collaboration through employing the concept of a social network is a potential fruitful research in trust and reputation for mobile ad hoc networks.

Also, the work in (Govindan & Mohapatra, 2012) presented a survey on trust management in mobile ad hoc networks based on the subjective nature of belief for behavior prediction. The survey explored different techniques for trust value calculation and trust propagation, and the concept of trust value can be used to promote confidence between the devices in the network. The survey also presents a summary of different related work on trust metrics, trust dynamics and how trust can change with time in the processes of trust propagation, trust aggregation and trust prediction. The study further reports different methods for trust measurements including fuzzy probability and trust scales.

Also, the work of (Bhuiyan, 2013) presented a general survey on the relationship between interest similarity and trust based on collecting information about the users' view regarding the relationship between trust and interest similarity. The survey reported that there is a positive relationship between trust and interest similarity which means that, the more two people are similar, the greater the trust relationships between them.

Implications of Trust-Based Routing Protocols in IoT for Smart Cities

Trust-based routing protocols in IoT for smart cities are an important routing scheme for efficient collaborative routing between smart devices in the network (Guo, 2018) (Bao & Chen, 2012). Thus, it is possible to hypothesize that collaborative routing schemes in IoT for smart cities may not efficiently work if the devices are not able to determine the corresponding trustworthy routing partners for data forwarding and collaborative tasks in a large IoT network. Therefore, ascribable to the considerable damages and adverse effects of the untrustworthy or unreliable device in the routing function, which by extension can affect the quality and reliability of data and routing protocols, analyzing the trust level of a device has a positive influence on the confidence with which a device conducts transactions in the networks (Chen, Bao, & Guo, 2016). A considerable amount of literature has been published on the importance and motivations of a trust-based routing protocol in collaborative routing schemes, see for example (A. B. Usman & J. Gutierrez, 2018) (Sicari, Rizzardi, Grieco, & Coen-Porisini, 2015), (Lize, Jingpei, & Bin, 2014). The following is a summary of the importance of trust-based routing protocols for IoT networks in a smart city environment.

1.   IoT Requirements for Smart Cities

A future Smart City environment is expected to consist of a huge number of autonomous IoT devices capable of providing services upon request. Subsequently, IoT-based solutions for smart cities are expected to have low cost, wider coverage, low energy consumption, high quality of service (QoS), high security and privacy, ultra-dense deployments, and multivendor interoperability. To fulfill the above requirements, several new techniques, algorithms and solutions are expected to be adopted to improve the cooperative communications and security considerations between IoT devices in smart city environment. Trust-based routing protocols and algorithms have been proved to play essential roles in cooperative routing between IoT devices, network traffic modeling for handling massive IoT traffic and mitigate security issues at a routing layer (Liu et al., 2018), (Mehmood et al., 2017).

2.   Detecting Selfish and Dynamic Behavior of Devices in the Network

The next generation IoT network for smart cities is expected to utilize a wide variety of sensing devices with significant sensing capabilities. However, it is possible for sensing devices to face challenges in the process of data transmission due to the devices dynamic routing behaviors. Data packets can be dropped by selfish malicious devices, causing serious damage to the network. Conventional security mechanisms cannot address the problem of selfish and dynamic changes of device behavior from honest to malicious and vice-versa. However, the concept of trust-based protocols have proved to be useful for different dynamic behavior analyses and providing a mechanism for detecting selfish behavior and adaptive dynamic decision processes (Govindan & Mohapatra, 2012).

3.   Generalize and Modular Solution Approach and Collusion Detection

To evaluate a device's cooperativeness/selfishness behavior in a heterogeneous IoT environment, global reports from other devices from IoT multi-hop networking elements may be required to strengthen the judgment made using local observations. For example a group of devices might collude to falsely accuse/praise a particular node to gain communication benefits at the cost of other nodes' resources. Also, evaluations of devices' behavior can be achieved either through local or global network observations by networking devices. Although, other forms of routing protocols use local observation/monitoring techniques such as overhearing, packet acknowledgment to prove the routing behavior of networking devices. However, trust-based protocols have been proved to combine both local and global observations for network monitoring. In addition the trust-based system provides a mechanism to the devices to fight against any attacks collectively in deciding about device behavior in the processes of rating a device, punishing a misbehaving device or rewarding a behaving device (Samian & Seah, 2017),(Pathan, 2016).

4.   Less Complexity

It is obvious that conventional security solutions require lots of energy and processing ability for the processes of key generation, key distribution, key management and both encryption and decryption tasks which are usually complex and computationally expensive and require significant memory and processing overheads. However, the use of trust and reputation do not necessarily involve such computational techniques as in the case of public key infrastructure (Chang & Chen, 2012) .

Requirements and Challenges of Trust-Based Routing Protocols in IoT for Smart Cities

As presented in the above section, the trust-based protocol is a promising solution to the problem of collaborative routing schemes for IoT. Difficulties arise however when an attempt is made to propose a trust-based routing scheme that can satisfy the desired properties of trust-based routing protocols. Summarized below are some of the desired properties of trust-based routing schemes in IoT.

1.   Scalability, Portability, and Anonymity

The process of trust management in IoT for smart city environments should be able to suit a large range of networks with a high level of accuracy and security. The devices should be able to join and leave the network securely at any time. The trust management process between the devices should be anonymous.

2.   Self-Organization and Decentralization

The IoT devices in smart cities should be able to organize themselves for trust-based decision making process. In other words, the trust establishment processes should be highly decentralized, so that the trust management can be implemented in highly dynamic and distributed network like wireless sensor networks.

3.   Integrated Confidence and Measure Credibility

The inaccuracy and uncertainty of many trust models increase with incomplete rating and recommendation information available in the network, thus it helps to add confidence measure that will minimize the trust rating sparsity problem.

4.   Behavior Prediction

A good trust-based protocols for IoT should incorporate the learning methods of the devices' previous behavior patterns for the prediction of device behavior. Relationship, location, similarities, social influence and behavioral attributes of the devices: A good trust and reputation model should take into account the relationship, behavior and attributes between the devices such as similarity preference, and friendship preference.

Research Challenges for Trust-Based Protocols in IoT for Smart Cities

Different proposed trust-based routing schemes for IoT have been vigorously challenged and contested by related studies. The purpose of presenting these challenges is to give research directions to new researchers in this domain.

1.   Due to the time and dynamical environment constraint of a smart city environment, building a trusted network of IoT for smart cities might be especially challenging with the presence of cheater devices that provide dishonest information, thus, a dynamic trust metric that will take into consideration the dynamic, context-aware and complex nature of distributed networks of IoT systems is essentially needed.
2.   Further, there is a need for a comprehensive risk assessment of adopting trust-based mechanisms in IoT devices for smart cities. There is also the need to explore how the trust mechanism can mitigate newly emerging smart cities network attacks and challenges based on vulnerabilities and threats analysis.
3.   Also there is a lack of standard orientation and generalized trust-based IoT framework and models from the view of service oriented architecture that is applicable to distributed, decentralized, and hybrid IoT for secure smart cities environments.

4. It will be interesting to have an integrated secure framework for trust-based protocols for IoT and cloud service to create an IoT service-community cloud utility for smart cities environment.

5. Additionally, in many distributed implementations of trust-based protocols for IoT there are some devices in the network that are known to be trustworthy (pre-trusted devices) in the initial stage of protocols implementations. The free, trusted devices (devices that are given initial trust values) are randomly selected with no prior experience of their reliability for data forwarding and collaborative routing tasks. In the event where the selected pre-trusted devices are non-capable devices, or they have limited resources, the network performance can be affected.

6. Also, while the distributed trust-based routing algorithms were purposely proposed to overcome the challenges of a) the computing and storing of the global trust vector, and b) the problem of central computation storage and message overhead, many trust and reputation models still face some challenges concerning exploring device reliability and capability for routing decisions.

7. Also, queries in many trust-based routings schemes are flooded throughout the network, with each device responding to a query if it possesses the file. The devices can then choose a corresponding device to communicate based on their reputation value alone; however, the devices cannot instantaneously consider the dynamic changes and intermittent connectivity between the devices and device resources which are all important factors for improving the quality of communication between the devices in the network.

8. Trust matrix sparsity arises from the phenomenon that the devices have no enough information of rating other device trust behavior or when the devices only rate or give a recommendation for the few devices in the network. This is very common in the implementation of trust and reputation algorithms in a sparse D2D network. This problem can affect the reputation integrity due to the limited information to judge and predict device trust behavior.

9. Most currently existing trust-based models for IoT have a principal drawback of not being able to detect non-cooperative (selfish) behavior. As a result, many attacks such as Sybil attack, DoS or colluding attacks, bad recommendation attacks, bad mouthing attacks, ballot stuffing attacks and on-off attacks are likely to succeed when using trust and reputation models. Therefore, simple, lightweight, and efficient security solutions should be designed to ensure data authenticity and integrity, and to provide secure communication between IoT devices.

10. The ability of an IoT device to predict another IoT device future behavior and expectancy is essential in trust-based models and algorithms in IoT for smart cities. However, this is difficult to be achieved without formulating an expectancy of the behavior and the dynamic network environment. Most of the existing trust-based protocols models do not take such aspects into consideration which limits their ability to promote self-organization principles in decision-making processes.

CONCLUSION

The Internet of Things brings Smart cities to life with seamless connectivity that improves technological, economic and environmental growth. The IoT is expected to play a vital role in the realization of Smart Cities characterized by heterogeneous network devices and communication protocols. Subsequently, the requirement and effectiveness of making cities smarter demands suitable provision of secure and efficient communication networks between IoT networking devices. This chapter collects the combined findings and proposed models for trust-based routing protocols and algorithms in a style that focuses on principles that are likely to be valuable for the design of future trust-based routing protocols for secure IoT in smart cities environments. Initially, the chapter presents the systematic foundation of trust-based models and algorithms toward trust-based protocols for the Internet of Things' routing protocols and gives an overview of the foundation of trust and reputation in IoT for smart cities; communication protocols and technologies that support trust-based routing protocols in IoT networks. The chapter also presents some interesting theoretical foundations of trust from social science to IoT networks and subsequently outlines the key ideas, notions, and theories of building trust-based protocols for IoT in smart cities environment. Finally, the chapter highlighted some requirements and challenges; implications of trust-based routing protocols in IoT for smart cities and open research problems.

REFERENCES

Airehrour, D., Gutierrez, J., & Ray, S. K. (2017). A trust-aware RPL routing protocol to detect blackhole and selective forwarding attacks. *Australian Journal of Telecommunications and the Digital Economy*, 5(1), 50. doi:10.18080/ajtde.v5n1.88

Bao, F., & Chen, I.-R. (2012). Dynamic trust management for internet of things applications. *Proceedings of the 2012 international workshop on Self-aware internet of things*. 10.1145/2378023.2378025

Bello, A., Liu, W., Bai, Q., & Narayanan, A. (2015a). *Exploring the Role of Structural Similarity in Securing Smart Metering Infrastructure*. Paper presented at the Data Science and Data Intensive Systems (DSDIS), 2015 IEEE International Conference on. 10.1109/DSDIS.2015.95

Bello, A., Liu, W., Bai, Q., & Narayanan, A. (2015b). Revealing the Role of Topological Transitivity in Efficient Trust and Reputation System in Smart Metering Network. *Proceedings of the 2015 IEEE International Conference on Data Science and Data Intensive Systems (DSDIS)*. 10.1109/DSDIS.2015.114

Bhuiyan, T. (2013). *Online Survey on Trust and Interest Similarity. In Trust for Intelligent Recommendation* (pp. 53–61). Springer. doi:10.1007/978-1-4614-6895-0_4

Chang, K.-D., & Chen, J.-L. (2012). A survey of trust management in WSNs, internet of things and future internet. *Transactions on Internet and Information Systems (Seoul)*, 6(1).

Chen, R., Bao, F., & Guo, J. (2016). Trust-based service management for social internet of things systems. *IEEE Transactions on Dependable and Secure Computing*, 13(6), 684–696. doi:10.1109/TDSC.2015.2420552

Cho, J.-H., Swami, A., & Chen, R. (2011). A survey on trust management for mobile ad hoc networks. *IEEE Communications Surveys and Tutorials*, 13(4), 562–583. doi:10.1109/SURV.2011.092110.00088

Chze, P. L. R., & Leong, K. S. (2014). *A secure multi-hop routing for IoT communication*. Paper presented at the Internet of Things (WF-IoT), 2014 IEEE World Forum on. 10.1109/WF-IoT.2014.6803204

Coleman, J. S. (2000). *Social capital in the creation of human capital. In Knowledge and social capital* (pp. 17–41). Elsevier. doi:10.1016/B978-0-7506-7222-1.50005-2

Dai, H., Jia, Z., & Qin, Z. (2009). Trust Evaluation and Dynamic Routing Decision Based on Fuzzy Theory for MANETs. *JSW*, *4*(10), 1091–1101. doi:10.4304/jsw.4.10.1091-1101

Engle, S. L. (2001). *Structural holes and Simmelian ties: Exploring social capital, task interdependence, and individual effectiveness*. Academic Press.

Govindan, K., & Mohapatra, P. (2012). Trust computations and trust dynamics in mobile adhoc networks: A survey. *IEEE Communications Surveys and Tutorials*, *14*(2), 279–298. doi:10.1109/SURV.2011.042711.00083

Gubbi, J., Buyya, R., Marusic, S., & Palaniswami, M. (2013). Internet of Things (IoT): A vision, architectural elements, and future directions. *Future Generation Computer Systems*, *29*(7), 1645–1660. doi:10.1016/j.future.2013.01.010

Guo, J. (2018). *Trust-based Service Management of Internet of Things Systems and Its Applications*. Virginia Tech.

Handfield, R. B., & Bechtel, C. (2002). The role of trust and relationship structure in improving supply chain responsiveness. *Industrial Marketing Management*, *31*(4), 367–382. doi:10.1016/S0019-8501(01)00169-9

Ho, J.-W., Wright, M., & Das, S. K. (2012). ZoneTrust: Fast zone-based node compromise detection and revocation in wireless sensor networks using sequential hypothesis testing. *IEEE Transactions on Dependable and Secure Computing*, *9*(4), 494–511. doi:10.1109/TDSC.2011.65

Hoffman, K., Zage, D., & Nita-Rotaru, C. (2009). A survey of attack and defense techniques for reputation systems. *ACM Computing Surveys*, *42*(1), 1–31. doi:10.1145/1592451.1592452

Kamvar, S. D., Schlosser, M. T., & Garcia-Molina, H. (2003). The eigentrust algorithm for reputation management in p2p networks. *Proceedings of the 12th international conference on World Wide Web*.

Kraijak, S., & Tuwanut, P. (2015). *A survey on IoT architectures, protocols, applications, security, privacy, real-world implementation and future trends*. Academic Press.

Liu, X., Xiong, N., Zhang, N., Liu, A., Shen, H., & Huang, C. (2018). A trust with abstract information verified routing scheme for cyber-physical network. *IEEE Access: Practical Innovations, Open Solutions*, *6*, 3882–3898. doi:10.1109/ACCESS.2018.2799681

Lize, G., Jingpei, W., & Bin, S. (2014). Trust management mechanism for Internet of Things. *China Communications*, *11*(2), 148–156. doi:10.1109/CC.2014.6821746

Mayer, R. C., Davis, J. H., & Schoorman, F. D. (1995). An integrative model of organizational trust. *Academy of Management Review*, *20*(3), 709–734. doi:10.5465/amr.1995.9508080335

Mehmood, Y., Ahmad, F., Yaqoob, I., Adnane, A., Imran, M., & Guizani, S. J. I. C. M. (2017). *Internet-of-things-based smart cities: Recent advances and challenges.* Academic Press.

Moraru, L., Leone, P., Nikoletseas, S., & Rolim, J. D. (2007). Near optimal geographic routing with obstacle avoidance in wireless sensor networks by fast-converging trust-based algorithms. *Proceedings of the 3rd ACM workshop on QoS and security for wireless and mobile networks*. 10.1145/1298239.1298246

Mui, L., Mohtashemi, M., & Halberstadt, A. (2002). A computational model of trust and reputation. *System Sciences, 2002. HICSS. Proceedings of the 35th Annual Hawaii International Conference on*. 10.1109/HICSS.2002.994181

Pathan, A.-S. K. (2016). *Security of self-organizing networks: MANET, WSN, WMN, VANET*. CRC Press.

Samian, N., & Seah, W. K. (2017). Trust-based Scheme for Cheating and Collusion Detection in Wireless Multihop Networks. *Proceedings of the 14th EAI International Conference on Mobile and Ubiquitous Systems: Computing, Networking and Services (MobiQuitous)*. 10.1145/3144457.3144486

Sarkar, S., & Datta, R. (2012). *A trust based protocol for energy-efficient routing in self-organized MANETs*. Paper presented at the India Conference (INDICON), 2012 Annual IEEE. 10.1109/INDCON.2012.6420778

Sen, J. (2010). *Reputation-and trust-based systems for wireless self-organizing networks*. Aurbach Publications, CRC Press.

Sicari, S., Rizzardi, A., Grieco, L. A., & Coen-Porisini, A. (2015). Security, privacy and trust in Internet of Things: The road ahead. *Computer Networks*, *76*, 146–164. doi:10.1016/j.comnet.2014.11.008

Simmel, G. (2011). *Georg Simmel on individuality and social forms*. University of Chicago Press.

Srinivasan, A., Li, F., & Wu, J. (2008). *A novel CDS-based reputation monitoring system for wireless sensor networks*. Paper presented at the Distributed computing systems workshops, 2008. ICDCS'08. 28th international conference on. 10.1109/ICDCS.Workshops.2008.17

Tajeddine, A., Kayssi, A., Chehab, A., Elhajj, I., & Itani, W. (2015). CENTERA: A centralized trust-based efficient routing protocol with authentication for wireless sensor networks. *Sensors (Basel)*, *15*(2), 3299–3333. doi:10.3390150203299 PMID:25648712

Tavakolifard, M., & Almeroth, K. C. (2012). Social computing: An intersection of recommender systems, trust/reputation systems, and social networks. *IEEE Network*, *26*(4), 53–58. doi:10.1109/MNET.2012.6246753

Thenmozhi, T., & Somasundaram, R. (2016). Towards modelling a trusted and secured centralised reputation system for VANET's. *Proceedings of the International Conference on Soft Computing Systems*. 10.1007/978-81-322-2674-1_64

Usman, A. B. (2018). *Trust-based protocols for secure collaborative routing in wireless mobile networks*. Auckland University of Technology.

Usman, A. B., & Gutierrez, J. (2016). A Reliability-Based Trust Model for Efficient Collaborative Routing in Wireless Networks. *Proceedings of the 11th International Conference on Queueing Theory and Network Applications*. 10.1145/3016032.3016057

Usman, A. B., & Gutierrez, J. (2018). Toward Trust Based Protocols in a Pervasive and Mobile Computing: A Survey. *Ad Hoc Networks*.

Usman, A. B., & Gutierrez, J. J. A. o. O. R. (2018). *DATM: a dynamic attribute trust model for efficient collaborative routing*. Academic Press.

Wang, Y., & Vassileva, J. (2007). Toward trust and reputation based web service selection: A survey. *International Transactions on Systems Science and Applications*, *3*(2), 118–132.

Yan, Z., & Holtmanns, S. (2008). Trust modeling and management: from social trust to digital trust. IGI Global.

Yan, Z., Zhang, P., & Vasilakos, A. V. (2014). A survey on trust management for Internet of Things. *Journal of Network and Computer Applications*, *42*, 120–134. doi:10.1016/j.jnca.2014.01.014

Yu, H., Shen, Z., Miao, C., Leung, C., & Niyato, D. (2010). A survey of trust and reputation management systems in wireless communications. *Proceedings of the IEEE*, *98*(10), 1755–1772. doi:10.1109/JPROC.2010.2059690

Zahariadis, T., Leligou, H. C., Trakadas, P., & Voliotis, S. (2010). Trust management in wireless sensor networks. *Transactions on Emerging Telecommunications Technologies*, *21*(4), 386–395.

Zhan, G., Shi, W., & Deng, J. (2012). Design and implementation of TARF: A trust-aware routing framework for WSNs. *IEEE Transactions on Dependable and Secure Computing*, *9*(2), 184–197. doi:10.1109/TDSC.2011.58

Zhou, X., Zou, Z., Song, R., Wang, Y., & Yu, Z. (2016). *Cooperative caching strategies for mobile peer-to-peer networks: A survey. In Information Science and Applications (ICISA) 2016* (pp. 279–287). Springer. doi:10.1007/978-981-10-0557-2_28

Chapter 6
Efficient Cyber Security Framework for Smart Cities

Amtul Waheed
Prince Sattam Bin Abdul Aziz University, Saudi Arabia

Jana Shafi
Prince Sattam Bin Abdul Aziz University, Saudi Arabia

ABSTRACT

Smart cities are established on some smart components such as smart governances, smart economy, science and technology, smart politics, smart transportation, and smart life. Each and every smart object is interconnected through the internet, challenging the security and privacy of citizen's sensitive information. A secure framework for smart cities is the only solution for better and smart living. This can be achieved through IoT infrastructure and cloud computing. The combination of IoT and Cloud also increases the storage capacity and computational power and make services pervasive, cost-effective, and accessed from anywhere and any device. This chapter will discuss security issues and challenges of smart city along with cyber security framework and architecture of smart cities for smart infrastructures and smart applications. It also presents a general study about security mechanism for smart city applications and security protection methodology using IOT service to stand against cyber-attacks.

DOI: 10.4018/978-1-5225-7189-6.ch006

INTRODUCTION

The concept of "Smart City "has stimulated highly in past two decades increasing demand in almost all fields in urban world since it has strong real-world background and realistic prerequisites. According to the latest study conducted by United Nations Population Fund above half of the world population resides in urban zone and it is anticipated by 2050 above 66 percent of the world's population will live in urban area ("Department of Economic and Social Affairs", 2014).

By the development of Information and communication technologies (ICT), cellular phones, global positioning system (GPS), Bluetooth, Wi-Fi etc. human beings are connected via numerous smart devices such as smartphones and smart gadgets (Bogatinoska, Malekian, Trengoska, & Nyako, 2016).

Advancements in high connectivity using Internet of Things, connectivity between vehicles, public places, institutions, homes and other social systems are connected with smart appliances, security devices and smart energy meters are being used in many cities improving the quality of life.

City infrastructures and services are refining by new updated systems of automating, controlling and monitoring to take full benefit from them. For instance integrated systems will assist an emergency responders and public safety in disaster recovery. Smart transport will help both public and private vehicles to access the updates from GPS location about weather and traffic.

Pillars of Smart Cities

1. **Institutional Infrastructure:** Information and Communications Technologies (ICT) delivered smart system designed for citizens for performing activities like smart governance, planning and management of city in effective, efficient, transparent and accountable.
2. **Physical Infrastructure:** Physical infrastructure states to smart infrastructure integrated with technologies such as high speed broadband infrastructure, energy system, water supply system, sewerage system, drainage system, solid waste management system etc.
3. **Social Infrastructure:** Social infrastructure pertain to the development of social values such as education, healthcare, entertainment, creativities, arts, sports, gardening, open spaces and children's parks.
4. **Economic Infrastructure:** It relates to emerging of economic infrastructure that can generate income, investments and employment.

5. **Smart Analysis:** With emergence of smart cities interaction, information gathering and interpretation has increased. How to act and react in critical conditions are provided by smart analysis.
6. **Transmission Analysis:** Analysis for information security at all stages of data storage and transmission in big data will be triggered.

Characteristics of Smart Cities

Success of smart city is determined by its supporting infrastructures and economic developments with assurance of environment protection and social inclusions. Smart city tools and characteristic are for both organizations and citizens for transforming city into intelligent city (Song & Wu, 2012), (Piro, Cianci, Grieco, Boggia, & Camarda, 2014). Table 1 shows brief characteristic of smart cities.

Table 1. Characteristic of smart cities

S. No	Characteristics of Smart Cities	Description
1.	Heterogeneity	Vast, distributed, independent data store used by various user
2.	Resource efficiency	Constrained network interfaces with unlimited memory, battery capacity and processing capabilities.
3.	Scalability	Fast emergence of smart city from small to large, results in rapid growth in both data and network traffic this is controlled by scalable system mechanism.
4.	Mobility	Wireless communication, well equipped infrastructure, monitoring traffic flow in the smart cities
5.	Connectivity	Provides connectivity between smart devices and smart cities
6.	Consumer contribution	Accomplishment, improvement and quality of smart city with all smart services depend upon infrastructure and technologies along with human involvement.
7.	Openness	The wireless sensor networks are open in air so most of the networks are open or in sharing network.
8.	Centralization	Smart city data storage is based on cloud computing technologies which are flexible due to centralized resource sharing method, but are very prone to risks.
9.	Collaboration	All most all information is open for collaboration, excluding some personal or secret information.
10	Permeability	Penetration level for network to human society will greatly enhance in the space or in time dimension of smart city.

Components of Smart Cities

ISO standard evaluated 17 key components in 37120/2014 titled "Sustainable Development of Communities". Components are education, energy, environment, finance, economy, fire and emergency response, health, safety, recreation, shelter, waste water, water and sanitation, solid waste, transportation, urban planning, telecommunication and innovation for smart cities based on their evaluation of performances for developing urban services and quality of life (Angelakis, Tragos, Pöhls, Kapovits, & Bassi, 2017). Figure 1 shows the components of smart cities

1. Smart Energy: Intelligent Energy Storage, Digital Management of Energy, Smart Meters, Smart Grids (Portela et al., 2013)..
2. Smart Buildings: Lighting Equipment, Automated Intelligent Buildings, Advanced Heating Ventilation And Air Conditioning Systems (HVAC) (Gomez & Paradells, 2010).
3. Smart Mobility: ITS-Enabled Transportation Pricing System, Intelligent Mobility, Advanced Traffic Management System (ATMS), Parking Management (Carlsen, 2014).
4. Smart Technology: Super Broadband, Free Wi-Fi, Seamless Connectivity, 4G Connectivity (Cimmino et al., 2014).
5. Smart Infrastructure: Digital Water and Waste Management, Digital Management Of Infrastructure, Sensor Networks (Burange & Misalkar, 2015), (Baldini, Kounelis, Fovino, & Neisse, 2013) .
6. Smart Governance and Smart Education: Disaster Management Solutions, Government-On-The-Go, E-Government, E-Education ("Stockholm City Executive Office", 2010).

Figure 1. Components of smart city

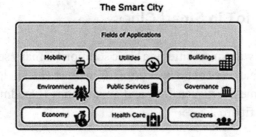

7. Smart Healthcare: Wearable Technology, Intelligent Healthcare Technology, Use Of E-Health And M-Health Systems (Solanas et al., 2014), (Patel, Park, Bonato, Chan, & Rodgers, 2012).
8. Smart Citizen: Civic Digital Natives; Use Of Green Mobility Options (Martínez-Ballesté, Pérez-Martínez, & Solanas, 2013).
9. Smart Security: Advanced Proactive Antivirus Protection Intelligent Threat Detection, Surveillance, Biometrics, Simulation Modelling and Crime Protection (Hoepman, 2014).
10. Smart Industry: Industry Preventive Maintenance, Smart Machines Communication
11. Smart Transport: Traffic Lane Monitoring, Congestion Reduction, Traffic Optimization, Remotely Finding Park Spots, Accident Prevention and Disaster Containment (Geronimo, Lopez, Sappa, & Graf, 2010).
12. Smart Home: Patterns, Schedule and Temperature Preference (Greveler, Glösekötterz, Justusy, & Loehr, 2012).
13. Smart Utilities: Water Consumption, Gas Consumption, Smart Meters, Utility Applications, Smart Grid (Dhungana, Engelbrecht, Parreira, Schuster, & Valerio, 2015).

IoT in Smart Cities

Advancement in ICT, mobile and social network technologies are exploding with the development of internet services (IoT) which touches almost all prospects of human life. More than 50 billion IoT devices are expected by the end of 2020.However the concept of IoT is presently overestimated in the areas of public sector, consumer sector and industrial sector. The main interest of Consumer sector is to deal with consumer applications and home area networks, whereas industrial sector are for improving business management, outcomes, manufacturing. And for public sector IoT is a key concept for developing the smart city services.

Architecture of IoT in Smart Cities

Composition of IoT with smart cities is a basic solution for many service applications delivered by smart cities. Basic architecture of IoT in smart cities helps in understanding operational deployment between IOT and smart cities (Goodman & Rapporteur, 2015), as shown in figure 2.

Figure 2. Architecture of IOT in smart cities

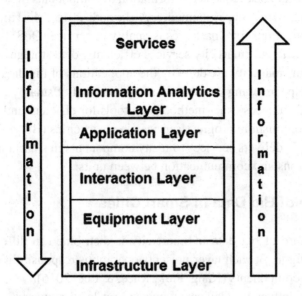

- Service Layer is responsible for making decisions. This layer obtains information from data analytics layer by using information it makes decisions.
- Information Analytics Layer gathers information from other layer, then it stores, analyses and process data.
- Interaction Layer is responsible for connectivity between hardware layer and other components for the flow of information between layers and hardware components.
- Equipment Layer is responsible for interaction between hardware elements (Such as chips, radios, sensors, actuators etc.) and environment or with other hardware elements or with the users.

Big Data in Smart Cities

Information gathered for smart services like transportation, healthcare, governance, grids etc. are collected into huge information which are autonomous, distributed, heterogeneous which are called as big data.

Smartness of city is defined by information extracted from these resources. Smart network are required for connecting smart components such as cars, smart phones and smart gadgets in almost all big data applications. These networks should be efficient in transferring the collected information from the resources to big data, then store, process and transfer back requested information to components of smart cities as it required (Nuaimi, Al Neyadi, Mohamed, & Al-Jaroodi, 2015).Most important aspect for big data applications for smart cites is quality of service (QoS) in a network.

To enhance different smart city services efficiently data storage and processing of information in smart city are done by the applications of big data technologies. Big data also helps in taking decisions for improvement of smart city services and resources. Big data needs exact methods and tools for efficient and effective data analysis for accomplishing progress services in smart cities (Borgia, 2014). Exact methods and tools delivers services to many components in smart city, in addition it boost interactions and communication between units.

Architecture of Big Data in Smart Cities

- The structure of big data in smart city is distributed in different layers as shown in figure 3; each layer presents operations responsibilities of big data smart city components (Kang, Park, Rhee, & Lee, 2016).
- First layer defines the devices connected on local networks. These devices generally generate a large amount of unstructured data per second.
- The second layer collects all the unstructured data and stores in databases of smart city data center or at big data storage. Then the data is processed according to request placed by smart city services using processing engines of big data.
- The last layer is application layer; it deals with user and machine interaction directly for making smart decisions. These are useful for many services such as web display analysis, fraud detection, intelligent traffic management, recommendation, sentiment analysis, etc.

Cloud Computing in Smart Cities

The importance of smart city is to connect, manage and process data. By creating real time and context specific information intelligence and analytics, from a complex set of objects such as people, devices software, sensors (Mitchel, Villa, & Stewart-Weeks, et al. 2013). High storage capacity and performance computing power is required for managing huge amount of heterogeneous data (Fu, Jia, & Hao, 2015). Therefore advance expansions and growth in the field of IoT and cloud computing is deployed with smart cities. Development of cloud computing enables big data storage and big data integration, analysis and processing in adequate time frames.

Figure 3. Architecture of big Data in Smart city

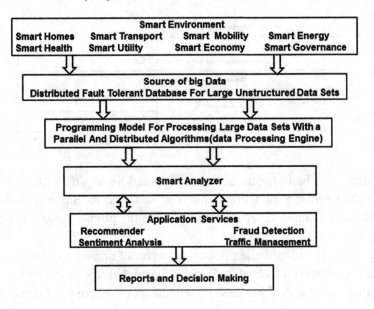

Architecture of Cloud Computing in Smart Cities

As shown in figure 4 the layer architecture of cloud computing in smart cites,

- Hardware layer deals with physical resources
- Operating System uses Linux distribution that supports OpenStack
- IaaS Layer is executed by means of OpenStack
- Data Service Layer holds the database and the file servers.
- Application VMs Layer Comprises the Virtual Machines that host the Smart City Applications.
- Access Layer is the front-end for the Smart City services.
- Layer It is implemented using Cloud applications
- Administration Layer Provides to the platform's administrators and to the application owners

BACKGROUND

In 2005 the former US president through his charitable organization the Clinton foundation challenged Cisco network equipment maker to purse technically how to make cites more sustainable. To achieve this connected urban development program, Cisco dedicated over five years for research on this topic by spending $25 million,

Figure 4. Architecture of cloud computing in smart city

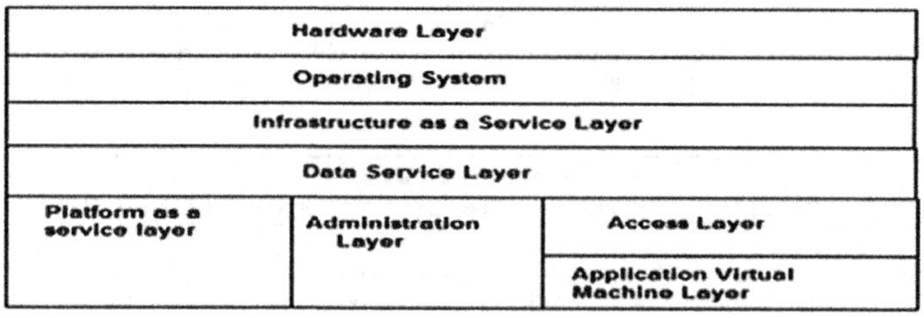

Hardware Layer		
Operating System		
Infrastructure as a Service Layer		
Data Service Layer		
Platform as a service layer	Administration Layer	Access Layer
		Application Virtual Machine Layer

which started with San Francisco, Amsterdam and Seoul. In 2010, when Cisco's pledge to the Clinton Foundation expired, it launched its Smart and Connected Communities division in order to commercialize the products and services that it had developed.

According to [Gordon] Feller, [Cisco's director of urban innovations], much of the technological work involved in Cisco's smart city projects is focused on older cities. Unlike the world of business communications where the Internet protocol has become standard, the telecommunications networks that underpin cities are typically based on incompatible legacy protocols.

With introduction of numerous smart applications several cyber security issues are also generating because of exposure in each layer of smart system. Attacks can be of any type from unauthorized access to denial of service to Sybil attack which can destroy or damage the quality of smart services or attack can cause physical interruptions in service availability (Zhang, et al., 2017).

To resolve security issue of smart services in smart cities both government and big data corporations are effectively establishing with robust advanced technologies. Further these types of cyber security issues cannot be single handed by technology itself. Effective polices and genuine business setup has to be introduced for successful smart city with all supported smart applications and without data breaches and security issues (NIST 800-122, 2010).

National Institute of Standard and Technology (NIST) formed an international team with above 100 organizations, universities and companies that advances and operates networked technologies. "Cybersecurity for Smart City Infrastructure" NIST planned to host workshop every year ("Cybersecurity for smart city architecture", 2015), ("City Science", 2015).

Intelligent cities are complex, wide, continuous and distributed systems. Encompassing data that has to be secured at each and every end. These cities are IoT-dependent and IoT-enabled where as their security rely on IoT protocols (Suo, Wan, Zou, & Liu, 2012).

Smart city IoT networks functions across multiple devices over heterogeneous technologies. The main aim is to provide secure interaction and authentication across these heterogeneous technologies (Pohls, et al., 2014).

A local cloud is required for communication of components it can be a central system for gathering information through which systems will communicate. Local cloud is central storage system which can quickly administrate to authorized requesters. In this system cloud is a tool for data analysis and connection rather than component

TOWARDS THE DESIGNING OF SECURE CYBER SMART CITY

Smart cities are responsible for producing, investigating and distributing of bulk of data. Data driven cities are the goal of the smart city technologies. City systems and services to be responsive allow to act upon timeline.

Observation and intelligence gathering is the foremost security stage of intelligence. This required tools such as CCTV, surveillance, hardware and software Biometrics in order to collect raw unprocessed data. Secure networks should be employed for protecting data, transmission must have Secured network to ensure data non-tempering.

Analyzing collected Data helps digest, decode and terabytes amount of information and data collected, via secured storage, analysis and forensic tools. Effective prevention is good to Change byte-sized to bite-sized to protect from threats or calamity reaction as well as alert situational awareness.

The human intervention in any security installations including physical security, perimeter protection apparatus, communication devices to the personnel are on the move. To get effective people and equipment mobilization is vital to the whole infrastructure of a resolute and safe location.

Smart City Cyber Security Challenges

Providing security to personal information and sensitive information is done by national laws in various ways as it is a basic right of every human. Smart city gathers data about people and places and gradually multiplying the range, volume and granularity of data, at this point concern of data privacy and security jolts.

Privacy and security can be breached and threatened in various ways which are normally unaccepted; still it is a part of smart city operational system which can be monitored by the following:

1.   **Observation:** Watching, Surveillance, following, audio or video recording of a person's action

2. **Accumulation:** Grouping of various characteristics of information regarding of the person to identify a pattern of actions.
3. **Improper Access:** Improper data protection policies can lead to improper access of information.
4. **Prolonged Usage:** Use of information gathered for prolonged practice for other purposes without owners consent.
5. **Exposed Devices:** Sensor and equipment are insured and exposed to hacker due to lack of standardized IoT devices, equipment and sensors. Hackers can disturb data; hack sensors feed wrong data which can cause signal failures and system shutdowns.
6. **Attacking Entire System:** ICT infrastructure allows building complex smart city structure which can manage various services. Even if a single device is hacked it is possible entire network or system can effect. This is due to weak security policies, encryption technique and poor maintenance or it can be a human error.
7. **Data Trafficking:** The bandwidth consumption of many devices to a single server can create data trafficking in network which can bring down the server or there are possibilities of security lapses due to strain on spectrum.
8. **Apps Hazard:** Smart cities are connected with smart devices from mapping apps, productive tools, to play games, social networking. In corporate environment with the increase in use of apps increases the risk of assisting Bring Your Own Device (BYOD).causing two security risks: first are malicious apps containing malicious code or security trash heap, second are developed by organization for company purpose containing security flaw.
9. **Software Bugs:** Simple Software glitch can cause a major shutdown of entire smart city, as it runs on many smart systems interconnected with each other handling different services.

Smart City Cyber Security Components

1. **Surveillance Management:** Smart city provides collective security measures by gathering data such as images, videos which are monitored from central location using cameras, CCTV etc. for surveillance which helps to act immediately in case of emergency situations.
2. **Analyzing Video Management:** These Videos observes identification of behavior, temporal and spatial of certain events. These video analyses are used in different fields like entertainment, home automation, health care, surveillance etc. for many purposes as active alarms to signal instant response in case of emergency and proactive monitoring tool.

3. **Centralized Data Storage Management:** Information gathered from all sensor devices in a network are stored and managed in centralized storage space, data center are designed according to applications executed in the smart city. These data centers offer real time monitoring for effective operations such as traffic control, analytics and video management.

4. **Central Command System:** Central command system is an infrastructure that provides data analysis for decision making, by gathering and providing integrated data from centralized data storage management.

5. **Efficient Services Management:** For smooth operation and implementations of services in smart city it is necessary to acquire knowledge and skills. Instruct the staff for operating and providing services efficiently for successful outputs.

Secure Architecture of Smart City

Smart cities IoT networks are across multiple device types over heterogeneous network and technologies. Providing security for such heterogeneous technologies is a challenging process. There are many different structures for representing secure architecture models of smart cities ("Fog Computing and the Internet of Things", 2015) (Jiong, Jayavardhana, Slaven, & Marimuthu) (Cerqueira Ferreira, Dias Canedo, & Timóteo de Sousa Junior) (Soliman, Abiodun, Hamouda, Zhou1, & Lung, 2013) (Zhan, 2013). Figure 5 represents all the components of secure smart city that are networks, software defined network controllers, secure gateways, data storage or cloud storage and smart city components. Architecture represents four major building blocks of secure smart city that are data privacy, confidentiality, integrity and authentication.

Security Framework of Smart City

Smart city cyber security system has to provide complete secure system by taking measurements for data protection, network security, evade illegal users, hackers, attackers to use and theft data, destructing information resources [37]. For full filling all security measures criteria security mechanisms has to be applied on each and every domain of smart environment such as equipment, technologies, laws and policies management. The smart city cyber security system consist of many areas of concerns like physical equipment security, data security mechanism, data management regulations, smart city security laws and public information security presented in the figure 6.

Figure 5. Components of secure smart city architecture

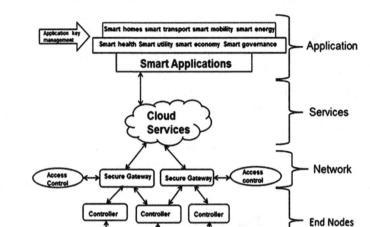

Figure 6. Presents the domains of secure smart city framework

Recommended Cyber Security Framework for Smart City

For the improvement of cyber security, ontology based security mechanisms are monitored, analyzed and classified in this recommended cyber security framework. The framework is divided into three layers which deal with security i.e. design time, run time, integration layer as shown in figure-7. First is design time, which provides security through methodology of service design and adaptation. Second is run time, which involves IoT environment through network and process monitoring regarding threats and vulnerabilities. Last layer is integration layer which allows both run time and design time to deals with the data or knowledge about IoT security ontology, it is used for reasoning mechanisms for providing suitable security services which can be varied at design time, adapting and actuating with IoT environments (Ducq, Agostinho, Chen, Zacharewicz, & Goncalves).

Smart City Security Mechanisms

This segment represents different mechanisms used to provide security to smart city.

Blockchain

In 2016, study conducted on applying blockchain to IOT domain. They realized and verified its significance and applications in development IoT ecosystem by increasing interest (Christidis & Devetsikiotis, 2016).

The key motive behind the increasing acceptance of blockchain – based IoT applications are, it enables applications to operate in distributed custom. Blockchain framework assures security for smart devices in smart city as well as it also supports newly developed framework features like confidentiality, integrity and availability (Dorri, Kanhere, Jurdak, & Gauravaram, 2017).New requirements of scalable IoT

Figure 7. Presents Recommended Cyber Security Framework

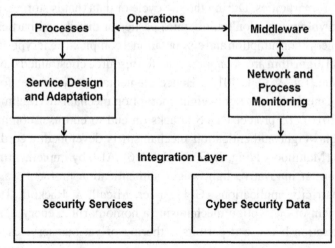

networks are not satisfied with the existing clouds (Sharma, Chen, & Park, 2017).To satisfy the required principles such as adaptability, security, scalability, efficiency and resilience, software defined networking and fog computing technology are developed with use of blockchain.

However blockchain technology resulted in convenient and reliable applications, still it in its early stage of development in IoT domain. To resolve the issues of privacy and security blockchain technology can be develop to its advances.

Biometrics

In IoT-based systems, biometrics is widely for authentication purpose. Biometric technology is based upon human behavioral and biological features such as fingerprints, retina detections, faces, voices, signatures etc. This technology is generally used for authentication purpose. Brainwavebased authentication technique gives assurance of high degree of authentication accuracy and efficiency (Zhou, Su, Chiu, & Yeh).

Mutual authentication protocol and a key negotiation are proposed to protect the confidential information of users in storage devices (Amin, Sherratt, Giri, Islam, & Khan, 2017). Risk of privacy and security hacking is high if bio based methods are not used properly. Developments in privacy preserving biometric (PPBSs) are under advancements (Wang, Wan, Guo, Cheung, & Yuen, 2017).

Cryptography

Cryptography is a smart and promising technology for providing security to smart applications in smart cities. During the life cycle of data that is storing, processing and sharing cryptography don't allow third parties or distributed parties to access data. Due to energy consumption and computational complexity encryption standards and traditional algorithm are not appropriate for resource constrained devices (Jing, Vasilakos, Wan, Lu, & Qiu, 2014). Hence elementary requirement for applying cryptographic mechanism is lightweight encryption mechanism (Mahmood, Ning, & Ghafoor, 2016). To protect DDoS attacks on end to end user communication, in 2016 a lightweight authentication mechanism is developed according to IoT requirements (Mahmood, Ning, & Ghafoor, 2016). Also by implementing a public key encryption an innovative lightweight authentication protocol was proposed to secure smart city applications (Goldwasser, Micali, & Rackoff, 1989). With the advancement of encryption mechanisms a homomorphic encryption (HE)has increasing attention because it permits gathering of various services and allows computations on encrypted data without revealing confidentiality of data. HE is used for many smart services such as protect electricity consumption in smart grid system, protect privacy for healthcare monitoring and to solve computing security issues (Talpur, Bhuiyan, & Wang, 2014) (Jabbar & Najim, 2016).Zero knowledge proofs are used as cryptographic effective authentication mechanism which handles authentication issues of smart devices such as smart cards. In this method, cryptographic mechanism is applied to one party to verify itself to other party without exposing information (Goldwasser, et al., 1989).

Ontology

Ontology is a division of philosophy and a successful tool to solve numerous issues such as knowledge patterns and unstructured data, to search for new knowledge and understand, define and reusing some formal represented knowledge. Innovations in ontology based security mechanism are also resolving security issues of IoT domain such as security risk management and cyber-attack detection (Razzaq, Anwar, Ahmad, Latif, & Munir, 2014) (Mozzaquatro, Jardim-Goncalves, & Agostinho, 2015).

Ontology based security mechanism model is developed in order to provide security to smart homes that interacts with smart devices in more efficient way by organizing security system (Tao, Zuo, Liu, Castiglione, & Palmieri, 2016).

QoPI Ontology base model is designed to represent, characterize, manage user personalized and active privacy control configurations under computing situations (Kim, Ko, & Kim, 2017).Use of ontology base security mechanism In IoT mostly focus on limited applications according to requirements instead of focusing on unified model which affects their application value. To overcome with this problem, one of the semantic-ontology-based situations reasoning mechanism has been designed to offer complete view of security state as well as refining the capability for emergency reply. Whoever this mechanism pays attention only on network layer of IoT architecture without addressing the overall security issues (Xu, Cao, Ren, Li, & Feng, 2017).

Game Theory

One of the powerful mathematical tool for field of smart city cyber security and privacy protection domain is game theory which is efficiently applied in different situations (Yu, 2016). According survey reports the attributes of game theory mechanism is more successful when compared to traditional defense mechanisms such as distributed solutions, reliable defense, proven mathematics and timely action (Do et al., 2017).

With the increasing interest of using game theory in the field of IOT smart city cyber security mechanism developments are also increasing such as analyzing schemes for cloud storage by evolutionary game theory (Abass, Xiao, Mandayam, & Gajic, 2017). Reducing energy consumption and assuring the accuracy both can be obtained by low powered devices and lightweight detection technique by means of game theory (Sedjelmaci, Senouci, & Taleb, 2017). Honeypot is another game theory model for solving security issues such as the attack and defense problems. This model supports in solving security issue of many smart applications such as smart healthcare, smart sensor networks, smart buildings etc (La, Quek, Lee, Jin, & Zhu, 2016). Another method of detecting spoofing attacks in cyber security is wireless network is zero sum game (Xiao, Li, Han, Liu, & Zhuang, 2016).Various security mechanism like K-anonymity and differential privacy are introduced by merging game theory with other privacy protection technologies (Liu, Liu, Guo, Li, & Fang, 2013).

With rapid growth in smart cities game theory will play a vital role in solving security and privacy issue of smart era.

Machine Learning and Data Mining

To protect network from attacks most commonly used security infrastructure is machine learning technologies, which is hired to improve effectiveness of intrusion detection systems. According to

Survey for securing Wireless sensor networks, it should implement machine learning technologies and its Machine learning algorithms (Alsheikh, Lin, Niyato, & Tan, 2014). For securing fusion and data sensing in wireless sensor networks, machine based scheme are introduced (Luo, Zhang, Yang, Liu, Chang, & Ning, 2016). Attacks in wi-fi networks can be detected by applying unique characteristic of extraction and selection model, which is having high detection rate (Aminanto, Choi, Tanuwidjaja, Yoo, & Kim, 2018).With fast growth in smartphone, smart devices and smart sensors networks privacy and security issues are also increasing. User centric machine learning technologies are used to analyze, predict and take decisions. Multisensor based authentication systems are used to study user behavior patterns and environmental structures (W.-H. Lee & R. B. Lee, 2015). Machine learning technology is used to develop a novel permission mechanism for mobile platform (Olejnik, et al., 2017).

By collaborating with defense mechanism, machine learning technology can be reinforced. Introduction of game theory model with machine learning can detect and prevent intrusions in wireless sensor networks (Shamshirband, Patel, Anuar, Kiah, & Abraham, 2014). Association of machine learning and biometric security system can provide secure wireless sensor networks systems (Biggio, Fumera, Russu, Didaci, & Roli, 2015).

Data Mining Technologies

According to survey conducted in the field of data mining, wide range of information is gathered by various devices and sensors about users are mined to provide better service (Tsai, Lai, Chiang, & Yang, 2014).

As a result some security and privacy issues will be triggered in data mining mechanism due to sensitive information about user such as location and behavior can be revealed. To overcome this issue privacy preserving data mining (PPDM) technology is introduced (Xing, Hu, Yu, Cheng, & Zhang, 2017).

Non-Technical Add-Ons

Following policy, governance, education and regulation is also an important aspect to provide cyber security with technical solutions (Kitchin, 2016).Government has authority to open, control and access the data. Government is responsible for data protection by implementing some data protection mechanism model to smart city (Walravens, 2012) (Batty et al., 2012).To improve skills of users, manufacturers, service providers and application designers educating the importance of security is necessary. For instance application users though have knowledge of privacy leakage and its destruction; still they ignore prevention method just for the sake of convenience (Ducq, Agostinho, Chen, Zacharewicz, & Goncalves). Further education develops the knowledge of citizens about using of smart devices and smart networks applying all security measures. Smart device manufacturers should improve standard of quality and safety measure. Vendors are obliged for updating firewalls and related software susceptibilities. Application designers should be trained in such a way that they should develop stable and strong coding (Misra, M. Maheswaran, and S. Hashmi, 2017) (Hurst. et al., 2017). Table 2 represents different mechanisms used to provide security to smart city.

Trends of Cybersecurity in Smart Cities

Increased Security Screening

Security screening technology plays a vital role in reducing risks and errors associated with security issues. Special algorithms are applied to recognizing facial and behavioral patterns which helps in reducing the potential behavioral threats.

Table 2. Different mechanisms used to provide security to smart city

Security Mechanisms	Smart Applications	Technologies Used
A. Blockchain	1. Smart homes 2. Smart transportation 3. Iot architecture	1. Block chain-based smart home architecture 2. Network topology and decentralized blockchain based framework 3. Distributed architecture based on blockchain technique and fog computing
B. Biometrics	1. Mobile sensors 2. Storage devices	1. Cascading bandpass filter for noise cancellation 2. Biometric based authentication and key negotiation protocol
C. Cryptography	1. Smart transportation 2. Smart grid 3. Smart shopping 4. Smart card	1. Two level authentication key exchange schema 2. Homomorphic encryption 3. RFID 4. Zero-knowledge proofs
D. Ontology	1. Smart home 2. Mobile computing 3. Iot architecture	1. Layers cloud architectural mode based on ontology 2. Context-aware and personalized privacy control 3. Semantic ontology based situation reasoning method
E. Game theory	1. Low resource Iot Devices 2. Honeypot-enabled networks 3. Wireless networks	1. Nash Equilibrium 2. Bayesian game of incomplete information 3. Zero-sum game
F. Machine learning and Data mining	1. Wi-Fi networks 2. Smartphone 3. Mobile devices 4. Social networking	1. Deep Feature learning 2. SVM based authentication system 3. Bayesian linear regression model 4. Privacy preserving K-means clustering

Emergency Apps and Crowdsourcing

Emergency app allows user to send alerts, request for assistance or services during emergency situations, and then the app detects user's location and instantly notify the nearest assistance center such as police station, medical personnel for services. Crowdsourcing of data on incidence, degree and nature supports to create large databases that can be used to identify areas meriting greater security.

Data-Based Crime Prevention and Predictive Policing

Big data analysis helps determine the most likely causes of new or emerging crime trends in different areas of the city by data analytics combined with real-time facial recognition, CCTV video linkages, and license plate scanning; analyze where a crime is most likely to take place on a specific date and time. Law enforcement agencies can use these insights to monitor specific neighborhoods showing increased crime, identify causes that have been determined to affect crime rates, identify individuals that have a higher risk of recidivism and increase officer patrols in areas with a higher likelihood of crime.

Drones for Risk Assessment

Minimizing risk for uncertain or dangerous situations is critical in a smart city environment. Drone or unmanned aerial vehicle (UAV) technology can pull together images to assess situations or possible dangers before sending in human beings. Drones can also help find fires, identify and prevent police ambushes, quickly search accident and crime scenes, and even detect heat from threats that may be concealed. Hence, drones can act as first responders before human intervention can take place.

FUTURE DIRECTIONS OF SECURE CYBER SMART CITY

Advancements and development are in progress in smart cities; with this increasing progress privacy and security of smart cities become a challenge. In this chapter we have discussed current cyber security scenario of smart city. With rapid growth of smart city more effective security mechanism has to be developed and implemented. The following areas of cyber security mechanisms have many prospects in further research directions.

Smart cities are almost everywhere in the world, smart project are under development in every country. There is no specific definition, concept or architecture of smart city. Therefore there is no specific security mechanism for smart city. Most of the developments are done on specific areas of smart city that cannot be shared with entire smart environment. For this reason detail study are required for reducing hurdles caused while securing smart cities.

Many security mechanisms have been developed but still only few of them are realistic for direct application. This is due to restricted energy resources, weak preserving algorithms and limited processing ability. Further development are required for lightweight counter measures, strong mobility, minimize overhead, flexibility, with assures security and dynamic low cost requirements.

Data mining task should be done with guarantee that data gathered, used and stored by IoT applications are minimized, and also knowledge discovered should be minimized without mining sensitive information of smart city user without prior authorization.

Security mechanism must be developed in such a way that it must be user friendly with improved security measures and comfort of numerous smart applications in smart cities.

IOT is heterogeneous networks which are interconnected and integrated such as internet, social networks, and industrial networks etc. To deal with the security issues of such latest complex networks effective technologies are needed.

SUMMARY

Interaction and sharing of information between smart devices, smart applications and smart users are the basic task of smart cities. With the widespread of smart environments has caused many security issues. To overcome this problem smart city should boost with advanced security mechanism without compromising on data privacy and security. This chapter discussed an overview of smart city, its security issues and challenges. It is necessary to have a robust security framework, architecture and policies for securing smart city and smart users. With advancement in technology different security mechanisms and strategies are developed. Still there is a long way to fulfill all security requirements and challenges of these smart applications. Cyber security as a model is dynamic; it is realistic and genuine to mitigate the current security challenges in further developments and related studies.

REFERENCES

Abass, Xiao, Mandayam, & Gajic. (2017). Evolutionary game theoretic analysis of advanced persistent threats against cloud storage. *IEEE Access*, *5*, 8482-8491.

Al Nuaimi, E., Al Neyadi, H., Mohamed, N., & Al-Jaroodi, J. (2015). Applications of big data to smart cities. *Journal of Internet Services and Applications*, *6*(1), 1–15. doi:10.118613174-015-0041-5

Alsheikh, M. A., Lin, S., Niyato, D., & Tan, H. P. (2018). Machine learning in wireless sensor networks: Algorithms, strategies, and applications. IEEE Commun. Surveys Tuts., 16(4), 1996-2018.

Amin, Sherratt, Giri, Islam, & Khan. (2017). A software agent enabled biometric security algorithm for secure _le access in consumer storage devices. *IEEE Trans. Consum. Electron.*, *63*(1), 53-61.

Aminanto, Choi, Tanuwidjaja, Yoo, & Kim. (2018). Deep abstraction and weighted feature selection forWi-Fi impersonation detection. *IEEE Trans. Inf. Forensics Security*, *13*(3), 621-636.

Angelakis, V., Tragos, E., Pöhls, H. C., Kapovits, A., & Bassi, A. (2017). *Designing, Developing, and Facilitating Smart Cities: Urban Design to IoT*. Springer. doi:10.1007/978-3-319-44924-1

Baldini, G., Kounelis, I., Fovino, I. N., & Neisse, R. (2013). A framework for privacy protection and usage control of personal data in a smart city scenario. In *Critical Information Infrastructures Security (LNCS 8328)* (pp. 212–217). Cham, Switzerland: Springer. doi:10.1007/978-3-319-03964-0_20

Batty. (2012). Smart cities of the future. *Eur. Phys. J. Special Topics*, *214*(1), 481-518.

Biggio, Fumera, Russu, Didaci, & Roli. (2015). Adversarial biometric recognition: A review on biometric system security from the adversarial machine-learning perspective. *IEEE Signal Process. Mag.*, *32*(5), 31-41.

Bogatinoska, D. C., Malekian, R., Trengoska, J., & Nyako, W. A. (2016). Advanced sensing and internet of things in smart cities. *2016 39th International Convention on Information and Communication Technology, Electronics and Microelectronics (MIPRO)*, 632-637.

Borgia, E. (2014). The Internet of Things vision: Key features, applications and open issues. *Computer Communications*, *54*, 1–31. doi:10.1016/j.comcom.2014.09.008

Burange, A. W., & Misalkar, H. D. (2015). Review of Internet of Things in development of smart cities with data management & privacy. *Proc. Int. Conf. Adv. Comput. Eng. Appl. (ICACEA)*, 189–195. 10.1109/ICACEA.2015.7164693

Carlsen, L. H. (2014). *The location of privacy—A case study of Copenhagen connecting's smart city* (M.S. thesis). Dept. Commun. Bus. Inf. Technol., Roskilde Univ., Roskilde, Denmark.

Christidis & Devetsikiotis. (2016). Blockchains and smart contracts for the Internet of Things. *IEEE Access*, *4*, 2292-2303.

Cimmino, A., Pecorella, T., Fantacci, R., Granelli, F., Rahman, T. F., Sacchi, C., ... Harsh, P. (2014). The role of small cell technology in future smart city applications. *Trans. Emerg. Telecommun. Technol. Special Issue Smart Cities*, *25*(1), 11–20. doi:10.1002/ett.2766

Cybersecurity for smart city architecture. (2015). National Institute of Standard and Technology (NIST). Retrieved from http://nist.gov/cps/cybersec_smartcities.cfm

Department of Economic and Social Affairs. (2014). *World Urbanization Prospects: The 2014 Revision, Highlights*. New York: United Nations Population Division.

Dhungana, D., Engelbrecht, G., Parreira, J. X., Schuster, A., & Valerio, D. (2015). Aspern smart ICT: Data analytics and privacy challenges in a smart city. *Proc. IEEE 2nd World Forum Internet Things (WF IoT)*, 447–452. 10.1109/WF-IoT.2015.7389096

Do, C. T., Tran, N. H., Hong, C., Kamhoua, C. A., Kwiat, K. A., Blasch, E., ... Iyengar, S. S. (2017). Game theory for cyber security and privacy. *ACM Computing Surveys*, *50*(2), 30. doi:10.1145/3057268

Dorri, A., Kanhere, S. S., Jurdak, R., & Gauravaram, P. (2017). Blockchain for IoT security and privacy: The case study of a smart home. *Proc. IEEE Int. Conf. Pervasive Comput. Commun. Workshops (PerCom Workshops)*, 618-623. 10.1109/PERCOMW.2017.7917634

Ducq, Y., Agostinho, C., Chen, D., Zacharewicz, G., & Goncalves, R. (n.d.). Generic methodology for service engineering based on service modelling and model transformation. In Manufacturing Service Ecosystem. Achievements of the European 7th FP FoF-ICT Project MSEE: Manufacturing SErvice Ecosystem (Grant No. 284860). Academic Press.

Fog Computing and the Internet of Things: Extend the Cloud to Where the Things Are. (2015). Available: http://www.cisco.com/c/dam/en_us/solutions/trends/iot/docs/computingoverview.pdf

Fu, Y., Jia, S., & Hao, J. (2015). A scalable cloud for the Internet of Things in smart cities. *Journal of Computers*, *26*(3), 63–75.

Geronimo, D., Lopez, A. M., Sappa, A. D., & Graf, T. (2010). Survey of pedestrian detection for advanced driver assistance systems. *IEEE Transactions on Pattern Analysis and Machine Intelligence*, *32*(7), 1239–1258. doi:10.1109/TPAMI.2009.122 PMID:20489227

Goldwasser, Micali, & Rackoff. (1989). The knowledge complexity of interactive proof systems. *SIAM J. Comput.*, *18*(1), 186-208.

Goldwasser, Micali, & Rackoff. (1989). The knowledge complexity of interactive proof systems. *SIAM J. Comput.*, *18*(1), 186-208.

Gomez, C., & Paradells, J. (2010). Wireless home automation networks: A survey of architectures and technologies. *IEEE Communications Magazine*, *48*(6), 92–101. doi:10.1109/MCOM.2010.5473869

Goodman, E. P. (2015). *Rapporteur. In The Atomic Age of Data: Policies for the Internet of Things* (p. 5). The Aspen Institute.

Greveler, U., Glösekötterz, P., Justusy, B., & Loehr, D. (2012). Multimedia content identification through smart meter power usage profiles. *Proc. Int. Conf. Inf. Knowl. Eng. (IKE)*, 383–390.

Hoepman, J.-H. (2014). Privacy design strategies. In *ICT Systems Security and Privacy Protection (SEC)* (pp. 446–459). Heidelberg, Germany: Springer. doi:10.1007/978-3-642-55415-5_38

Hurst, W., Shone, N., El Rhalibi, A., Happe, A., Kotze, B., & Duncan, B. (2017). Advancing the micro-CI testbed for IoT cyber-security research and education. *Proc. CLOUD Comput.*, 139.

Jabbar & Najim. (2016). Using fully homomorphic encryption to secure cloud computing. *Internet Things Cloud Comput.*, *4*(2), 13-18.

Jin, Gubbi, Marusic, & Palaniswami. (n.d.). An Information Framework for Creating a Smart City Through Internet of Things. In *IoT Architecture to Enable Intercommunication Through REST API and UPnP Using IP, ZigBee and Arduino*. Academic Press.

Jing, Vasilakos, Wan, Lu, & Qiu. (2014). Security of the Internet of Things: Perspectives and challenges. *Wireless Netw.*, *20*(8), 2481-2501.

Kang, Y.-S., Park, I.-H., Rhee, J., & Lee, Y.-H. (2016). MongoDB-Based Repository Design for IoT-Generated RFID/Sensor Big Data. *IEEE Sensors Journal*, *16*(2), 485–497. doi:10.1109/JSEN.2015.2483499

Kim, Ko, & Kim. (2017). Quality of private information (QoPI) model for effective representation and prediction of privacy controls in mobile computing. *Comput. Secur.*, *66*, 1-19.

Kitchin. (2016). *Getting smarter about smart cities: Improving data privacy and data security*. Dept. Taoiseach, Data Protection Unit, Dublin, Ireland, Tech. Rep.

La, Quek, Lee, Jin, & Zhu. (2016). Deceptive attack and defense game in honeypot-enabled networks for the Internet of Things. *IEEE Internet Things J.*, *3*(6), 1025-1035.

Lee, W.-H., & Lee, R. B. (2015). Multi-sensor authentication to improve smartphone security. *Proc. Int. Conf. Inf. Syst. Secur. Privacy (ICISSP)*, 1-11.

Liu, X., Liu, K., Guo, L., Li, X., & Fang, Y. (2013). A game-theoretic approach for achieving k-anonymity in location based services. *Proc. IEEE INFOCOM*, 2985-2993. 10.1109/INFCOM.2013.6567110

Luo, Zhang, Yang, Liu, Chang, & Ning. (2016). A kernel machine-based secure data sensing and fusion scheme in wireless sensor networks for the cyber-physical systems. *Future Gener. Comput. Syst.*, *61*, 85-96.

Mahmood, Z., Ning, H., & Ghafoor, A. (2016). Lightweight two-level session key management for end user authentication in Internet of Things. *Proc. IEEE Int. Conf. Internet Things (iThings) IEEE Green Comput. Com-mun. (GreenCom) IEEE Cyber, Phys. Social Comput. (CPSCom) IEEE Smart Data (SmartData)*, 323-327. 10.1109/iThings-GreenCom-CPSCom-SmartData.2016.78

Manadhata, P. K., & Wing, J. M. (2011). An Attack Surface Metric. *IEEE Transactions on Software Engineering, 37*(3), 371–386. doi:10.1109/TSE.2010.60

Martínez-Ballesté, A., Pérez-Martínez, P. A., & Solanas, A. (2013, June). The pursuit of citizens' privacy: A privacy-aware smart city is possible. *IEEE Communications Magazine, 51*(6), 136–141. doi:10.1109/MCOM.2013.6525606

Misra, S., Maheswaran, M., & Hashmi, S. (2017). *Security Challenges an Approaches in Internet of Things.* Springer. doi:10.1007/978-3-319-44230-3

Mitchel, S., Villa, N., & Stewart-Weeks, M. (2013). *The Internet of Everything for Cities: Connecting People, Process, Data, and Things to Improve the 'Livability' of Cities and Communities.* Point of View, Cisco.

Mozzaquatro, B. A., Jardim-Goncalves, R., & Agostinho, C. (2015). Towards a reference ontology for security in the Internet of Things. *Proc. IEEE Int. Workshop Meas. Netw. (M&N)*, 1-6. 10.1109/IWMN.2015.7322984

NIST 800-122. (2010). *Guide to Protecting the Confidentiality of Personally Identifiable Information (PII).* National Institute of Standards and Technology.

Olejnik, K., Dacosta, I., Machado, J. S., Huguenin, K., Khan, M. E., & Hubaux, J.-P. (2017). SmarPer: Context-aware and automatic run time permissions for mobile devices. *Proc. 38th IEEE Symp. Secur. Privacy (SP)*, 1058-1076.

Patel, S., Park, H., Bonato, P., Chan, L., & Rodgers, M. (2012). A review of wearable sensors and systems with application in rehabilitation. *Journal of Neuroengineering and Rehabilitation, 9*(1), 21. doi:10.1186/1743-0003-9-21 PMID:22520559

Piro, G., Cianci, I., Grieco, L. A., Boggia, G., & Camarda, P. (2014). Information centric services in smart cities. *Journal of Systems and Software, 88*, 169–188. doi:10.1016/j.jss.2013.10.029

Pohls, Angelakis, Suppan, Fischer, Oikonomou, Tragos, … Mouroutis. (2014). Rerum: Building a reliable iot upon privacyand security-enabled smart objects. In Wireless Communications and Networking Conference Workshops (WCNCW) (pp. 122–127). IEEE.

Portela, C. M. (2013). A flexible, privacy enhanced and secured ICT architecture for a smart grid project with active consumers in the city of Zwolle—NL. *Proc. 22nd Int. Conf. Electricity Distrib. (CIRED)*, 1–4. 10.1049/cp.2013.0628

Razzaq, Anwar, Ahmad, Latif, & Munir. (2014). Ontology for attack detection: An intelligent approach to Web application security. *Comput. Secur.*, *45*, 124-146.

Science, C. (2015). Massachusetts Institute of Technology (MIT). Retrieved from http://cities.media.mit.edu/

Sedjelmaci, Senouci, & Taleb. (2017). An accurate security game for low-resource IoT devices. *IEEE Trans. Veh. Technol.*, *66*(10), 9381-9393.

Shamshirband, Patel, Anuar, Kiah, & Abraham. (2014). Cooperative game theoretic approach using fuzzy Q-learning for detecting and preventing intrusions in wireless sensor networks. *Eng. Appl. Artif. Intell.*, *32*, 228-241.

Sharma, Chen, & Park. (2017). A software de_ned fog node based distributed blockchain cloud architecture for IoT. *IEEE Access*, *6*, 115-124.

Solanas, A., Patsakis, C., Conti, M., Vlachos, I., Ramos, V., Falcone, F., ... Martinez-Balleste, A. (2014, August). Smart health: A context-aware health paradigm within smart cities. *IEEE Communications Magazine*, *52*(8), 74–81. doi:10.1109/MCOM.2014.6871673

Soliman, M., Abiodun, T., Hamouda, T., Zhou, J., & Lung, C. (2013). Smart Home: Integrating Internet of Things with Web Services and Cloud Computing. *IEEE 5th International Conference on Cloud Computing Technology and Science*.

Song, G., & Wu, L. (2012). Smart city in perspective of innovation 2.0. City Management, 9, 53-60.

Stockholm City Executive Office. (2010). *Living in Stockholm Should Be e-asy.* Available: http://international.stockholm.se/globalassets/ovriga-bilder-och-filer/e-tjanster_broschyr-16-sid_4.pdf

Suo, H., Wan, J., Zou, C., & Liu, J. (2012). Security in the internet of things: a review. In *Computer Science and Electronics Engineering (ICCSEE), 2012 International Conference on* (vol. *3*, pp. 648–651). Academic Press. 10.1109/ICCSEE.2012.373

Talpur, Bhuiyan, & Wang. (2014). Shared-node IoT network architecture with ubiquitous homomorphic encryption for healthcare monitoring. *Int. J. Embedded Syst.*, *7*(1), 43-54.

Tao, Zuo, Liu, Castiglione, & Palmieri. (2016). Multi-layer cloud architectural model and ontology-based security service framework for IoT-based smart homes. *Future Gener. Comput. Syst.*, *78*, 1040-1051.

Tsai, Lai, Chiang, & Yang. (2014). Data mining for Internet of Things: A survey. *IEEE Commun. Surveys Tuts.*, *16*(1), 77-97.

Walravens. (2012). Mobile business and the smart city: Developing a business model framework to include public design parameters for mobile city services. *J. Theor. Appl. Electron. Commerce Res.*, *7*(3), 121-135.

Wang, Wan, Guo, Cheung, & Yuen. (2017). Inference-based similarity search in randomized Montgomery domains for privacy preserving biometric identification. *IEEE Trans. Pattern Anal. Mach. Intell.*, *40*(7), 1611-1624.

Xiao, Li, Han, Liu, & Zhuang. (2016). PHY-layer spoofing detection with reinforcement learning in wireless networks. *IEEE Trans. Veh. Technol.*, *65*(12), 10037-10047.

Xing, Hu, Yu, Cheng, & Zhang. (2017). Mutual privacy preserving k-means clustering in social participatory sensing. *IEEE Trans. Ind. Informat.*, *13*(4), 2066-2076.

Xu, Cao, Ren, Li, & Feng. (2017). Network security situation awareness based on semantic ontology and user-defined rules for Internet of Things. *IEEE Access*, *5*, 21046-21056.

Yu. (2016). Big privacy: Challenges and opportunities of privacy study in the age of big data. *IEEE Access*, *4*, 2751-2763.

Zhan, X. S. (2013). Promoting the construction of information security system. Information Security and Communications Privacy, 5, 9-12.

Zhang, Ni, Yang, Liang, Ren, & Shen. (2017). Security and privacy in smart city applications: Challenges and solutions. *IEEE Communication Magazine, 55*(1), 122-129.

Zhou, L., Su, C., Chiu, W., & Yeh, K.-H. (n.d.). You think, therefore you are: Transparent authentication system with brainwave-oriented bio-features for IoT networks. *IEEE Trans. Emerg. Topics Comput.* Available: https://ieeexplore.ieee.org/abstract/document/8057810/

Chapter 7
Privacy Preservation in Smart Grid Environment

Muhammad Aminu Lawal
King Abdulaziz University, Saudi Arabia

Syed Raheel Hassan
King Abdulaziz University, Saudi Arabia

ABSTRACT

Smart grids are conceived to ensure smarter generation, transmission, distribution, and consumption of electricity. It integrates the traditional electricity grid with information and communication technology. This enables a two-way communication among the smart grid entities, which translates to exchange of information about fine-grained user energy consumption between the smart grid entities. However, the flow of user energy consumption data may lead to the violation of user privacy. Inference on such data can expose the daily habits and types of appliances of users. Thus, several privacy preservation schemes have been proposed in the literature to ensure the privacy and security requirements of smart grid users. This chapter provides a review of some privacy preservation schemes. The schemes operational procedure, strengths, and weaknesses are discussed. A taxonomy, comparison table, and comparative analysis are also presented. The comparative analysis gives an insight on open research issues in privacy preservation schemes.

DOI: 10.4018/978-1-5225-7189-6.ch007

INTRODUCTION

The integration of information and communication technology (ICT) and traditional electricity grid gives rise to what is referred to as smart grid. The notion of smartness comes from the ability of a two-way communication among the smart grid entities. This allows for better management through effective monitoring and controlling of energy generation and consumption. Overall, the smart grid is envisioned to achieve the following objectives (Şimşek, Okay, Mert, & Özdemir, 2018):

- Integration of green/renewable power sources into the existing energy distribution.
- Billing through effective control and observation of measurements.
- Load balancing of energy consumption.
- Bi-directional communication among stakeholders.
- Enhancing resistance against attacks from malicious users.
- Autonomous management to increase reliability.
- Optimal use of assets to increase efficiency.

Despite the benefits of the smart grid, the two-way communication paves way for collection of fine grain information of user consumption through the smart meter. This introduces a major challenge of privacy violation because the information collected can be used to infer user habits, types of appliances, total energy consumption as shown in figure 1. This fine-grained information can be obtained through the use of "Non-intrusive Appliance Load Monitoring" (NALM) technique (Hart, 1992). Hence privacy preservation is very vital in the smart grid environment. Several schemes have been proposed in the literature to ensure user security and privacy preservation in smart grid. The relationship between privacy preservation and security is very close, hence effective privacy preservation is nearly difficult without security. However, this chapter focuses on privacy preservation in smart grid environment.

This chapter intends to discuss the background of smart grid and privacy with a review of some privacy preservation schemes in smart grids. The operational procedure, strength and weakness of each scheme is highlighted. In addition, a taxonomy and comparative table based on types and techniques employed in the schemes are presented at the end of the chapter.

The rest of the chapter is organised as follows: section 2 discusses the background of smart grid and privacy in smart grid, section 3 presents the review of the privacy preservation schemes, taxonomy, comparison table, and the comparative analysis. Finally, section 4 concludes the chapter.

Figure 1. Energy consumption information inference
(Efthymiou & Kalogridis, 2010)

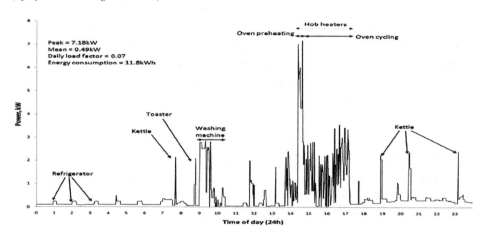

BACKGROUND OF SMART GRID

The blackout that occurred in North America in August of 2003 left several power stations affected and also incapacitated millions of lives. Revelations from investigations carried out showed that the blackout was because of unbalanced load consumption and lack of real-time monitoring amid other reasons (Lu, Liang, Member, & Li, 2012). Events like this make innovations such as smart grid obligatory. With investments on smart grid expected to reach $133.7 billion over the next 10 years in Western Europe (Northeastgroupllc, 2017).

As mentioned earlier, smart grid is an improvement on the traditional electricity system. However, it does not have a universal definition. The European Technology Platform ("European Commission (2006) European smart grids technology platform: vision and strategy for Europe's electricity,") Defines the Smart Grid as:

"A Smart Grid is an electricity network that can intelligently integrate the actions of all users connected to it—generators, consumers and those that do both—in order to efficiently deliver sustainable, economic and secure electricity supplies."

Another definition given by the U.S. Department of Energy (Department of Energy, 2009) Is:

"A smart grid uses digital technology to improve reliability, security, and efficiency (both economic and energy) of the electrical system from large generation, through

the delivery systems to electricity consumers and a growing number of distributed-generation and storage resources."

From the above definition, smart grid tends to provide smarter generation, transmission, distribution and consumption of electricity. Figure 2 shows a conceptual model provided by the NIST (*National Institute of Standards and Technology (2013): NIST framework and roadmap for smart grid interoperability standards, Release 2.0. smart grid interoperability panel (SGIP)*,). It has seven domains: generation, distribution, transmission, operations, service providers and customers.

The generation involves the traditional electricity (non-renewable) and renewable sources as well, which are further integrated into the transmission and distribution network. The model also incorporates open markets as third parties services in order to improve customer experience as well as competition between the service providers. The most vital part of this model in terms of privacy is the distribution stage. It is the interface between the smart grid and the customers, all possible privacy violations take place at that interface (*Uludag, Zeadally, & Badra, 2015*). The smart meter which is an enabling component of the smart grid is located at customer side i.e. the home area network (HAN), building area network (BAN) or industrial area network (IAN). It provides the fine grained energy consumption and all related customer information from the customer side to distribution stage utilities for necessary processing and analysis. Hence, the privacy of the customers can be violated when the information shared between the smart meter and the distribution stage utilities is compromised. This can lead to the following possible consequences according to a non-profit group called electronic privacy information centre (EPIC) (Electronic Privacy Information Center, 2011):

Figure 2. NIST Smart grid conceptual model

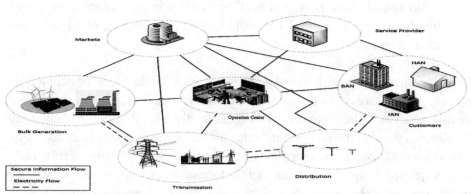

- Inference on personal behaviour patterns.
- Determine precise types of appliances used.
- Perform real-time surveillance.
- Profiling based on energy consumption.

Thus, preserving privacy at the distribution stage is important.

Privacy and Smart Grids

The term privacy varies among different individuals, cultures or institutions, hence it does not have a universal definition. However, Roger Clarke (Clarke, n.d.) has defined it in four dimensions:

1. Privacy of personal data or information is often referred to as data privacy or information privacy. It dictates that personal information about a person should not be disclosed to unauthorised parties (individuals or organisation). Privacy of personal data or information encompasses the right on how, when, whom, where and what level a person discloses personal information. Furthermore, in a situation where a third party has the data or information, a person should have a degree of control over the data and its usage.
2. Privacy of a person involves the privacy of a person's body in relation to integrity. It could refer to the right of a person over his body. It covers areas such as health problems, physical requirements and use of medical devices.
3. Privacy of personal behaviour often referred to as media privacy. It involves a person's right in terms of personal activities such as religious affiliations, political views and personal habit. A person should have the right over choosing to disclose or share such information.
4. Privacy of personal communication involves a person's right to free communication. A person should be able to communicate with anyone by utilizing various forms of media without unwarranted observation and monitoring.

In the smart grid perspective, the privacy of personal data or information is the most important aspect. As it relates to information concerning persons which can be used to identify the persons directly or indirectly. This type of information can be obtained through the fine-grained information of energy usage recorded by the smart meter. Although, the other dimensions can be related to the smart grid because of new types of energy usage. Example: the location of an electric vehicle (EV) can be known from a charging station, if the same charging station is constantly used.

This shows that the owner of the EV frequently visits the location for a purpose. This can fit into privacy of personal behaviour.

System Model

The system model considered by the majority of the privacy preservation schemes in smart grids consist of three major components, which are: smart meters (SM), data collector and third party systems as depicted in figure 3. The smart meter is located at the customer side i.e. HAN, BAN or IAN. The data collector collects data from a group of SMs for onward transmission to the third party systems for analysis and processing. The data collection function is performed by entities named differently by various schemes. Some of the names used to represent the data collector are: base stations, data collection units, aggregator, or gateway among others. Also the third party systems in the privacy preservation schemes are represented with different terms based on their functionality. Some third party systems includes: operation center or control center and trusted authority among others.

REVIEW OF PRIVACY PRESERVATION SCHEMES IN SMART GRID

Privacy preservation in smart grids is achieved by employing numerous techniques such as data aggregation, anonymization, and perturbation. Data aggregation is one of the popular methods used for periodic collection of fine-grained user

Figure 3. System model of privacy preservation schemes in smart grids

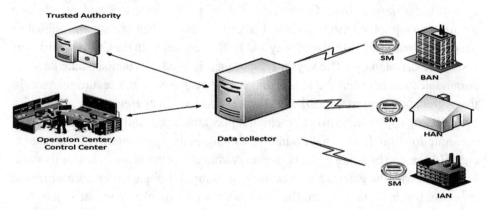

energy consumption from the smart meters. To ensure privacy preservation and security requirements during data aggregation, multiparty computation (MPC) or homomorphic encryption is applied.

The MPC based techniques allow multiple parties to jointly compute a value according to individually owned data without sharing the content of the data with other parties involved. While the homomorphic encryption based techniques allows arithmetic computations to be carried out on cipher text which matches the same operation on plaintext. This feature is important in the smart grid because entities can perform required computations without knowing the data.

Another popular technique used to ensure privacy in smart grid is anonymization, it hides the true identity of users by using pseudonyms, hashing value among others. Hence, associating real users with their energy consumption data becomes a difficult task. In addition some schemes are hybrid based i.e. combing more than one technique in ensuring privacy. Such techniques include time perturbation, Shamir's secret sharing, secret key and bit rotation techniques. Several researchers have proposed privacy preserving schemes in a smart environment, below is a review and categorization of some schemes based on types of techniques used:

Aggregation-Based Schemes

Khalid and Xiaondong (Alharbi & Lin, 2012) proposed a lightweight privacy-preserving data aggregation (LPDA) scheme for the smart grid. The LPDA employs the bilinear pairing technique and uses one-time masking method to shield users' privacy while ensuring a lightweight aggregation. The scheme involves three parties: a control center (CC), a building area network gateway (BAN-Gateway), and several home area network (HAN) within the BAN. It also has three stages: system initialization, aggregation request, and aggregation response stages. In the system initialization, the CC generates the bilinear parameters and selects two secure cryptographic hash functions. It keeps the master key private and publishes the public parameters. BAN–gateway and HANs register with the CC to obtain their respective private keys. The keys are subsequently used to establish static keys for communication between the HANs and the BAN gateway in a secure manner. In the aggregation request stage, the BAN gateway sends an aggregation request (with a time stamp appended to avoid potential replay attack) about the respective energy consumption to all HANs in the building. Finally, in the aggregation response stage, each HAN in the building collects its energy consumption message. It uses its static key shared with the gateway and one-time masking technique to generate a masked message, which is authenticated and forwarded to the BAN gateway in a multi-hop manner. The BAN gateway recovers the message and verify its correctness, process the message again and send it to the CC using a secure channel, the CC also verifies the correctness of the message. The scheme was evaluated using a custom simulator developed with java. The LPDA was able to prevent security vulnerabilities in smart grid in an efficient manner. However, the scheme incurs high burden due to key management (Bao & Chen, 2016) and communication overhead because of the multi-hop communication.

An In-network aggregation using homomorphic encryption scheme is proposed in (F. Li, Luo, & Liu, 2010) to guard user privacy in smart grids. The scheme utilizes a spanning tree created using a breadth-first transversal to gather user energy usage messages in a distributed manner with the collector as the root node and homomorphic encryption to encrypt intermediate messages between the nodes. Firstly, an aggregation route which covers all the smart meters in the area of interest is constructed. Using an operation plan, the energy usage information is aggregated in the bottom-up approach. Each parent node (SM) gathers energy usage messages from its children nodes (SMs) and sum it up to its own energy usage message. The process will continue until all the aggregated message reaches the root (collector), where all messages will be decrypted. The scheme ensures user privacy during aggregation because intermediate SMs are unable to know the content of messages. However, the scheme is vulnerable to forgery of data due to lack of auditing of messages.

An efficient and privacy-preserving aggregation (EPPA) scheme (Lu et al., 2012) was proposed to ensure secure smart grid communication. The scheme employs a bilinear pairing, homomorphic Paillier cryptosystems, and a superincreasing sequence to organize multidimensional energy consumption data. EPPA engages three entities: smart meters (SM), residential area gateway (GW) and the operational authority (OA). It also has four phases: system initialization, user report generation, privacy-preserving report aggregation, and secure report reading & response phases. At the system initialization phase, the OA generates all system parameter as well as two secure cryptographic hash functions. It publishes the hash functions and selects a superincreasing sequence. Similarly, the SM and GW generate their respective public and private keys. During the user report generation phase, each SM prepares its energy consumption report. It uses its private key to sign, append a time stamp, encrypt and send to the GW. At the privacy-preserving report aggregation phase, the GW receives all encrypted energy consumption report from the SMs. It utilizes a batch verification method to verify the reports and calculate the sums of all the encrypted energy consumption reports. It also appends a timestamp uses its private key to sign the aggregated report and sent to the OA. Finally, the OA verifies the aggregated report and decrypts using its master key. The OA analyses the report and responds to the SMs in a secure manner with their respective usage cost in order to enable them to use the electricity efficiently. The scheme was evaluated using PBC and MIRACL libraries. EPPA satisfies security requirements with low computational and communication overhead but vulnerable to an internal attacker.

An efficient privacy-preserving demand response (EPPDR) Scheme (H. Li et al., 2014) was proposed for privacy-preserving demand aggregation and efficient response. The scheme uses bilinear pairing, a homomorphic encryption technique, and adaptive key evolution to ensure forward secrecy of session keys. It operates in a similar way to EPPA (Lu et al., 2012) with a modification at the initialization phase. The OA and GW employ an identity-based signature to generate private keys and session keys. These keys are shared in a non-interactive manner between the smart grid entities (OA and GW share a session key, GW and SM also share a session key). EPPDR also uses three secure cryptographic hash functions instead of two used in EPPA. In addition, EPPDR has a key evolution phase, which it utilizes to generate new session keys at the end of a predefined period. The scheme was evaluated using PBC and MIRACL libraries. EPPDR achieves efficiency in terms of computational and communication overhead. Additionally, it controls key evolution adaptively to ensure balance trade-off in terms of security and communication efficiency. However, the scheme suffers from performance degradation when the number of smart meters increases (A. R. Abdallah & Shen, 2017).

An efficient and privacy-preserving data aggregation scheme (He, Kumar, Zeadally, Vinel, & Yang, 2017) was proposed for smart grid against internal adversaries. The scheme employs Boneh–Goh–Nissim public key cryptography. The scheme comprises three entities, a trusted third party (TTP), an aggregator (Agg) and a smart meter (SM). It also has three stages: initialization stage, the registration stage, and aggregation stage. During initialization, the aggregator employs the Boneh–Goh–Nissim public key cryptography technique to generate the system parameters used in the scheme. The blinding factors are generated by Agg and TTP. At the registration stage, the smart meter registers with the aggregator to obtain its secret key and finally, the aggregator collects the electricity consumption data from the SM and extracts the sum. The scheme was evaluated using the MIRACL library (Scott., 2011). The scheme is provably secure and can meet the security requirements (privacy, authentication, and integrity) but has a high communication and computation cost.

A lightweight lattice-based homomorphic privacy-preserving data aggregation scheme was proposed for the residential electricity consumers in a smart grid (A. Abdallah & Shen, 2018). The scheme utilizes the lightweight lattice-based homomorphic encryption scheme (C. A. Melchor, Castagnos, & Gaborit, 2008), it uses a vector space structure to encrypt messages as noisy lattices. Hence it assures the security of messages with minimal computation complexity because it executes simple multiplication and addition in vector spaces. The scheme consists of initialization and aggregation stages. The initialization stage establishes a secure connection between the smart home appliances (APs) and the CC through the SM and base station (BS) using APs unique identifiers and public keys of the SM, BS, and CC provided by trusted authority TA. The aggregation stage allows the household APs to collect their respective energy consumption data devoid of the smart meter (SM) knowledge. The SM obtains the encrypted and aggregated consumption reading of all the APs in the home area network from the assigned aggregator AP at each round and relay it to the BS for onward transmission to the control center. The SM and BS are ignorant of the total consumption reading but have the ability to verify the authenticity of the messages. The scheme was evaluated using MATLAB. The scheme assures consumers' privacy and security requirements i.e. confidentiality and integrity of messages with lightweight communication and acceptable computation overhead. Furthermore, it fits devices with low computational resources such as APs and SMs but communication overhead increases with increase in APs.

A practical privacy-preserving data aggregation (3PDA) Scheme (Liu, Guo, Fan, Chang, & Cheng, 2018) was proposed for Smart Grid. The 3PDA employs the lifted EC-ElGamal cryptosystem (ElGamal, 1985)(Desmedt, 1994), which is developed based on elliptic curve group and the Camenisch–Lysyanskaya (CL) signature scheme (J. Camenisch, Hohenberger, & Pedersen, 2012). The scheme involves three entities: smart meters (SM), data collection unit (DCU) and operational center (OC). It also consists of five stage: system setup, aggregation area creation, ciphertext generation, ciphertext aggregation, and distributed decryption. In the system setup stage, digital certificates are issued to SM and DCU by the OC. In the aggregation area creation stage, the SMs create an aggregation area by generating a group key. The energy consumption messages are encrypted by the SM and send to the DCU at the ciphertext generation stage. At the ciphertext aggregation stage, the DCU collects all the encrypted aggregated messages and send to the OC. Finally, at the distributed decryption stage, the OC demands for a distributed decryption which involves the SMs to obtain the aggregated sum. The 3PDA scheme was evaluated using PBC and OpenSSL library. The scheme was able to achieve its design goals (privacy, authentication, and integrity) but it is not feasible on limited resource smart meters due to computationally inefficient operations of the cryptosystem used (Gope, Sikdar, & Member, 2018).

Multiparty Computation Based Schemes

A secure multiparty computation based privacy preserving smart metering scheme (Thoma, Cui, & Franchetti, 2012) was proposed for smart meter based load management. The scheme employs secure multiparty computation and homomorphic encryption (Paillier cryptosystem) techniques. It utilizes the secure summation and comparison properties of the secure multiparty computation for ensuring the privacy of the energy consumption data. The secure summation is done via Paillier cryptosystem in four steps: setup, start, encryption addition and stop. During setup, the Paillier cryptosystem is used to generate the public key and private key. The public key is passed to the users and the utility while the private key is given to the utility only. At the start step, the user encrypts his energy usage message and pass to the next user. At the encryption addition step, the user encrypts his own energy consumption message and add up with previous users' message, this continues until the last user. Finally, at the stop step, the last user sends the encrypted message to the utility, where the utility will decrypt the message using the private key. The secure comparison is accomplished using the Yao's millionaire example. During the secure comparison, the predefined consumption limits and the users energy consumption at real-time (normal and high demand periods) are compared between the user and utility for verification of the consumed energy and generation of the consumption bill. The scheme was implemented as a GUI using Java and Apache MINA. The implementation shows that the scheme is feasible and can be adopted in real systems.

Mustafa et al. (Mustafa, Cleemput, Aly, Abidin, & Leuven, 2017)proposed a secure and privacy-preserving protocol for smart metering in smart grids. The protocol employs a multiparty computation model (MPC) (Canetti, 2000), which enables numerous data receivers (grid operators and suppliers) to aggregate readings from the smart meter at the users end for computing distribution, transmission, and the balance fees as well as unbalance fines. The protocol allows users to swap suppliers at any time, it comprises of three entities: dealers (smart meters(SM)), computational parties (delegated by the data communications company (DCC)) and output parties (distribution network operators (DNO), transmission system operator (TSO), and suppliers). It also consists of four stages: generation and distribution of input data, region-based data aggregation, grid-based data aggregation and output data distribution. At the generation and distribution of input data stage, the smart meters generate data on contracted suppliers, consumed and generated electricity. This data is forwarded to computational parties. At the region-based data aggregation stage, the computational parties aggregate the production and consumption of electricity of all regions and utilize the oblivious equality test per loop operation to associate each supplier with the consumption and production data. At the grid-based data aggregation stage, the aggregate grid-based data is obtained by summing up the regional based aggregation data by the computational party. Finally, the output data distribution stage distributes the computed aggregations to DNO, TSO, and suppliers to enable them to reconstruct their respective results by adding corresponding shares. To ensure security, the authors assumed that all entities have unique identifiers, time synchronized and all communications are encrypted and authenticated also the smart meters are tamper-proof. Further security in the protocol against malicious DNOs, TSO, and suppliers as well as a semi-honest (honest-but-curious) DCC is ensured by MPC (Ben-Or, Goldwasser, & Wigderson, 1988)(D. Chaum, Cŕepeau, & Damg°ard, 1988). The protocol was evaluated using synthetically generated electricity data and the smart metering architecture proposed in the UK (Aly, 2015). The protocol's security and complexity analysis showed its practical feasibility but may incur high communication cost due to communication among several parties.

Anonymity-Based Schemes

An escrow-based anonymization (Efthymiou & Kalogridis, 2010) was proposed for smart grid metering. The scheme utilizes two types of identities (IDs) hardcoded in the smart meters (SM) for transmitting readings to the utility: anonymous ID (high-frequency ID (HFID)) and attributable ID (low-frequency ID (LFID)). These IDs are contained in the personally identifiable SM profile (PISM) and anonymous SM profile (ANSM) of the SM and can be linked by an escrow service (manufacturer or trusted third party) only. The HFID is used to transmit electricity usage messages more frequently e.g. 5mins interval while the LFID is used to send account management and billing messages at longer intervals e.g. weekly or monthly. By employing a secure communication protocols PISM and ANSM are used to further create Client Data Profile (CDP) and Anonymous Data Profile (ADP) at random intervals. Their associated parameters are also created at the smart grid entities for the onward exchange of earlier defined messages. Security analysis indicates that security is obtainable through the employed trusted escrow service and set up of the CDP and ADP at random intervals at the SM. However, the scheme fails to provide sufficient privacy when the anonymity set is small e.g. when the SM needs to be replaced due to faults.

A credential-based privacy-preserving power request scheme was proposed in (Hui & Li, 2011) for the smart grid network. The scheme employs a blind signature technique to ensure user information privacy to third parties and the operator. It allows a customer to create credentials and blinding factors and asks the control center to sign them using its private key. This enables the customer to subsequently communicate with the control center anonymously. The scheme comprises of four stages: setup, registration, power requesting and reconciliation stages. During the setup stage, the control center generates public and private keys based on RSA and the SM will be allotted with a unique identity and a secret value for the authentication process. At the registration stage (done at the beginning of the month), the SMs register by providing their identities and secret values for authentication. Each SM generates credentials (to be used once) and request the control center to sign, the credential consists of ID, date, and value (amount of power that can be requested). The control center verifies the signatures by utilizing a scheme used in (David Chaum, 1982) and store necessary details about each signature in its database. The SM also stores the credentials for subsequent power request at any time during the month. During the power request stage, to request for power increment the SM takes a credential randomly and send to the control center. The control center verifies its signature on the credential and grants the request if all verification is successful otherwise another credential will be requested. Finally, at the reconciliation stage (end of the month), the SM authenticates its self at the control center and send unused credentials. The control center verifies them and reconcile the value as well as the consumed power and generate the bill for the customer. The credential signing of the scheme is evaluated using a test program developed with java. Analysis indicates that the scheme has low communication overhead. However, credential collision is unavoidable due to two or more users registering with identical credential identity (Zeadally, Pathan, Alcaraz, & Badra, 2013).

Samaneh and Mohammad (Hajy, Zargar, & Yaghmaee, 2013) proposed a privacy-preserving anonymization scheme via group signature scheme (PPGS) for the smart grid. The scheme has three entities: group manager (it controls group members' i.e. SMs), gateway (GW) and the control center. The scheme also has three stages: Bootstrapping, Smart Meter data generation and Smart Meter data aggregation stages. At the bootstrapping stage, the GM generates parameters (group public key, master opening key, master issuing key and master linking key) by executing the setup stage of Short Group signatures with Controllable Linkability (CL-GS) scheme (J.Hwang, S.Lee, B.Chung, & H.Cho, 2013). It sends group public key and master linking key to the GW through a secure channel. A smart meter can join the group by running the user join stage of CL-GS scheme to obtain the user signing key. The control center uses Paillier's Cryptosystem to compute the public key and private key to ensure confidentiality of smart meter data request. During the Smart Meter data generation stage, all SMs sends the energy consumption reading appended with a timestamp to the GW via a secure channel, which is provided by the group signature and if the channel is insecure, the paillier's cryptosystem will be used. Finally, at the Smart Meter data aggregation stages, the GW verifies the timestamp and the signature using the batch verification scheme (Ferrara, A. L., Green, M., Hohenberger, S., & Pedersen, 2009). It calculates the gathered energy consumption data in plaintext and forwards to the control center through a secure channel. The gateway subsequently groups the signatures based on their link indices and store them by utilizing the link algorithm in CL-GS for impending usage. It also checks for data duplicity before aggregation. This scheme has minimal computation and communication overhead but vulnerable to internal attack due to difficulty in locating illegal smart meter (Wang, And, & Chen, 2016).

A privacy-preserving smart metering scheme (Diao, Zhang, & Cheng, 2015) using linkable anonymous credential (PSMLAC) was proposed to ensure privacy protection, security properties as well as message authentication and traceability of fault smart metering. The PSMLAC utilizes the Camenisch and Lysyanskaya signature (Camenisch & Lysyanskaya, 2003) to create the linkable anonymous credential protocol. The protocol also works devoid of a trust-third party, allows a dynamic enrolment of SMs and revocation as well as performs complex statistical analysis of energy usage. The PSMLAC involves two entities: smart meter and collector and execute five algorithms: setup, join, data upload, link, verify and trace algorithms. The necessary protocol credentials are generated when the setup algorithm executes and a user list is initialized which will hold the SMs credentials. During join algorithm execution, an SM can join the system by interacting with the collector. The SM uses the zero proof knowledge to prove its credentials to the collector. The collector adds the SM to the user list and after verifying the credentials, the SM subsequently computes its private signing key. In the data upload, the SMs signatures (independent at different intervals) are further updated using a time stamp. At the link algorithm execution, the collector verifies the time stamps of the signatures. If the timestamps are all different, it runs the verify algorithm otherwise the trace algorithm will be executed. The verify algorithm will be executed to further verify the signature. Otherwise, the collector runs the trace algorithm. It sends the time stamps of the linked signatures that fail the test in the link algorithm stage to the SMs to prove the ownership of the signature. Subsequently, it runs the verify algorithm on the unlinked signatures and perform the complex statistical analysis of the data. The protocol was simulated using Maple 12. The PSMLAC achieves its goals in an efficient manner but the scheme is vulnerable to forgery attack (Qu, Shang, Lin, & Sun, 2015).

Hybrid-Based Schemes

A light-weight secret key-based privacy-preserving scheme for home area networks in Smart grid was proposed (Ullah, Khan, Ullah, & Ali, 2017). The scheme employs a secret key technique using XOR and bit rotation processes to ensure the privacy of data between the smart home appliances (SHA) and the smart meter (SM). The secret key is shared between the SHA and SM at the initialization stage during installation of the SHA. At the SHA end, the recorded energy consumption and the established key undergo the bit rotation, XOR operation, and hashing in order to conceal information from the adversaries. The received data at the SM end undergoes the reverse process to obtain the original message. The scheme was evaluated through simulation on a MATLAB simulator. The scheme assures customers' privacy and security requirements i.e. confidentiality and integrity of messages with lightweight communication and acceptable computation overhead. In addition, it fits devices with low computational resources. However, the scheme suffers from key repetition.

A privacy-preserving data collection protocol (TPS3) (Şimşek et al., 2018) for smart grids was proposed to guarantee the privacy of smart grid consumer data. The protocol employs temporal perturbation and Shamir's secret sharing (SSS) to achieve its objective. TPS3 involves three entities: smart meters, data collection servers and third-party systems and four phases: generating message, perturbing time information, fragmenting and sending message and re-generating and storing message phases. During the generating message phase, the SMs gathers the energy consumption message from all home appliances, indicates their respective IDs in the message, appends a time stamp and a seed using a hash function to prevent replay attacks and ensure integrity respectively. At the perturbing time information phase, the SMs uses the Laplace distribution to perturb the time information of the message in order to prevent attackers from understanding the time series of the messages. The message is encrypted and held for a time interval. After the expiration of the time interval, the fragmenting and sending message phase employs SSS to fragment the message and send to DCS using different routes. Finally, the regenerating and storing message phase collects all the messages from the SMs and recover them putting privacy into consideration. The messages are subsequently sent to the third party systems in a secure manner at request. The protocol was evaluated using a simulator developed with C#. TPS3 enhances the privacy of smart grid consumers from honest but curious HBC attackers. However, the protocol incurs an increase in time delay due to time perturbation and communication overhead because of the Shamir's secret sharing scheme employed to send messages via different paths.

The taxonomy and comparison table are presented in figure 4 and table 1 respectively. The taxonomy is based on the techniques employed in the privacy preservation. While in the table, performance metrics evaluated, strengths and weakness of the privacy preservation techniques are included.

Comparative Analysis

Table 1 presents the comparison of the reviewed privacy preservation schemes in the smart grid based on the techniques employed, performance metrics evaluated, strengths and weaknesses. The schemes assure privacy preservation and security requirements

Figure 4. Taxonomy of privacy preservation schemes in smart grid environment

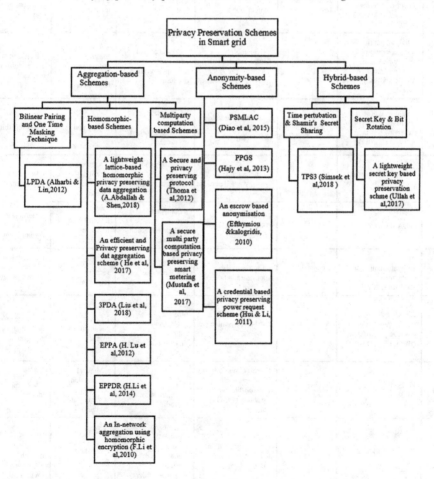

Table 1. Comparison of privacy preservation schemes in smart grid environment

S.No.	Name of Schemes	Types	Techniques	Performance Metrics Evaluated	Strengths	Weaknesses
1	LPDA (Alharbi & Lin, 2012)	Aggregation based	i. Bilinear pairing ii. One time masking	i. Average aggregation delay (AAD).	i. Prevent security vulnerabilities	i. Incurs high burden due to key management. ii. Communication overhead because of the multi-hop communication.
2	An efficient and privacy-preserving data aggregation scheme (He et al., 2017)	Aggregation based	i. Homomorphic (Boneh–Goh–Nissim public key cryptography)	i. Runtime. ii. Computational cost. iii. Communication overhead.	i. Meet security requirements	i. High computational cost.
3	A secure and privacy-preserving protocol for smart metering in smart grids (Mustafa et al., 2017)	Aggregation based	i. Multiparty computation	i. Computational cost.	i. Practically feasible	i. High communication cost
4	A lightweight lattice-based homomorphic privacy-preserving data aggregation (A. Abdallah & Shen, 2018)	Aggregation based	i. Homomorphic (lattice-based homomorphic encryption)	i. Computational cost. ii. Communication overhead.	i. Assures consumers' privacy and security requirements.	i. Increased communication overhead.
5	A light-weight secret key-based privacy-preserving scheme (Ullah et al., 2017)	Hybrid based	i. Secret key technique using XOR ii. Bit rotation	i. Computational cost. ii. Communication overhead.	i. Assures customers' privacy and security requirements.	i. Key repetition.
6	PSMLAC (Diao et al., 2015)	Anonymity based	i. Camenisch and Lysyanskaya signature	i. Computational cost. ii. Communication overhead.	i. Achieves security requirements efficiently.	i. Vulnerable to forgery attacks.
7	3PDA (Liu et al., 2018)	Aggregation based	i. Homomorphic (Lifted EC-ElGamal cryptosystem)	i. Computational cost. ii. Communication overhead.	i. Fulfill security requirements.	i. Inefficient operations of the cryptosystem
8	An escrow-based anonymization (Efthymiou & Kalogridis, 2010)	Anonymity based	i. Escrow-based anonymization	Not applicable	i. Achieve security requirements.	i. Insufficient security when anonymity set is small.
9	A credential-based privacy-preserving power request scheme (Hui & Li, 2011)	Anonymity based	i. A blind signature technique	i. Credential signing time. ii. Credential collision iii. Communication overhead	i. Low communication overhead.	i. Unavoidable credential collision.
10	PPGS (Hajy et al., 2013)	Anonymity based	i. Short Group signatures with Controllable Likability (CL-GS) scheme	i. Computational cost. ii. Communication overhead	i. Minimal computation ii. Minimal communication overhead	i. Vulnerable to internal attack.
11	TPS3 (Şimşek et al., 2018)	Hybrid based	i. Time perturbation ii. Shamir's secret sharing	i. Number of total messages. ii. Number of compromised messages. iii. Order (Order of report in time series.)	i. Enhances privacy.	i. Increase in time delay. ii. High communication overhead.

continued on following page

Table 1. Continued

S.No.	Name of Schemes	Types	Techniques	Performance Metrics Evaluated	Strengths	Weaknesses
12	A secure multiparty computation based privacy preserving smart metering (Thoma et al., 2012)	Aggregation based	i. Multiparty computation	Not applicable	i. Practically feasible	No performance evaluation presented.
13	An In-network aggregation using homomorphic encryption (F. Li et al., 2010)	Aggregation based	i. Homomorphic encryption.	i. Computational overhead.	i. Ensures user privacy.	i. Vulnerable to forgery attack.
14	EPPA (Lu et al., 2012)	Aggregation based	i. Homomorphic (Paillier cryptosystems)	i. Computational cost. ii. Communication overhead	i. Low communication overhead. ii. Low computational cost.	i. Vulnerable to internal attackers.
15	EPPDR (H. Li et al., 2014)	Aggregation based	i. Homomorphic encryption	i. Computational cost. ii. Communication overhead	Efficient in terms of i. Communication overhead. ii. Computational cost	i. Performance degradation under large number of smart meters.

(confidentiality, integrity, and authentication). However, the proposed schemes also have some limitations such as high computational cost, high communication overhead, vulnerable to internal attacks, and poor performance under a large number of users.

The authors of MPC based schemes (Mustafa et al., 2017),(Thoma et al., 2012) stated that their schemes are practically feasible but due to the need for much communication between the smart grid entities, the scheme may incur high communication overhead. Similarly, TPS3 (Şimşek et al., 2018), LPDA (Alharbi & Lin, 2012) and lightweight lattice-based homomorphic privacy-preserving data aggregation scheme (A. Abdallah & Shen, 2018) incur high communication overhead due to the SSS technique applied, the multi-hop operation applied and increase in appliances respectively.

To ensure privacy preservation with minimal communication cost, some schemes based on anonymization (Diao et al., 2015),(Efthymiou & Kalogridis, 2010),(Hui & Li, 2011),(Hajy et al., 2013) and homomorphic encryption (He et al., 2017),(Liu et al., 2018),(F. Li et al., 2010),(Lu et al., 2012),(H. Li et al., 2014) are proposed. However, these schemes suffer from high computational cost. Additionally, some of the schemes (Diao et al., 2015),(Hajy et al., 2013),(F. Li et al., 2010),(Lu et al., 2012) are vulnerable to internal attacks and forgery attacks.

Towards mitigating internal adversaries attacks, the scheme (He et al., 2017) achieves privacy preservation and security requirements as well as protect smart grid users from internal attacks but also has high communication cost. To guarantee privacy preservation with minimal computational cost and communication overhead, the scheme (Ullah et al., 2017) employed XOR Bit Rotation and Secret Key. However, the scheme suffers from key repetition due to the utilization of the same key obtained during initialization. In addition, it provides privacy preservation in the home area network only. Similarly, the scheme (H. Li et al., 2014) is efficient in terms of computational cost and communication overhead but experience poor performance under a large number of users.

CONCLUSION AND FUTURE WORK

The privacy preservation in smart grid cannot be overemphasized. The fine-grained information about the energy consumption of users can be used to infer lifestyle, types of appliances and total energy consumption of the users, which translates to violation of user's privacy and may have negative consequences. In this chapter, a review of some privacy preservation schemes has been discussed in details. The operational procedure, strengths, and weaknesses of the schemes are discussed. A hierarchal taxonomy of the different techniques has been explained with their relevance. Furthermore, a comparison table and comparative analysis are also presented. The comparison table is based on techniques employed, performance metrics evaluated, strength and weaknesses of the schemes. The comparative analysis gives an insight on open research issues in privacy preservation schemes.

Although several schemes have been proposed to ensure privacy and security requirements. There is still a need for more robust privacy preservation schemes. These schemes should be:

- Computationally inexpensive.
- Have Minimal communication overhead.
- Resistant to internal attacks.
- Assure privacy from the HAN to the operation center.
- Have Stable performance under a large number of customers.

REFERENCES

Abdallah, A., & Shen, X. S. (2018). A Lightweight Lattice-Based Homomorphic Privacy-Preserving Data Aggregation Scheme for Smart Grid. *IEEE Transactions on Smart Grid*, 9(1), 396–405. doi:10.1109/TSG.2016.2553647

Abdallah, A. R., & Shen, X. S. (2017). A Lightweight Lattice-based Security and Privacy- Preserving Scheme for Smart Grid Customer side Networks. *IEEE Transactions on Smart Grid*, 8(3), 1064–1074. doi:10.1109/TSG.2015.2463742

Alharbi, K., & Lin, X. (2012). LPDA: A Lightweight Privacy-preserving Data Aggregation Scheme for Smart Grid. *2012 International Conference on Wireless Communications and Signal Processing (WCSP)*. 10.1109/WCSP.2012.6542936

Aly, A. (2015). *Network flow problems with secure multiparty computation*. Universt´e catholique de Louvain.

Bao, H., & Chen, L. (2016). A lightweight privacy-preserving scheme with data integrity for smart grid communications. *Concurrency and Computation: Practice and Experience, 28*, 1094–1110. doi:10.1002/cpe

Ben-Or, M., Goldwasser, S., & Wigderson, A. (1988). Completeness theorems for non-cryptographic fault-tolerant distributed computation. In *STOC* (pp. 1–10). Chicago: ACM. doi:10.1145/62212.62213

Camenisch, J., Hohenberger, S., & Pedersen, M. Ø. (2012). Batch verification of short signatures. *Journal of Cryptology*, 25(4), 723–747. doi:10.100700145-011-9108-z

Camenisch, J., & Lysyanskaya, A. (2003). A signature scheme with efficient protocols. In *3rd Int. Conf. Security Commun. Netw.* (pp. 268–289). Academic Press. 10.1007/3-540-36413-7_20

Canetti, R. (2000). Security and composition of multiparty cryptographic protocols. *Journal of Cryptology*, 13(1), 143–202. doi:10.1007001459910006

Chaum, D. (1982). Blind signatures for untraceable payments. In Advances in Cryptology - Crypto '82 (pp. 199–203). Springer-Verlag.

Chaum, D., Cr´epeau, C., & Damgard, I. (1988). Multiparty unconditionally secure protocols. In *STOC. ACM* (pp. 11–19). ACM.

Clarke, R. (n.d.). *What's Privacy?* Retrieved from www.rogerclarke.com/DV/Privacy.html

Desmedt, Y. G. (1994). Threshold cryptography. *European Transactions on Telecommunications*, *5*(4), 449–458. doi:10.1002/ett.4460050407

Diao, F., Zhang, F., & Cheng, X. (2015). A Privacy-Preserving Smart Metering Scheme Using Linkable Anonymous Credential. *IEEE Transactions on Smart Grid*, *6*(1), 461–467. doi:10.1109/TSG.2014.2358225

Efthymiou, C., & Kalogridis, G. (2010). Smart Grid Privacy via Anonymization of Smart Metering Data. In *2010 First IEEE International Conference on Smart Grid Communications* (pp. 238–243). IEEE. 10.1109/SMARTGRID.2010.5622050

Electronic Privacy Information Center. (2011). *The Smart Grid and Privacy*. Retrieved from https://epic.org/privacy/smartgrid/smartgrid.html

ElGamal, T. (1985). A public key cryptosystem and a signature scheme based on discrete logarithms. *IEEE Transactions on Information Theory*, *31*(4), 469–472. doi:10.1109/TIT.1985.1057074

European Commission. (2006). *European smart grids technology platform: vision and strategy for Europe's electricity*. Retrieved from http://www.ec.europa.eu/

Ferrara, A. L., Green, M., Hohenberger, S., & Pedersen, M. Ø. (2009). Practical Short Signature Batch Verification. In Proc.2009 CT-RSA (pp. 309–324). Academic Press. doi:10.1007/978-3-642-00862-7_21

Gope, P., Sikdar, B., & Member, S. (2018). An Efficient Data Aggregation Scheme for Privacy-Friendly Dynamic Pricing-based Billing and Demand-Response Management in Smart Grids. *IEEE Internet of Things Journal*. doi:10.1109/JIOT.2018.2846299

Hajy, S., Zargar, M., & Yaghmaee, M. H. (2013). Privacy Preserving via Group Signature in Smart Grid. In *1st Electric Industry Automation Congress, EIAC 2013* (pp. 1–5). Academic Press.

Hart, G. (1992). Nonintrusive appliance load monitoring. *IEEE*, *80*(12), 1870–1891.

He, D., Kumar, N., Zeadally, S., Vinel, A., & Yang, L. T. (2017). Efficient and Privacy-Preserving Data Aggregation Scheme for Smart Grid Against Internal Adversaries. *IEEE Transactions on Smart Grid*, *8*(5), 2411–2419. doi:10.1109/TSG.2017.2720159

Hui, L. C. K., & Li, V. O. K. (2011). Credential-based Privacy-preserving Power Request Scheme for Smart Grid Network. In *2011 IEEE Global Telecommunications Conference - GLOBECOM 2011*. Kathmandu, Nepal: IEEE.

Hwang, J., Lee, S., Chung, B., Cho, H., & Nyang, D. H. (2013). Group signatures with controllable linkability for dynamic membership. *Information Sciences*, *222*, 761–778. doi:10.1016/j.ins.2012.07.065

Li, F., Luo, B., & Liu, P. (2010). Secure Information Aggregation for Smart Grids Using Homomorphic Encryption. In *2010 First IEEE International Conference on Smart Grid Communications* (pp. 327–332). IEEE. 10.1109/SMARTGRID.2010.5622064

Li, H., Lin, X., Yang, H., Liang, X., Lu, R., & Shen, X. (2014). EPPDR: An efficient privacy-preserving demand response scheme with adaptive key evolution in smart grid. *IEEE Transactions on Parallel and Distributed Systems*, *25*(8), 2053–2064. doi:10.1109/TPDS.2013.124

Liu, Y., Guo, W., Fan, C., Chang, L., & Cheng, C. (2018, February). A Practical Privacy-Preserving Data Aggregation (3PDA) Scheme for Smart Grid. *IEEE Transactions on Industrial Informatics*, 1–1.

Lu, R., Liang, X., Member, S., & Li, X. (2012). EPPA: An Efficient and Privacy-Preserving Aggregation Scheme for Secure Smart Grid Communications. *IEEE Transactions on Parallel and Distributed Systems*, *23*(9), 1621–1632. doi:10.1109/TPDS.2012.86

Melchor, C. A., Castagnos, G., & Gaborit, P. (2008). Lattice-based homomorphic encryption of vector. In *2008 IEEE International Symposium on Information Theory* (pp. 1858–1862). Toronto: IEEE. 10.1109/ISIT.2008.4595310

Mustafa, M. A., Cleemput, S., Aly, A., Abidin, A., & Leuven, K. U. (2017). An MPC-based Protocol for Secure and Privacy-Preserving Smart Metering. In 2017 IEEE PES Innovative Smart Grid Technologies Conference Europe (ISGT-Europe) (pp. 1–6). IEEE. doi:10.1109/ISGTEurope.2017.8260202

National Institute of Standards and Technology. (2013). *National Institute of Standards and Technology (2013): NIST framework and roadmap for smart grid interoperability standards, Release 2.0. smart grid interoperability panel (SGIP).* Retrieved from http:// j.mp/1rs1tKs http://collaborate.nist.gov/twiki- sggrid/pub/ SmartGrid/IKBFramework/ NIST_Framework_Release_2-0_corr.pdf

North-east group llc. (2017). *Western Europe Smart Grid : Market Forecast*. Retrieved from http://www.northeast-group.com/

Qu, H., Shang, P., Lin, X. J., & Sun, L. (2015). Cryptanalysis of A Privacy-Preserving Smart Metering Scheme Using Linkable Anonymous Credential. *IACR Cryptology EPrint Archive, 2015*, 1066.

Scott, M. (2011). *Miracl Library*. Retrieved from http://www.shamus.ie

Şimşek, M. U., Okay, F. Y., Mert, D., & Özdemir, S. (2018). TPS3 : A privacy preserving data collection protocol for smart grids. *Information Security Journal: A Global Perspective, 27*(2), 102–118.

Thoma, C., Cui, T., & Franchetti, F. (2012). Secure Multiparty Computation Based Privacy Preserving Smart Metering System. In *44th North American Power Symposium (NAPS)* (pp. 1–6). Academic Press. 10.1109/NAPS.2012.6336415

Ullah, S., Khan, E., Ullah, S., & Ali, W. (2017). A Light-Weight Secret Key-Based Privacy Preserving Technique for Home Area Networks in Smart Grid. In *2017 13th International Conference on Natural Computation, Fuzzy Systems and Knowledge Discovery (ICNC-FSKD)* (pp. 895–899). IEEE. 10.1109/FSKD.2017.8393395

Uludag, S., Zeadally, S., & Badra, M. (2015). *Techniques, Taxonomy, and Challenges of Privacy Protection in the Smart Grid: Privacy in Digital, Networked World*. Springer. doi:10.1007/978-3-319-08470-1_15

US Department of Energy. (2009). *Smart grid system report*. Retrieved from http://www.doe.energy.gov/

Wang, X.-F., And, Y. M., & Chen, R.-M. (2016). An efficient privacy-preserving aggregation and billing protocol for smart grid. *Security and Communication Networks, 9*(17), 4536–4547. doi:10.1002ec.1645

Zeadally, S., Pathan, A.-S. K., Alcaraz, C., & Badra, M. (2013). Towards Privacy Protection in Smart Grid. *Wireless Personal Communications, 73*(1), 23–50. doi:10.100711277-012-0939-1

ADDITIONAL READING

Bennati, S., & Pournaras, E. (2018). Privacy-enhancing aggregation of Internet of Things data via sensors grouping. *Sustainable Cities and Society*, *39*(February), 387–400. doi:10.1016/j.scs.2018.02.013

Braun, T., Fung, B. C. M., Iqbal, F., & Shah, B. (2018). Security and privacy challenges in smart cities. *Sustainable Cities and Society*, *39*(February), 499–507. doi:10.1016/j.scs.2018.02.039

de Oliveira, F. B. (2016). *On privacy-preserving protocols for smart metering systems: Security and privacy in smart grids. On Privacy-Preserving Protocols for Smart Metering Systems: Security and Privacy in Smart Grids*. Springer.

Eckhoff, D., & Wagner, I. (2017). Privacy in the Smart City - Application, Technologies, Challenges and Solutions. *IEEE Communications Surveys and Tutorials*, *20*(1), 1–28.

Ferrag, M. A., Maglaras, L. A., Janicke, H., Jiang, J., & Shu, L. (2018). A systematic review of data protection and privacy preservation schemes for smart grid communications. *Sustainable Cities and Society*, *38*(December 2017), 806–835.

Finster, S., & Baumgart, I. (2015). Privacy-aware smart metering: A survey. *IEEE Communications Surveys and Tutorials*, *17*(2), 1088–1101. doi:10.1109/COMST.2015.2425958

Leszczyna, R. (2018). Cybersecurity and privacy in standards for smart grids – A comprehensive survey. *Computer Standards and Interfaces*, *56*(April 2017), 62–73.

Li, S., Xue, K., Yang, Q., & Hong, P. (2018). PPMA: Privacy-preserving multisubset data aggregation in smart grid. *IEEE Transactions on Industrial Informatics*, *14*(2), 462–471. doi:10.1109/TII.2017.2721542

Lyu, L., Nandakumar, K., Rubinstein, B., Jin, J., Bedo, J., & Palaniswami, M. (2018). PPFA: Privacy preserving fog-enabled aggregation in smart grid. *IEEE Transactions on Industrial Informatics*, *14*(8), 3733–3744. doi:10.1109/TII.2018.2803782

Tan, S., De, D., Song, W.-Z., Yang, J., & Das, S. K. (2017). Survey of Security Advances in Smart Grid: A Data Driven Approach. *IEEE Communications Surveys and Tutorials*, *19*(1), 397–422. doi:10.1109/COMST.2016.2616442

KEY TERMS AND DEFINITIONS

Aggregation: A process of gathering information together from different sources.

Anonymization: The process of hiding a person's true identity.

Bits Rotation: The process of shifting or rearranging bits by employing the same pattern (number of positions and direction) over the whole set of bits.

Building Area Network (BAN): A network that combines several local area networks (LANs) together to cover an entire building.

Hashing: The process of transforming a set of characters into a smaller fixed-length key or parameter that uniquely represents the original set of characters.

Home Area Network (HAN): A network limited within a user's home that connects all appliances.

Industrial Area Network (IAN): A network that covers an entire industrial environment and connects its machines to control and monitoring systems.

Masking Technique: A process of concealing authentic data with arbitrary an input.

Pseudonyms: A set of character used to hide person's true identity.

Secret Key: This is a parameter utilized for encryption and decryption of messages in a secret-key or symmetric encryption technique.

Shamir's Secret Sharing: The process of dividing a secret message into a number of segments before sending it and to reconstruct the original message a certain number (threshold) of the segments is required.

Smart Meter: An internet enabled device that records and send energy consumption information of buildings to the utility companies for accurate billing and monitoring.

Time Perturbation: The process of obfuscating the real time interval of recording an information by introducing random time delay.

Chapter 8
A Review on Cyberattacks:
Security Threats and Solution Techniques for Different Applications

Gaganjot Kaur Saini
Charles Sturt University, Australia

Malka Halgamuge
ⓘD https://orcid.org/0000-0001-9994-3778
Charles Sturt University, Australia

Pallavi Sharma
Charles Sturt University, Australia

James Stephen Purkis
Charles Sturt University, Australia

ABSTRACT

Research questions remain to be answered in terms of discovering how security could be provided for different resources, such as data, devices, and networks. Most organizations compromise their security measures due to high budgets despite its primary importance in today's highly dependent cyber world and as such there are always some loopholes in security systems, which cybercriminals take advantage of. In this chapter, the authors have completed an analysis of data obtained from 31 peer-reviewed scientific research studies (2009-2017) describing cybersecurity issues and solutions. The results demonstrated that the majority of applications in this area are from the government and the public sector (17%) whereas transportation and other areas have a minor percentage (6%). This study determined that the government sector is the main application area in cybersecurity and is more susceptible to cyber-attacks whereas the wireless sensor network and healthcare areas are less exposed to attack.

DOI: 10.4018/978-1-5225-7189-6.ch008

INTRODUCTION

The term "Cyber Security" refers to protecting all networks, computing/smart devices and data/information from vulnerabilities and hackers. This is needed because all security techniques, including firewalls and anti-virus measures, are still vulnerable in terms of protecting cyber security systems from attacker (Kumar, Yadav, Sharma, & Singh, 2016). Attackers can be categorized as intentional (attackers who have some kind of personal motive and benefit by hacking or attacking devices and networks) or non-intentional (attacker that may mistakenly or unknowingly send some kind of malicious code to the devices or network). There are three categories of attacks; (i) Network attacks (Intrusions, Web defacement, Denial-of-Service attacks), (ii) Network abuse (Phishing, Forgery, SPAM) and (iii) Malicious codes (Viruses, Worms, Trojen horse, Spyware, Key loggers, BOTs). Cyber threats can take the form of cyber war, cyber terrorism and cyber-crime. Firstly, cyber war occurs when one country aims to destroy the networks and computing devices of another nation, for example Denial of Service attacks and viruses (Kumar et al., 2016). Secondly, cyber terrorism occurs when terrorist organizations or groups organize activities using cyber means that cause or spread terror, for example network attacks and hacking systems (Kumar et al., 2016). Lastly, cyber-crime is motivated by data theft, monetary gain or wicked- hacking, for example, Debit/credit card data, crashing website (Kumar et al., 2016).

Cyber security is very crucial when it comes to protecting devices, data and networks in this field. This field includes banking, finance, insurance, education, telecommunication, taxation, and accounting. Moreover, privacy and security remain significant areas of study because there are always different types of issues in protecting data, networks and devices. In terms of security, it is important to consider what, who and how is involved. That is, what devices to protect, who is authorized to apply these measures and how the devices should be protected. Importantly, every sector depends on Information Technology and the internet for doing its operations (P. Chauhan, N. Singh, & N. Chandra, 2013). Consequently, privacy and security are important topics, which every company needs to be concerned about. Almost everything in our daily life is connected to technology such as of communication, transportation and health care (P. Chauhan, N. Singh, & N. Chandra, 2013). Moreover, human beings are completely dependent on technology such as people using phones and computers for communication as well as the internet for online shopping and e-billing. Hence, attackers have more options to hack these systems than ever before. Furthermore, this dependence on technology and the internet has led to more potential vulnerabilities for attackers to exploit (H. Iguer, Medromi,

Sayouti, Elhasnaoui, & Faris, 2014). Cyber security in transportation systems is also very crucial otherwise attackers can take advantages of the in-vehicle or other vulnerabilities in GPS, security cameras, emergency communication (two-way radios) and other related systems (Bowen et al., 2015).

Smart City is a new concept and there is no proper definition yet. However, a smart city can be described as a city, which includes ICT for improving the superiority as well as execution of metropolitan services, e.g. transportation, electricity and other conveniences for diminishing wastage, and overall cost consumption of resources ("Smart City,"). The main objective of the smart city is to improve the quality and life style of the citizens by using smart technology ("Smart City,"). A smart city includes all the facilities for its citizens such as proper supply of water, electricity, sanitation, management of wastes, transportation services, security and safety, education, health services, sustainable environment, housing and other digital services for making life easier than ever before ("Features of Smart Cities,").

Different authors have proposed different solutions for different issues in cyber security. Some suggest preventative measures while others seek to identify attacks. For instance, there are the following techniques for protecting networks of businesses, industries, government sectors from attacks: 'kHIVE', 'IDS', 'IronKey', 'MOZART' and 'CAULDRON' (S. C. Paladino & J. E. Fingerman, 2009).

Researchers are trying to figure out different types of techniques to prevent, detect, and mitigate different types of cyber-attacks in different interdisciplinary fields. They are working on solutions for cyber-attacks and cyber security from different perspectives (H. Iguer et al., 2014). For example, researchers (H. Iguer et al., 2014) have proposed the prototype 'CySeMoL' as a security model. According to (W. Yang & Zhao, 2014), security researchers there has been more attention on control logic, Firmware and services of PLC in ICS in industries. Furthermore, researchers in the area of cyber security are looking for deficiency of vulnerabilities in security mechanisms, tools, technologies, and laws. Significantly, as almost everything is stored in cyberspace and due to the ubiquity of computing and smart devices (H. Iguer et al., 2014), cyber security issues as this study aims to seek solutions in the following fields by reviewed literature: public and private sectors, industries, smart grid, health care, Wireless Sensor Networks, Internet of Things etc.). The next section will briefly describe Smart Grid, Wireless Sensor Networks and the Internet of Things as these are some of these key areas.

Firstly, Smart Grid is a new and rising technology, which can satisfy the need for renewable energy resources and the finite availability of fossil fuels by integrating it with digital technologies (J. Hu & A. V. Vasilakos, 2016). Modern efficient digital power systems that distribute electricity (energy) to the consumers consist of different measures of power, technology, resources (M. M. Pour, A. Anzalchi, & A. Sarwat, 2017). However, this area has turned into a main area of cyber security, especially regarding the software and networks in the grid (S. Shapsough, F. Qatan, R. Aburukba, F. Aloul, & A. R. Al Ali, 2015). There are also numerous interlinked automatic devices, such as smart meters, GPS, computers and other monitoring devices, wireless routers, in the smart grid, (M. M. Pour et al., 2017). For example, recently there has been a national power grid cyber-attack reported (Ly, Sun, & Jin, 2016) as a consequence of this incident it is crucial to secure the communication channels in smart power grids (Ly et al., 2016).

Secondly, Wireless Sensor Networks is the inter-connections of devices of limited computing abilities (A. Tyagi, J. Kushwah, & M. Bhalla, 2017). Wireless sensor networks are able to detect any changes or conditions, including pressure, temperature, and movement, within their surroundings. These wireless sensor networks can configure themselves automatically according to the changes detected and it can also be used to transfer information, regarding these changes, to a central location through the wireless network. As this includes a number of devices connected to each other, they can be easily targeted by the attackers (A. Tyagi et al., 2017).

Thirdly, the 'Internet of Things' is also an area where security is the main issue. The major concern in this area is to maintain the availability, integrity and confidentiality of electronic data on communication systems (M. Radovan & B. Golub, 2017). With the growth in digital devices, there are higher chances of cyber-attacks therefore securing systems is becoming a pivotal task for almost all organizations (Heikkilä, Rättyä, Pieskä, & Jämsä, 2016). Security related concerns need to be addressed urgently due to the latest technology dependency and challenges that come with it (Ly et al., 2016). Researchers are particularly concerned about national and economic security (Ly et al., 2016) hence public-sector applications are important. There are already new challenges that are encountered by merging huge smart/embedded devices with networking devices and topologies in the Cyber Physical System establishment (Ly et al., 2016).

Cyber-Physical Systems (CPS) is the combination of networking, computation, and physical processes ("Cyber-Physical Systems,"). The networks and embedded computers observe and direct (control) the physical processes ("Cyber-Physical Systems,"). The financial and public capabilities of these systems are significantly superior to what has been recognized, and huge investments are being prepared at international level to build up the technology. CPS unites the activeness of the physical processes with those of the networking and software giving modeling and abstractions, analysis and design techniques for the integrated total ("Cyber-Physical Systems,").

Cyber security in every field is a necessity and of crucial concern to the public. Proper security of different assets is extremely costly, so as a result of this most organizations compromise their security systems. Furthermore, attackers find new tricks to launch attack as the technology grows and develops. There are still some need to identify various solutions, which would be cheaper, or at least of reasonable in cost to apply and give complete security to assets. Moreover, the challenges and security threats in cyber space are also increasing. These attacks are growing at a faster rate than ever in conjunction with technological development and despite various advanced security solutions (Duić, Cvrtila, & Ivanjko, 2017). Cyber security is the crucial concern of every nation, organization and individual. This situation has made it more problematic as there are no actionable international written document, which demonstrates proper and clear legitimate rules and regulations for cybercrimes (Duić et al., 2017).

The main purpose of this research is to examine cyber security issues and find appropriate solutions. Studies into cyber security concepts, applications, issues and solutions have discussed the definitions of cyber security and similar concepts as well as describe numerous types of solutions. Developers have created various alternatives to deal with a range of cyber security issues. However, technological advances often present new problems and one major concern is the protection of privacy and security of data, such as networks and smart/computing devices. As a result, this study undertook a content analysis of the literature published in regard to security and privacy issues in cyber security.

Graphical Abstract: Figure 1 represents the pictorial demonstration of the abstract of this chapter

Definitions of Cyber Security Solution Techniques and Tools

See Table 1.

Comparison of Different Key Factors

See Table 2.

There is a motive behind every attack whether it is money related, organizational reconnaissance, or related to national spying. Most targeted attacks have the motive of getting information about another nation's domestic, trading, and military operations. Attacks on these networks can be extremely useful for the attackers as vast quantity of information can be disclosed from it (Sood & Enbody, 2013).

Figure 1. Graphical Abstract: The data gathered from thirty-one peer-reviewed research papers from 2009 – 2017 are outlined. Some recent (2018) cyber security issues have been mentioned. Cyber security issues, threats, proposed solutions, applications and benefits of solutions by authors are studied to gather data. The analysis of this research conducted clearly shows that the Government Public sector and private enterprises along with organizations are the main targets for cyber-attacks. Research has mostly focused on government and public sector application areas followed by private enterprises and organization sectors that compared the issues to find solution techniques.

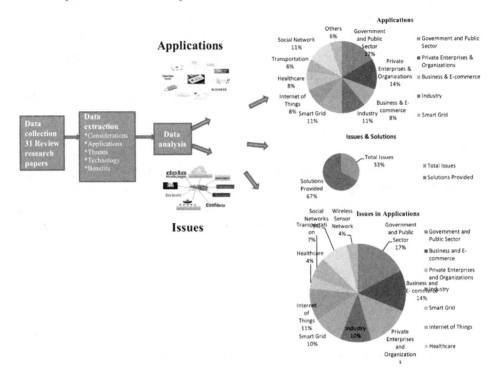

According to (Sood & Enbody, 2013) the United State Government Accountability Office's current report recommends that the motives of these attacks can also be to interfere with military grade electronic segments. This allows fraudulent vendors to sell inferior products to exploit authorized numbers from genuine parts. There can also be a scenario in which a targeted attacker can compromise and tamper with the website of a vendor to get information and offer inferior electronic parts that would consequently be sold to the military. These attacks present a major risk for the national security of each nation (Sood & Enbody, 2013).

Table 1. Definitions of cyber security solution techniques/tools

No.	Cyber Security Solution	Description
1.	kHIVE(Salvatore C Paladino & Jason E Fingerman, 2009)	kHIVE is a software-based host integrity engine for authentication. It is intended to identify the existence of sophisticated malware such as user or kernel rootkits. The program is able to observe running processes, memory, and the Windows Registry.
2.	MOZART(Salvatore C Paladino & Jason E Fingerman, 2009)	MOZART is computerized webpage evaluation tool to look for interior and exterior websites for a diversity of information as described by selected sets of rules. Privacy act, sensitive, and other processes security connected information are highlighted for evaluate to recognize the happening of irregularity as given in the sets of rules.
3.	CAULDRON (Combinatorial Analysis Utilizing Logical Dependencies Residing on Networks)(Salvatore C Paladino & Jason E Fingerman, 2009)	It is a hazard evaluation tool, which finds for series of interdependent loop holes spread across a network.
4.	IronKey(Salvatore C Paladino & Jason E Fingerman, 2009)	It is a Universal Serial Bus (USB) flash drive which offers improved security functions such as password-protected access control, hardware-level encryption, and a self-destruct characteristic to avoid leakage of data in the occasion of device theft or loss.
5.	CySeMoL (Cyber Security Modeling Language) (H. Iguer et al., 2014)	It is a software tool and modeling language, that can be utilized for analysis of cyber security of enterprises. The key purpose of CySeMoL is to permit users to make replica of their architectures and make computations on the probability of various cyber-attacks being victorious.
6.	Demilitarized zone(Marc Kowtko, 2011)	It is different network from local network. The key function is protecting the network parts from intrusion and its effects. These are used to guard the internal secure network from intrusion. it can be seen as a security's second layer.
7.	Honeypot(Marc Kowtko, 2011)	It is a computer that is a planned trap for hackers that permit network administrators to watch hacker actions and expose their strategies. Honeypots are for attracting hackers.
8.	ECDSA (Elliptic Curve Digital Signature Algorithm)(Zheng et al., 2015)	It is an asymmetric cryptographic algorithm. In this a digital signature is created with a private key and public key pair by sender, that is utilized for authentication, and a mixture of AES-CCM and Elliptic Curve Integrated Encryption Scheme are utilized for the encryption.
9.	eTRON(M. Fahim Ferdous Khan & Ken Sakamura, 2016)	It supports utilize tamper resistant devices and public key cryptography.
11.	PDS (Personal Data Stores) (Chaudhary & Kumar, 2015)	User's personal data can be reserved at single place. It can be put under the organization of user or security specialists. It makes it simpler to secure the data.
12.	DECENT (Decentralized) (Chaudhary & Kumar, 2015)	It is decentralized architecture of the online social network has been suggested in. It has main focal point on privacy and security. To save the user's data, it uses distributed hash table.
13.	Secure vault (Chaudhary & Kumar, 2015)	It is a Secure Social Networking Site. It entails combination of the encryption concepts, dislocation of data and false information. "False" information is given to illegal users. Secure vault utilizes encryption to secure private or receptive users' information.

Table 2. Comparison of different Key factors, such as applications issues, solutions and their benefits are surmised as data for this study taken from 31 peer-reviewed scientific publications (2009-2017) describing cyber security related issues

No.	Author	Key Factors	Application	Threats/Issues	Technique/ Solution	Benefits
1.	Paladino et al. (2009) (Salvatore C Paladino & Jason E Fingerman, 2009)	Confidentiality Integrity Availability	Government agencies and Private enterprises	Privacy/Data breach, Network attacks, Malware	1. kHIVE 2. MOZART 3. Wireless intrusion detection system 4. CAULDRON (Combinatorial Analysis Utilizing Logical Dependencies Residing on Networks) 5. IronKey	1. Provides sophisticated malware detection 2. Generate quality reports, allow apprentice analysts to support in forensic examinations of data loss or device theft 3. Identify instances of device misconfiguration and unauthorized access 4. Provides situational awareness for network vulnerabilities and security measures to exclude potential attack paths 5. Enhanced security features (hardware level encryption, password-protected access control, self-destruct features in case
2.	IGUER et al. (2014) (Hajar IGUER, MEDROMI, SAYOUTI, & 2014)	Confidentiality Integrity Availability, Accountability, Audit	Businesses	Phishing e-mails Malware Identity theft Intellectual property theft Hacking	CySeMol (cyber Security Modeling Language)	It permits cyber-security analyses of enterprise architectures without needing any major cyber-security experience of the modeler.
3.	Kowtko (2011) (Marc Kowtko, 2011)	Data/ information leakage, Confidentiality, Availability,	Private and Public (Organizations) sector	Intrusion Denial-of-Service attack	1. Demilitarized zones 2. Intrusion Prevention System 3. Intrusion Detection System 4. Dedicated firewalls 5. Honeypot 6. Denial-of-Service filters 7. Biometric system	1. Separate a company's web server, e-mail server or other servers accessed by the public (Prevent intrusion) 2. Prevent network attack 3. Detect network attack 4. Adequate protection 5. Intelligence, Distraction 6. Capture and block bad requests 7. Authentication
4.	Qian (2013)(Qian, 2013)	Security of data, devices and networks	Government Data Centre	Unauthorized access Hacking Virus attacks	1. Firewall 2. Intrusion Detection System 3. Intrusion Prevention System 4. Anti-Viruses	1. Controlling access 2. Real-time monitoring, detecting viruses and hacking 3. Filter network traffic 4. Scanning virus
5.	Yang et al. (2014) (Wen Yang & Qianchuan, 2014)	(Industrial Control Systems) a. Programmable logic Controller security b. Supervisory software security	Industries	Firmware modification Man-in-the-middle attack Stuxnet Password recovery Authentication bypass	Cryptography	Improved security

continued on following page

Table 2. Continued

No.	Author	Key Factors	Application	Threats/Issues	Technique/ Solution	Benefits
6.	Chauhan et al., (2013) (Pavitra Chauhan, Nikita Singh, & Nidhi Chandra, 2013)	Confidentiality Integrity Availability, Business transactions	Organizations	Unauthorized access and interception, Network integrity violations Denial of Service, Data theft, Espionage	1. Bioinformatics 2. SWOT (Strength, Weaknesses, Opportunities, Threats) 3. SLEPT (Social, Legal, Economic, Political, Technological)	1. Predict cyber security attacks by discovering rising threats based on computer contamination and incursions analysis based on human illness model.
7.	Chatterjee (2015) (Sheshadri Chatterjee, 2015)	Security Privacy Confidentiality Integrity Availability Auditing Non-repudiation Authentication	E - Commerce (Online shopping)	Unauthorized access, reuse of other data without permission (replay attack)	1. Firewall 2. Encryption software 3. Digital signature	1. Provide authenticity 2. Provide data security 3. Provide secure payment
8.	Pour, et al. (2017) (Maneli Malek Pour, Arash Anzalchi, & Arif Sarwat, 2017)	Devices and network Confidentiality Integrity	Smart Grid	Man-in-the-middle attack Distributed Denial-of-Service attack False data injection attack Jamming attack	1. IP Fast Hopping mechanism 2. Intrusion Detection System -based Technologies 3. Encryption Mechanisms	1. Hide actual IP address of servers between large network 2. Detect doubtful power failure, unusual log information and traffic 3. Ameliorate data confidentiality and Integrity
9.	Shapsough, et al. (2015) (Salsabeel Shapsough, Fatma Qatan, Raafat Aburukba, Fadi Aloul, & A R. Al Ali, 2015)	Confidentiality Integrity Availability Accountability Audit Non-repudiation	Smart Grid	Man-in-the-middle attack Distributed Denial-of-Service attack False data injection attack Jamming attack	1. DoS Mitigation (1a. Pushback 1b. Rate limiting 1c. Filtering 2. DoS detection 2a. Using Flow Entropy 2b. Using Signal Strength 2c. Using Transmission Failure Count 2d. Using Signatures 2e Using Sensing Time Measurement)	1a. Router obstructs whole traffic that matches (blocked) Characteristics 1b. Lessen effectiveness of attack 1c. Filtering of the source IP of suspicious packet with blacklist's detector and block it 2a. Probabilistic method to analyze all the traffic 2b. Empirical method (differentiate between legitimate data and jamming attack) 2c. Detecting jamming attack to keep track of transmission failure 2d. Using the previous kind of (known) characteristics and patterns of attacks. 2e. Transmitter acknowledge the channel to verify that is it free prior starting to send data
10.	Hu, et al. (2016) (Jiankun Hu & Athanasios V Vasilakos, 2016)	Data privacy/ security Energy-theft detection	Smart grid (Energy Big data)	Man-in-the-middle attack Distributed Denial-of-Service attack False data injection attack Jamming attack	1. Scalable key management (Identity-based key tree Cryptosystem) 2. Cloud service based hierarchical cryptosystem 3. Privacy-preserving range query scheme	1. Efficient for de-synchronization attack, secure communication 2. Confidentiality, Integrity, Authenticity, non-repudiation, scalability 3. Privacy preserving intelligent applications, scalability

continued on following page

Table 2. Continued

No.	Author	Key Factors	Application	Threats/Issues	Technique/ Solution	Benefits
11.	Hossain, et al. (2017) (Mahmud Hossain, Ragib Hasan, & Anthony Skjellum, 2017)	Devices, Networks	Internet of Things	Interruption Eavesdropping Modification Fabrication Message Replay User/Software/ Hardware compromise Denial of Service attack Internal External Physical Logical attack	1. End to End Network Security (Host Identity Protocol (HIP)-based Schemes) 2. End to End Transport Security (Constrained application protocol), 3. Access Control Mechanisms (Role-based Access Control and Capability-based Access Control)	1. IoT devices' authentication by considering by devices' mobility properties 2. End-to-end datagram transportation 3. a. Suitable for human-to-things communication b. Maps access rights like read/write
12.	Radovan et al. (2017) (M Radovan & B Golub, 2017)	Data Devices Networks Confidentiality Integrity Availability	Internet of Things	Unauthorized access radio connection hacking	1. Access control 2. Bidirectional gap 3. Application white-listing 4. Asset identification 5. Antivirus 6. Assessment and Audit 7. Continuous monitoring and log analysis Data forensics Anomaly detection tools	1. Authentication/ Authorization 2. Firewalls, between control systems and rest of network 3. To control which applications are permitted to install or execute on a specific host 4. Visibility of components within the control system equipment and network activity
13.	Tyagi, et al. (2017) (Akshat Tyagi, Juhi Kushwah, & MOnica Bhalla, 2017)	Confidentiality Integrity Availability, Accountability, Audit Self - Organization, Data Freshness, Non-repudiation	Wireless sensor networks	Jamming Exhaustion Tampering Wormhole attack Identity replication attack	1. Intrusion Detection 2. Secure Data aggregation 3. Secure Group Management	1. Intrusion detection (traffic and computation analysis) 2. Securing aggregated data in network 3. Securing group of nodes where important services are done by group of nodes
14.	Aledhari, et al. (2017) (Mohammed Aledhari, Ali Marhoon, Ali Hamad, & Fahad Saeed, 2017)	Patient security, Data privacy, Confidentiality, Nodes/Devices, Networks	Healthcare networks	Identity theft Intrusion	1. Genomic-based Cryptography 2. Deterministic Chaos Theory, 3. One-Time Pad Encryption Method	1. Information carrier and as a modern implementation tool in biological technology 2. Extreme sensitivity to initial conditions and some other interesting properties, such as pseudo-randomness, wide spectrum, and sound correlation, the chaotic signal looks like noise to unauthorized users, low-cost, secure, fast encryption 3. simple

continued on following page

Table 2. Continued

No.	Author	Key Factors	Application	Threats/Issues	Technique/ Solution	Benefits
15.	Chiuchisan1, et al. (2017) (I. Chiuchisan, D.-G. Balan, O. Geman, I. Chiuchisan, & I. Gordin, 2017)	Information security Privacy Quality of Service	Healthcare networks	Intrusion attack Viruses	1. Secure File Transfer Protocol 2. Hyper Text Transfer Protocol Secure 3. Encrypted Data Base tables 4. Virtual Private Network	1. More secure transfer 2. Web security 3. Authorized access only 4. End-to-end secure connection
16.	Zheng et al. (2015) (Zheng et al., 2015)	Intra-vehicle security Safety critical information Authentication	Underground railway system	DoS Spoofing Masquerade Replay attacks Message falsification Impersonation	1. Encryption/ Cryptography techniques 2. Elliptic Curve Digital Signature Algorithm (ECDSA)	1. Improved security 2. Secure authentication
17.	Pelargonio et al. (2014) (Pierluigi Pelargonio & Marco Pugliese, 2014)	Information security, storage processing Networks	Underground railway system	Intrusion	1. Intrusion Detection System 2. Object Counting 3. Changing scenario Monitoring	1. Able to do differentiation between human and other things, reduced false alarms 2. Output in graphical/tabular form 3. Inform about the suspicious approaches, burglary, unwatched stuff etc.
18.	Khan et al. (2016) (M. Fahim Ferdous Khan & Ken Sakamura, 2016)	Data security and privacy, Authentication, Access control	Healthcare	Man-in-the-middle-attack, Trojan horse attack	1. Role based access control (RBAC) 2. Discretionary access control (DAC) 3. Mandatory access control (MAC) 4. eTRON and eTNet architecture	1. It links the users to dissimilar roles, where each role is approved an appropriate group of authorizations compulsory and enough to do job roles of any person (employee) assuming that function. 2. It permits the users to allow access to any entity, which is in their control. 3. It limits the access to things based on the sensitivity of the information. 4. It uses the public key cryptography and interfere conflict devices.

continued on following page

Table 2. Continued

No.	Author	Key Factors	Application	Threats/Issues	Technique/ Solution	Benefits
19.	Khan et al. (2014) (Khan & Mashiane, 2014)	Information privacy and security	Social network	Identity theft and authentication, Intellectual property theft, Vandalism, Stalking and harassment, Malware and viruses	1. Controlled settings of social networking site account. 2. Test (by using own information and run the inquiry for verifying if any of the individual's private data/ information is visible to public) 3. Supervise (always observing own privacy settings of account)	1. Third person cannot look into your social networking/ data account. 2. It decides that the user is secure from possible dangerous queries of graph. 3. To guarantee that one's data is protected and detecting potential threats
20.	Yadav et al. (2016) (Suman Avdhesh Yadav, Shipra Ravi Kumar, Smita Sharma, & Akanksha Singh, 2016)	Service availability, Data confidentiality, Information integrity	Smart Grid	Threats of Physical layer (Sabotage, vandalism, Theft) Thefts of Logical layer (data confidentiality), Malware, Replay attacks, Unauthorized access, DoS attacks, Traffic analysis	Constant testing, Profiling	To keep constantly performing traffic analysis and security
21.	Nandhini et al. (2016) (Nandhini & Das, 2016)	Information and account privacy and security	Social networks	Malware download, Identity theft, Social scams	1. Methods based on Graph structure 2. Method of opinion mining 3. Image verification algorithms and Image processing methods	1. Sensing the fraud 2. To authenticate the originality of posts of wicked user and examines them if it is the genuine or bogus user posting posts 3. Verifying that pictures are actual or not
22.	DRIAS et al. (2015) (Zakarya DRIAS, Ahmed SERHROUCHNI, & Olivier VOGEL, 2015)	Performance, Availability, Safety, Flexibility, Integrity, Confidentiality,	Industrial Control Systems	Denial-of-Service, Eavesdropping, Man-in-the-middle attack, Breaking into a system	1. Intrusion detection and prevention system 2. Cryptographic counter measures	2. To guarantee truthfulness and secrecy of data and verification of origin of messages and the message itself and non-repudiation actions executed in the control system
23.	Chaudhary et al. (2015) (Chaudharya & Kumarb, 2015)	Confidentiality, Privacy, Integrity, Authenticity, Availability	Social networks	Identity theft, Personalized spam, Digital stalking	1. Personal data stores (PDS) 2. Data mining algorithms 3. Access control, 4. Social graph anonymization 5. DECENT (decentralized online social network architecture) 6. Secure vault	1. Simpler to shield the data, because entire data is residing at a single place 2. Protection of privacy 5. To store the user information, using a dispersed hash table, CIA user's contents uses Cryptographic methods 6. False data is supplied to users who are not authorized, for protection encryption is used

continued on following page

Table 2. Continued

No.	Author	Key Factors	Application	Threats/Issues	Technique/ Solution	Benefits
24.	Mahmood et al.(2013) (D. T. Mahmood & U. Afzal, 2013)	Information security, Passwords, Digital certificates, GPS, Locations	Organizations (Big data)	Spamming, Botnets, Malware, Search poisoning, Phishing, Denial of service	1. Threat detection a. Network traffic b. Web transactions c. Network server d. Network source e. User credentials 2. Corporate Security Analytics Solutions a. Analysis of root Cause b. Analysis of pathway c. Discovery of application d. Discovery of data leakage e. Analysis of insider threat	a. Sensing and forecast doubtful origin and end-place, along with traffic patterns, which is not normal. b. Sensing and forecasts user access patterns that are not normal, mainly in the practice of activities or vital resources. c. Sense and forecast server control patterns that are not normal e.g., unusual or unexpected changes in configuration, non-compliance with pre-defined policy etc. d. Sense and forecast unusual patterns of use of any machine, e.g., associated to type of information/data the origin broadcasts, process and collects. e. Sensing irregularity w.r.t. a user, or a user's group, not obeying with its inherent access behaviour, e.g., unusual access transaction amount or time 2. a. Capability to go reverse in time from any security incident to establish the source cause, to decide that something took place, what just occurred, and how it came into the network. b. Capability to go ahead in time from a security incident to decide the full claim of this happening in the upcoming time e.g., damage, predictable path etc c. Capability to find out doubtful applications on network, permitting security systems to obstruct applications based on the ID of the application. d. Capability to supplied the motives behind loss of data based on the framework of the missing information e. Capability to remain path of the actions of employees to predict and notice reasons of probable attacks carried out by them.

continued on following page

Table 2. Continued

No.	Author	Key Factors	Application	Threats/Issues	Technique/ Solution	Benefits
25.	Jayasingh et al. (2016) (Bipin Bihari Jayasingh, M. R. Patra, & D Bhanu Mahesh, 2016)	Privacy, Security, Integrity, Authenticity	Commercial banks Credit card Phone companies (Big data)	Fraud	1. Security Incident and Event Management (SIEM), 2. Encryption (Rijndeal, Blow fish, Transparent, Application), 3. API (Application Programming Interface) at application level, 4. Log analysis	1. Detection of fraud 2. Controls access at the file level by transparent encryption and specific columns are encrypted by application encryption in an application ahead of it writes fields to the database.
26.	Deore et al. (2016) (Deore & Waghmare, 2016)	Information security	Private and Government sectors (For distributed data)	Eavesdropping, Phishing attack, Spoofing, Man-in-the-middle attack, Denial-of-service attack, Virus attack	1. One-time password (OTP) 2. Firewalls, Anti-virus 3. Log analysis 4. Distributed Cyber Security Automation Framework for Experiments (DCAFE)	1. It is helpful to protect a system. It is authentic for only one single session. 2. Protect the private information/data and the computer networks 3. For sensing the threats 4. The application establishes DCAFE, a lightweight, realistic, dispersed structure for the computerization of cyber security investigation experiments, based on s/w agents, which can handle system tasks, computerize data gathering, examine results, and run novel research without human involvement. For a specified attack requirement, DCAFE has confirmed its capability to find out occasion log configurations holding compromise indicators, which can be successfully engaged by mechanisms of attack recognition.

continued on following page

Table 2. Continued

No.	Author	Key Factors	Application	Threats/Issues	Technique/ Solution	Benefits
27.	Fan et al. (2015) (Fan, Fan, Wang, & Zhou, 2015)	Information security, Functional security, Physical security,	Power plants, transportation industries etc. (Industrial Control System)	(Information threats) Unauthorized access to resources of system and change by chance, unauthorized or loss or damage (Functional threats) Device malfunction and breakdown (Physical threats) Fire, electric shock, mechanical hazards radiation, chemical risks and other aspects	1. Defense- in-depth model 2. Programmable Logic Controllers (PLC) 3. Authentication technology and Access Control using the dispersed firewall 4. Intrusion Detection Technology	It separated industrial control network into various zones of security. By setting up intrusion detection, firewalls, loopholes examining and other measures for security can create an important safeguard. 2. Corporations have been established to upload the PLC (Programmable Logic Controllers) which control the automation processes of industries. 3. It includes a defensive layer among inner subnet contrast with the traditional boundary firewall. This can build various configurations for every service item, and completely consider the running applications and network processing load while setting the configure. Firewall rule configuration utilizes the mechanism of white list that builds dynamic decisions of behaviour between information network and control network. 4. It is able for discriminating usual and aberrant traffic in ICS, noticing security events before time phase.
28.	Cashion et al. (2011) (Jeffrey Cashion & Bassiouni, 20111)	Security, Privacy	Social networking sites (Businesses)	Session -hijacking, Phishing attacks, Spamming	1. Self-Configuring Repeatable Hash Chains (SCRHC), 2. Rolling Code protocol,	1. This authentication protocol hinders the session hijacking cookies. It places two different types of the flavor of authentication: one for mobile devices using wireless connections and the other for high-end workstations using high-speed broadband connections. The core of the two flavors is the same, however, they differ in how they exercise different aspects of the protocol. 2. The protocol, called Rolling Code utilizes the initial secure HTTPS authentication to exchange a shared secret between the server and the user browser.

continued on following page

Table 2. Continued

No.	Author	Key Factors	Application	Threats/Issues	Technique/ Solution	Benefits
29.	Sharma et al. (2016) (D. P. Sharma, M. D. Doshi, & M. M. M. Prajapati, 2016)	Security and privacy of Sensitive data, electronic transaction protection,	Nation, e-governance, Businesses	Virus/warm, Malware, Financial crimes, Cyber pornography, Online gambling, Cyber defamation, Web jacking, Email spoofing, Data diddling Identity fraud and theft Criminal Activity and Money laundering Email spoofing Phishing Communal Violence and Fanning Tensions Hacking Cyber bullying Cyber stalking	Anti-virus	Defends against attacks of viruses
30.	Irshad (2016) (Irshad, 2016)	Data authentication, Data integrity, Data confidentiality	Internet of Things	Unauthorized access Data modification Information disclosure	1. Cisco security framework 2. Floodgate security framework 3. Constrained Application Protocol Frameworks (CoAP) 4. OSCAR: Object Security Framework for Internet of Thing	1. Threat Detection, Anomaly Detection, Predictive Analysis Contextual-Awareness 2. Identify the threats and Floodgate firewall IDS Support Compliance Assistance Security Evaluation 3. Best Fit in Application Security Framework 4. (a) Access Control Confidentiality, Authenticity, Availability (b) Analyzed and extracted risks of utilizing cloud computing by using the Risk Breakdown Structure (RBS) method.
31.	Razzaq et al.(2013) (Abdul Razzaq, Ali Hur, H Farooq Ahmad, & Muddassar Masood, 2013)	Security of Information and other industry's resources	Industry	Denial of service attack, Malware, Viruses, Fraud, Identity theft, Cyber stalking, Phishing scams, Information warfare	1. Vulnerability Scanners 2. Intrusion Prevention System 3. Intrusion Detection System 4. Network and application Firewall.	1. These the automated tools which initial crawls a web application, and then check its web pages to detect the vulnerabilities in the application using the passive technique. Here, the Scanners produces a probe inputs and then compare response upon these input for security vulnerabilities 2. These are the software designed to detect the unauthorized access to the resources, however, to prevent these resources from the unauthorized access. 3. These are the software designed to detect the illegal access to the system or resources. Signatures based Intrusion Detection Systems identify

MATERIALS AND METHODS

- **Collection of Raw Data:** The raw data was collected from thirty-one research papers focusing on different **cyber security** applications. The attributes were compared by key factors, applications, threats and issues, techniques and benefits.
- **Data Inclusion Criteria:** To assess data inclusion criteria a comparison table was constructed which included attributes such as Author, Key factors, Applications, Threats/Issues, Techniques and Benefits.
- **Analysis of Raw Data:** The analysis of raw applications was grouped according to the major categories such as government and public sector, businesses, smart grid, healthcare, industry, and transportation. Different types of issues were combined together according to the areas of application then the cyber security techniques were combined. Lastly, solution techniques proposed in the literature were analyzed.

The entire number of cyber security applications, threats/issues and techniques/ solutions were described, and their application percentages calculated. Percentage of issues in each applications and percentage of suggested solutions according to the nature of the issue were then calculated accordingly.

The research question on the topic is: what can be done to enhance cyber security and protect confidential information/data, devices and networks from the attackers or cyber/crimes without the need for big budgets? The data for this research was collected from the research studies on cyber security in different areas such as industries, smart grid, internet of things, transportation, social networks, business & e-commerce, government & public sector, healthcare, private enterprises & organizations published from the period 2009 to 2017.

The following tables summarise the information about different applications, threats/issues and the techniques used to solve these issues.

RESULT

Applications and Issues of Cyber Security Using the Data From the 31 Peer-Reviewed Scientific Publications

See Table 3.

Table 3. Applications and Issues of cyber security using the data from the 31 peer-reviewed scientific publications

No.	Application	Issues
1.	Government and public sector	• Privacy/Data(Salvatore C Paladino & Jason E Fingerman, 2009) • Network attacks (Salvatore C Paladino & Jason E Fingerman, 2009) • Malware(Salvatore C Paladino & Jason E Fingerman, 2009), (H. Iguer et al., 2014), (Khan & Mashiane, 2014), (Suman Avdhesh Yadav et al., 2016),(Nandhini & Das, 2016),(D. T. Mahmood & U. Afzal, 2013), (P. Sharma, D. Doshi, & M. M. Prajapati, 2016),(A. Razzaq, A. Hur, H. F. Ahmad, & M. Masood, 2013) • Unauthorized access (Qian, 2013), (P. Chauhan, N. Singh, & N. Chandra, 2013), (S. Chatterjee, 2015),(M Radovan & B Golub, 2017), (Suman Avdhesh Yadav et al., 2016),(Fan et al., 2015), (Irshad, 2016) • Hacking (H. Iguer et al., 2014),(M Radovan & B Golub, 2017), (P. Sharma et al., 2016) • Virus attacks (I. Chiuchisan, D. Balan, O. Geman, I. Chiuchisan, & I. Gordin, 2017), (M. F. F. Khan & K. Sakamura, 2016), (Deore & Waghmare, 2016), (P. Sharma et al., 2016), (A. Razzaq et al., 2013) • Intrusion(Marc Kowtko, 2011),(M. Aledhari, A. Marhoon, A. Hamad, & F. Saeed, 2017), (I. Chiuchisan et al., 2017), (P. Pelargonio & M. Pugliese, 2014) 1) Eavesdropping (M. Hossain, R. Hasan, & A. Skjellum, 2017), (Z. Drias, A. Serrhrouchni, & O. Vogel, 2015), (Deore & Waghmare, 2016) • Phishing attack (H. Iguer et al., 2014),(D. T. Mahmood & U. Afzal, 2013), (Deore & Waghmare, 2016), (J. Cashion & Bassiouni, 2011),(P. Sharma et al., 2016), (A. Razzaq et al., 2013) • Spoofing (Bowen et al., 2015) (Deore & Waghmare, 2016), (P. Sharma et al., 2016) • Man-in-the-middle attack (W. Yang & Zhao, 2014), (M. M. Pour et al., 2017)., (S. Shapsough et al., 2015), (J. Hu & A. V. Vasilakos, 2016), (M. F. F. Khan & K. Sakamura, 2016),(Z. Drias et al., 2015),(Deore & Waghmare, 2016) • Denial-of-service attack (Marc Kowtko, 2011),(P. Chauhan, N. Singh, & N. Chandra, 2013),(Bowen et al., 2015),(Suman Avdhesh Yadav et al., 2016), (Z. Drias et al., 2015),(D. T. Mahmood & U. Afzal, 2013), (Deore & Waghmare, 2016), (A. Razzaq et al., 2013) • Financial crimes (P. Sharma et al., 2016) • Cyber pornography (P. Sharma et al., 2016) • Online gambling (P. Sharma et al., 2016) • Cyber defamation (P. Sharma et al., 2016) • Web jacking (P. Sharma et al., 2016) • Data diddling (P. Sharma et al., 2016) • Identity fraud (P. Sharma et al., 2016) and theft (H. Iguer et al., 2014), (M. Aledhari et al., 2017), (M. F. F. Khan & K. Sakamura, 2016),(Nandhini & Das, 2016),(Chaudhary & Kumar, 2015),(P. Sharma et al., 2016), (A. Razzaq et al., 2013) • Criminal Activity and Money laundering (P. Sharma et al., 2016) • Communal Violence and Fanning Tensions (P. Sharma et al., 2016) • Cyber bullying (P. Sharma et al., 2016) • Cyber stalking (P. Sharma et al., 2016), (A. Razzaq et al., 2013)
2.	Businesses and E-commerce	• Phishing e-mails (H. Iguer et al., 2014),(D. T. Mahmood & U. Afzal, 2013), (Deore & Waghmare, 2016), (J. Cashion & Bassiouni, 2011),(P. Sharma et al., 2016), (A. Razzaq et al., 2013) • Malware(Salvatore C Paladino & Jason E Fingerman, 2009), (H. Iguer et al., 2014),(D. T. Mahmood & U. Afzal, 2013), (P. Sharma et al., 2016), (A. Razzaq et al., 2013) • Identity theft (H. Iguer et al., 2014), (M. Aledhari et al., 2017),(M. F. F. Khan & K. Sakamura, 2016), (Nandhini & Das, 2016), (Chaudhary & Kumar, 2015), (P. Sharma et al., 2016), (A. Razzaq et al., 2013) • Intellectual property theft (H. Iguer et al., 2014), (M. F. F. Khan & K. Sakamura, 2016) • Hacking (H. Iguer et al., 2014),(M Radovan & B Golub, 2017), (P. Sharma et al., 2016) • Fraud, (A. Razzaq et al., 2013) • Session -hijacking (J. Cashion & Bassiouni, 2011) • Spamming (Chaudhary & Kumar, 2015),(D. T. Mahmood & U. Afzal, 2013), (J. Cashion & Bassiouni, 2011) • Virus (I. Chiuchisan et al., 2017), (M. F. F. Khan & K. Sakamura, 2016), (Deore & Waghmare, 2016), (P. Sharma et al., 2016), (A. Razzaq et al., 2013)/warm • Financial crimes (P. Sharma et al., 2016) • Online gambling (P. Sharma et al., 2016) • Cyber defamation (P. Sharma et al., 2016) • Web jacking (P. Sharma et al., 2016) • Email spoofing (Bowen et al., 2015), (Deore & Waghmare, 2016), (P. Sharma et al., 2016) • Data diddling (P. Sharma et al., 2016) • Criminal Activity and Money laundering (P. Sharma et al., 2016) • Unauthorized access (Qian, 2013), (P. Chauhan, N. Singh, & N. Chandra, 2013), (S. Chatterjee, 2015),(M Radovan & B Golub, 2017), (Suman Avdhesh Yadav et al., 2016),(Fan et al., 2015),(Irshad, 2016) • Reuse of other data without permission (replay attack) (S. Chatterjee, 2015),(M. Hossain et al., 2017), (Bowen et al., 2015), (Suman Avdhesh Yadav et al., 2016)

continued on following page

Table 3. Continued

No.	Application	Issues
3.	Private enterprises and organizations	• Intrusion(Marc Kowtko, 2011),(M. Aledhari et al., 2017), (I. Chiuchisan et al., 2017), (P. Pelargonio & M. Pugliese, 2014) • Denial-of-Service attack(Marc Kowtko, 2011), (P. Chauhan, N. Singh, & N. Chandra, 2013), (M. M. Pour et al., 2017)., (S. Shapsough et al., 2015), (J. Hu & A. V. Vasilakos, 2016),(M. Hossain et al., 2017), (Bowen et al., 2015) • Privacy/Data breach (Salvatore C Paladino & Jason E Fingerman, 2009) • Network attacks (Salvatore C Paladino & Jason E Fingerman, 2009) • Malware(Salvatore C Paladino & Jason E Fingerman, 2009), (H. Iguer et al., 2014),(M. F. F. Khan & K. Sakamura, 2016), (Suman Avdhesh Yadav et al., 2016), (Nandhini & Das, 2016),(D. T. Mahmood & U. Afzal, 2013), (P. Sharma et al., 2016), (A. Razzaq et al., 2013) • Unauthorized access (Qian, 2013), (P. Chauhan, N. Singh, & N. Chandra, 2013), (S. Chatterjee, 2015),,(M Radovan & B Golub, 2017), (Suman Avdhesh Yadav et al., 2016),(Fan et al., 2015),(Irshad, 2016) • Hacking (H. Iguer et al., 2014), (M Radovan & B Golub, 2017), (P. Sharma et al., 2016) • Eavesdropping (M. Hossain et al., 2017), (Z. Drias et al., 2015), (Deore & Waghmare, 2016) • Phishing attack (H. Iguer et al., 2014),(D. T. Mahmood & U. Afzal, 2013), (Deore & Waghmare, 2016), (J. Cashion & Bassiouni, 2011),(P. Sharma et al., 2016), (A. Razzaq et al., 2013) • Spoofing (Bowen et al., 2015), (Deore & Waghmare, 2016), (P. Sharma et al., 2016) • Man-in-the-middle attack (W. Yang & Zhao, 2014), (M. M. Pour et al., 2017)., (S. Shapsough et al., 2015),(M. F. F. Khan & K. Sakamura, 2016),(Z. Drias et al., 2015),(Deore & Waghmare, 2016) • Virus attack (Qian, 2013), (I. Chiuchisan et al., 2017), (M. F. F. Khan & K. Sakamura, 2016), (Deore & Waghmare, 2016), (P. Sharma et al., 2016), (A. Razzaq et al., 2013) • Interception (P. Chauhan, N. Singh, & N. Chandra, 2013) • Network integrity violations (P. Chauhan, N. Singh, & N. Chandra, 2013) • Data theft (P. Chauhan, N. Singh, & N. Chandra, 2013) • Espionage (P. Chauhan, N. Singh, & N. Chandra, 2013) • Spamming (Chaudhary & Kumar, 2015),(D. T. Mahmood & U. Afzal, 2013), (J. Cashion & Bassiouni, 2011) • Botnets(D. T. Mahmood & U. Afzal, 2013) • Search poisoning(D. T. Mahmood & U. Afzal, 2013)
4.	Industries	• Firmware modification (W. Yang & Zhao, 2014) • Man-in-the-middle attack (W. Yang & Zhao, 2014),(M. F. F. Khan & K. Sakamura, 2016),(Z. Drias et al., 2015),(Deore & Waghmare, 2016) • Stuxnet (W. Yang & Zhao, 2014) • Password recovery (W. Yang & Zhao, 2014) • Authentication bypass (W. Yang & Zhao, 2014) • Eavesdropping (M. Hossain et al., 2017), (Z. Drias et al., 2015), (Deore & Waghmare, 2016) • Breaking into a system (Z. Drias et al., 2015) • Malware(Salvatore C Paladino & Jason E Fingerman, 2009), (H. Iguer et al., 2014),(M. F. F. Khan & K. Sakamura, 2016), (Suman Avdhesh Yadav et al., 2016),(Nandhini & Das, 2016),(D. T. Mahmood & U. Afzal, 2013), (P. Sharma et al., 2016), (A. Razzaq et al., 2013) • Viruses (Qian, 2013), (I. Chiuchisan et al., 2017), (M. F. F. Khan & K. Sakamura, 2016), (Deore & Waghmare, 2016), (P. Sharma et al., 2016), (A. Razzaq et al., 2013) • Denial of service attack(Marc Kowtko, 2011), (P. Chauhan, N. Singh, & N. Chandra, 2013), (S. Shapsough et al., 2015), (J. Hu & A. V. Vasilakos, 2016),(M. Hossain et al., 2017), (Bowen et al., 2015) • Fraud (B. B. Jayasingh, M. R. Patra, & D. B. Mahesh, 2016), (A. Razzaq et al., 2013) • Identity theft (H. Iguer et al., 2014), (M. Aledhari et al., 2017), (M. F. F. Khan & K. Sakamura, 2016), (Nandhini & Das, 2016),(Chaudhary & Kumar, 2015),(P. Sharma et al., 2016), (A. Razzaq et al., 2013) • Phishing scams (H. Iguer et al., 2014),(D. T. Mahmood & U. Afzal, 2013), (Deore & Waghmare, 2016), (J. Cashion & Bassiouni, 2011),(P. Sharma et al., 2016), (A. Razzaq et al., 2013) • Information warfare (A. Razzaq et al., 2013)

continued on following page

Table 3. Continued

No.	Application	Issues

5.	Smart grid	• Man-in-the-middle attack (W. Yang & Zhao, 2014), (M. M. Pour et al., 2017)., (S. Shapsough et al., 2015), (J. Hu & A. V. Vasilakos, 2016)
		• Distributed Denial-of-service attack (M. M. Pour et al., 2017)., (S. Shapsough et al., 2015), (J. Hu & A. V. Vasilakos, 2016)
		• False data injection attack (M. M. Pour et al., 2017)., (S. Shapsough et al., 2015), (J. Hu & A. V. Vasilakos, 2016)
		• Jamming attack (M. M. Pour et al., 2017)., (S. Shapsough et al., 2015), (J. Hu & A. V. Vasilakos, 2016), (A. Tyagi et al., 2017)
		• Physical layer threats (Sabotage, vandalism, Theft)(Suman Avdhesh Yadav et al., 2016)
		• Logical layer thefts (data confidentiality)
		• Malware(Salvatore C Paladino & Jason E Fingerman, 2009), (H. Iguer et al., 2014),(M. F. F. Khan & K. Sakamura, 2016), (Suman Avdhesh Yadav et al., 2016),(Nandhini & Das, 2016),(D. T. Mahmood & U. Afzal, 2013), (P. Sharma et al., 2016), (A. Razzaq et al., 2013)
		• Replay attacks (S. Chatterjee, 2015),, (M. Hossain et al., 2017), (Bowen et al., 2015), (Suman Avdhesh Yadav et al., 2016)
		• Traffic analysis(Suman Avdhesh Yadav et al., 2016)
		• Information threats (Fan et al., 2015)
		Unauthorized access to system resources and unauthorized or
		Accidental change damage or loss
		• Functional threats (Fan et al., 2015)
		Equipment failure and malfunction
		• Physical threat (Fan et al., 2015)
		    Electric shock
		    Fire
		    Radiation
		    Mechanical hazards
		    Chemical hazards and other factors

6.	Internet of Things	• Interruption (M. Hossain et al., 2017) • Eavesdropping (M. Hossain et al., 2017), (Z. Drias et al., 2015), (Deore & Waghmare, 2016) • Modification (M. Hossain et al., 2017) • Fabrication (M. Hossain et al., 2017) • Message Replay (S. Chatterjee, 2015),, (M. Hossain et al., 2017), (Bowen et al., 2015), (Suman Avdhesh Yadav et al., 2016) • User/Software/Hardware compromise (M. Hossain et al., 2017) • Dos attack(Marc Kowtko, 2011),(P. Chauhan, N. Singh, & N. Chandra, 2013),(M. M. Pour et al., 2017)., (S. Shapsough et al., 2015),(J. Hu & A. V. Vasilakos, 2016), (M. Hossain et al., 2017), (Bowen et al., 2015) • Internal (M. Hossain et al., 2017) • External (M. Hossain et al., 2017) • Physical (M. Hossain et al., 2017) • Logical attack (M. Hossain et al., 2017) • Unauthorized access (Qian, 2013), (P. Chauhan, N. Singh, & N. Chandra, 2013), (S. Chatterjee, 2015),,(M Radovan & B Golub, 2017) • Radio connection hacking(M Radovan & B Golub, 2017) • Information disclosure (Irshad, 2016)
7.	Wireless sensor networks	• Jamming (M. M. Pour et al., 2017)., (S. Shapsough et al., 2015), (J. Hu & A. V. Vasilakos, 2016), (A. Tyagi et al., 2017) • Exhaustion (A. Tyagi et al., 2017) • Tampering (A. Tyagi et al., 2017) • Wormhole attack (A. Tyagi et al., 2017) • Identity replication attack (A. Tyagi et al., 2017)
8.	Healthcare networks	• Identity theft 7, (M. Aledhari et al., 2017),(Nandhini & Das, 2016),(Chaudhary & Kumar, 2015),(P. Sharma et al., 2016), (A. Razzaq et al., 2013) • Intrusion attack(Marc Kowtko, 2011), (M. Aledhari et al., 2017), (I. Chiuchisan et al., 2017), (P. Pelargonio & M. Pugliese, 2014) • Viruses (Qian, 2013), (I. Chiuchisan et al., 2017), (M. F. F. Khan & K. Sakamura, 2016), (Deore & Waghmare, 2016), (P. Sharma et al., 2016), (A. Razzaq et al., 2013) • Man-in-the-middle-attack (W. Yang & Zhao, 2014), (M. M. Pour et al., 2017)., (S. Shapsough et al., 2015), (J. Hu & A. V. Vasilakos, 2016) • Trojan horse attack, (M. F. F. Khan & K. Sakamura, 2016)

continued on following page

Table 3. Continued

No.	Application	Issues
9.	Underground railway system (transportation system)	• Denial-of-Service(Marc Kowtko, 2011), (P. Chauhan, N. Singh, & N. Chandra, 2013), (M. M. Pour et al., 2017)., (S. Shapsough et al., 2015), (J. Hu & A. V. Vasilakos, 2016), (M. Hossain et al., 2017), (Bowen et al., 2015) • Spoofing (Bowen et al., 2015), (Deore & Waghmare, 2016), (P. Sharma et al., 2016) • Masquerade (Bowen et al., 2015) • Replay attacks (S. Chatterjee, 2015),, (M. Hossain et al., 2017), (Bowen et al., 2015), (Suman Avdhesh Yadav et al., 2016) • Message falsification Impersonation (Bowen et al., 2015) • Intrusion(Marc Kowtko, 2011), (M. Aledhari et al., 2017), (I. Chiuchisan et al., 2017), (P. Pelargonio & M. Pugliese, 2014) • Information threats (Fan et al., 2015) Unauthorized access to system resources and unauthorized or unplanned change, damage or loss • Functional threats (Fan et al., 2015) Equipment failure and malfunction, • Physical threats (Fan et al., 2015) Electric shock Fire Radiation Mechanical hazards Chemical hazards and other factors
10.	Social networks	• Identity theft (H. Iguer et al., 2014), (M. Aledhari et al., 2017), (Nandhini & Das, 2016), (Chaudhary & Kumar, 2015),(P. Sharma et al., 2016), (A. Razzaq et al., 2013) and authentication (M. F. F. Khan & K. Sakamura, 2016) • Intellectual property theft (H. Iguer et al., 2014), (M. F. F. Khan & K. Sakamura, 2016) • Vandalism (M. F. F. Khan & K. Sakamura, 2016), (Suman Avdhesh Yadav et al., 2016) • Stalking and harassment (M. F. F. Khan & K. Sakamura, 2016) • Malware(Salvatore C Paladino & Jason E Fingerman, 2009), (H. Iguer et al., 2014),(M. F. F. Khan & K. Sakamura, 2016), (Suman Avdhesh Yadav et al., 2016), (Nandhini & Das, 2016),(D. T. Mahmood & U. Afzal, 2013), (P. Sharma et al., 2016), (A. Razzaq et al., 2013) • Viruses (I. Chiuchisan et al., 2017), (M. F. F. Khan & K. Sakamura, 2016), (Deore & Waghmare, 2016), (P. Sharma et al., 2016), (A. Razzaq et al., 2013) • Social scams (Nandhini & Das, 2016) • Digital stalking (Chaudhary & Kumar, 2015) • Session-hijacking (J. Cashion & Bassiouni, 2011) • Phishing attacks (H. Iguer et al., 2014),(D. T. Mahmood & U. Afzal, 2013), (Deore & Waghmare, 2016), (J. Cashion & Bassiouni, 2011),(P. Sharma et al., 2016), (A. Razzaq et al., 2013) • Spamming (Chaudhary & Kumar, 2015),(D. T. Mahmood & U. Afzal, 2013), (J. Cashion & Bassiouni, 2011)

Issues and Solution Techniques for Cyber Security Using the Data From the 31 Peer-Reviewed Scientific Publications

See Table 4.

Analysis of Applications Used in Cyber Security

The data has been grouped together according to the following categories: government and public sector, business & e-commerce, healthcare, smart grid, internet of things, industries, private enterprises and organizations, transportation, and wireless sensor networks.

Table 4. Issues and solution techniques for Cyber Security using the data from the 31 peer-reviewed scientific publications

		Issues	Solution Techniques

1.		Privacy/Data breach Network attacks Malware	• Wireless Intrusion Detection System (S. C. Paladino & J. E. Fingerman, 2009)
			• MOZART (S. C. Paladino & J. E. Fingerman, 2009)
			• IronKey (S. C. Paladino & J. E. Fingerman, 2009)
			• Combinatorial Analysis Utilizing Logical (S. C. Paladino & J. E. Fingerman, 2009)
			• Dependencies Residing on Networks (CAULDRON) (S. C. Paladino & J. E. Fingerman, 2009)
			• kHIVE (S. C. Paladino & J. E. Fingerman, 2009)
			• CySeMoL (Cyber Security Modeling Language) (H. Iguer et al., 2014)
			• Restricted account settings of social networking site, (M. F. F. Khan & K. Sakamura, 2016)
			• Test (to use one's own data, run queries to identify if any personal information is publicly accessible), (M. F. F. Khan & K. Sakamura, 2016)
			• Monitor (continuously monitoring own account's privacy settings), (M. F. F. Khan & K. Sakamura, 2016)
			• Continuous testing (S. A. Yadav, S. R. Kumar, S. Sharma, & A. Singh, 2016)
			• Profiling (S. A. Yadav et al., 2016)
			• Threat detection (T. Mahmood & U. Afzal, 2013)
			  o Network traffic
			  o Web transactions
			  o Network server
			  o Network source
			  o User credentials
			• Corporate Security Analytics Solutions (T. Mahmood & U. Afzal, 2013)
			  o Root Cause Analysis
			  o Pathway analysis
			  o Application discovery
			  o Data leakage discovery
			  o Insider threat analysis
			• VulnerabilityScanners (A. Razzaq et al., 2013)
			• Intrusion Prevention System (M. Kowtko, 2011), (Qian, 2013), (Z. Drias et al., 2015), (A. Razzaq et al., 2013)
			• Intrusion Detection System (Marc Kowtko, 2011),(M. Kowtko, 2011), (M. M. Pour et al., 2017)., (A. Tyagi et al., 2017), (P. Pelargonio & M. Pugliese, 2014), (Z. Drias et al., 2015), (Fan et al., 2015), (A. Razzaq et al., 2013)
			• Network and application Firewall (A. Razzaq et al., 2013)

2.		Phishing e-mails Identity theft Intellectual property theft Hacking	• CySeMol (Cyber Security Modeling Language) (H. Iguer et al., 2014)
			• Firewall (M. Kowtko, 2011), (S. Chatterjee, 2015),, (Deore & Waghmare, 2016)
			• Intrusion Detection System (M. Kowtko, 2011), (Qian, 2013), (M. M. Pour et al., 2017)., (A. Tyagi et al., 2017),(P. Pelargonio & M. Pugliese, 2014),(Z. Drias et al., 2015),(Fan et al., 2015), (A. Razzaq et al., 2013)
			• Intrusion Prevention System (M. Kowtko, 2011), (Qian, 2013), (Z. Drias et al., 2015), (A. Razzaq et al., 2013)
			• Anti-Viruses (Qian, 2013), (M. Radovan & B. Golub, 2017), (Deore & Waghmare, 2016), (P. Sharma et al., 2016)
			• Restricted account settings of social networking site (M. F. F. Khan & K. Sakamura, 2016)
			• Test (to use one's own data and run queries to determine if any personal information is publicly accessible) (M. F. F. Khan & K. Sakamura, 2016)
			• Monitor (continuously monitoring own account's privacy settings) (M. F. F. Khan & K. Sakamura, 2016)
			• Threat detection (T. Mahmood & U. Afzal, 2013)
			  o Network traffic
			  o Web transactions
			  o Network server
			  o Network source
			  o User credentials
			• Corporate Security Analytics Solutions(D. T. Mahmood & U. Afzal, 2013) (T. Mahmood & U. Afzal, 2013)
			  o Root Cause Analysis
			  o Pathway analysis
			  o Application discovery
			  o Data leakage discovery
			  o Insider threat analysis
			• One-time password (OTP) (Deore & Waghmare, 2016)
			• Log analysis (Deore & Waghmare, 2016)
			• Distributed Cyber Security Automation Framework for Experiments (DCAFE) (Deore & Waghmare, 2016)
			• Self-Configuring Repeatable Hash Chains (SCRHC) (J. Cashion & Bassiouni, 2011)
			• Rolling Code protocol (J. Cashion & Bassiouni, 2011)
			• Vulnerability Scanners (A. Razzaq et al., 2013)

continued on following page

Table 4. Continued

	Issues	Solution Techniques
3.	Intrusion Denial-of-Service attack Distributed Denial-of-Service attack	• Dedicated firewalls (M. Kowtko, 2011) • Intrusion Detection System (M. Kowtko, 2011), (Qian, 2013), (M. M. Pour et al., 2017)., (P. Pelargonio & M. Pugliese, 2014), (A. Tyagi et al., 2017) • Intrusion Prevention System (M. Kowtko, 2011), (Qian, 2013), (Z. Drias et al., 2015), (A. Razzaq et al., 2013) • Demilitarized zones (M. Kowtko, 2011) • Honeypot (M. Kowtko, 2011) • Biometric system (M. Kowtko, 2011) • DoS filters (M. Kowtko, 2011) • Secure File Transfer Protocol (I. Chiuchisan et al., 2017) • HTTPS (I. Chiuchisan et al., 2017) • Encrypted DB tables (I. Chiuchisan et al., 2017) • Virtual Private Network (I. Chiuchisan et al., 2017) • Scalable key management (Identity-based key tree Cryptosystem) (J. Hu & A. V. Vasilakos, 2016) • Cloud service based hierarchical cryptosystem (J. Hu & A. V. Vasilakos, 2016) • Privacy-preserving range query scheme (J. Hu & A. V. Vasilakos, 2016) • Continuous testing, profiling (S. A. Yadav et al., 2016) • Cryptographic counter measures (Z. Drias et al., 2015) • Threat detection (M. M. Pour et al., 2017). o Network traffic o Web transactions o Network server o Network source o User credentials • Corporate Security Analytics Solutions (T. Mahmood & U. Afzal, 2013) o Root Cause Analysis o Pathway analysis o Application discovery o Data leakage discovery o Insider threat analysis • One-time password (OTP) (Deore & Waghmare, 2016) • Anti-virus (Qian, 2013), (M. Radovan & B. Golub, 2017), (Deore & Waghmare, 2016), (P. Sharma et al., 2016) • Log analysis (Deore & Waghmare, 2016) • Distributed Cyber Security Automation Framework for Experiments (DCAFE) (Deore & Waghmare, 2016) • Vulnerability Scanners (A. Razzaq et al., 2013)
4.	Unauthorized access Virus attacks	• Firewall (M. Kowtko, 2011), (Qian, 2013) (S. Chatterjee, 2015),, (Deore & Waghmare, 2016) • Intrusion Detection System (M. Kowtko, 2011), (Qian, 2013), (M. M. Pour et al., 2017)., (P. Pelargonio & M. Pugliese, 2014), (A. Tyagi et al., 2017) • Intrusion Prevention System (M. Kowtko, 2011), (Qian, 2013), (Z. Drias et al., 2015), (A. Razzaq et al., 2013) • Anti-Viruses (Qian, 2013), (M. Radovan & B. Golub, 2017), (Deore & Waghmare, 2016), (P. Sharma et al., 2016) • Restricted account settings of social networking site (M. F. F. Khan & K. Sakamura, 2016) • Test (to use one's own data and run queries to identify if any personal information is publicly accessible) (M. F. F. Khan & K. Sakamura, 2016) • Monitor (continuously monitoring own account's privacy settings) (M. F. F. Khan & K. Sakamura, 2016) • Continuous testing (S. A. Yadav et al., 2016) • Profiling (S. A. Yadav et al., 2016) • One-time password (OTP) (Deore & Waghmare, 2016) • Log analysis (Deore & Waghmare, 2016) • Distributed Cyber Security Automation Framework for Experiments (DCAFE) (Deore & Waghmare, 2016) • Cisco security framework (Irshad, 2016) • Floodgate security framework (Irshad, 2016) • CoAPConstrained Application Protocol Frameworks (Irshad, 2016) • OSCAR: Object Security Framework for Internet of Thing (Irshad, 2016) • Vulnerability Scanners (A. Razzaq et al., 2013)

continued on following page

Table 4. Continued

		Issues	Solution Techniques

5.		Firmware modification Man-in-the-middle attack Stuxnet, Password recovery Authentication bypass	• Cryptography (W. Yang & Zhao, 2014)
			• IP Fast Hopping mechanism (M. M. Pour et al., 2017).
			• Scalable key management (Identity-based key tree Cryptosystem) (J. Hu & A. V. Vasilakos, 2016)
			• Cloud service based hierarchical cryptosystem (J. Hu & A. V. Vasilakos, 2016)
			• Privacy-preserving range query scheme (J. Hu & A. V. Vasilakos, 2016)
			• Role based access control (RBAC), (M. F. F. Khan & K. Sakamura, 2016)
			• Discretionary access control (DAC), (M. F. F. Khan & K. Sakamura, 2016)
			• Mandatory access control (MAC), (M. F. F. Khan & K. Sakamura, 2016)
			• eTRON and eTNet architecture, (M. F. F. Khan & K. Sakamura, 2016)
			• Intrusion detection (M. Kowtko, 2011), (Qian, 2013), (M. M. Pour et al., 2017)., (P. Pelargonio & M. Pugliese, 2014), (A. Tyagi et al., 2017)
			    and prevention system (M. Kowtko, 2011), (Qian, 2013), (Z. Drias et al., 2015), (A. Razzaq et al., 2013)
			• Restricted account settings of social networking site (M. F. F. Khan & K. Sakamura, 2016)
			• Test (to use one's own data and run queries to identify if any personal information is publicly accessible) (M. F. F. Khan & K. Sakamura, 2016)
			• Monitor (continuously monitoring own account's privacy settings) (M. F. F. Khan & K. Sakamura, 2016)
			  • One-time password (OTP) (Deore & Waghmare, 2016)
			  • Firewalls(Qian, 2013),(S. Chatterjee, 2015),, (Deore & Waghmare, 2016)
			  • Anti-virus (Qian, 2013), (M. Radovan & B. Golub, 2017), (Deore & Waghmare, 2016), (P. Sharma et al., 2016)
			  • Log analysis (Deore & Waghmare, 2016)
			  • Distributed Cyber Security Automation Framework for Experiments (DCAFE) (Deore & Waghmare, 2016)

6.		Interception Network integrity violations Data theft Espionage	• Bioinformatics (P. Chauhan, N. Singh, & N. Chandra, 2013) • SLEPT (Social, Legal, Economic, Political, Technological) (P. Chauhan, N. Singh, & N. Chandra, 2013) • SWOT (Strength, Weaknesses, Opportunities, Threats) (P. Chauhan, N. Singh, & N. Chandra, 2013)
7.		Reuse of others' data without permission (replay attack)	• Digital signature (S. Chatterjee, 2015), • Firewall (M. Kowtko, 2011), (Qian, 2013) (S. Chatterjee, 2015),,(Deore & Waghmare, 2016) • Encryption software (S. Chatterjee, 2015), • Continuous testing, profiling (S. A. Yadav et al., 2016)

8.		False data injection attack Jamming attack	• Scalable key management (Identity-based key tree Cryptosystem) (J. Hu & A. V. Vasilakos, 2016)
			• Cloud service based hierarchical cryptosystem (J. Hu & A. V. Vasilakos, 2016)
			• Privacy-preserving range query scheme (J. Hu & A. V. Vasilakos, 2016)
			• Encryption Mechanisms (M. M. Pour et al., 2017).
			• IDS-based Technologies (M. M. Pour et al., 2017).
			• DoS detection (S. Shapsough et al., 2015)
			  o Using Flow Entropy
			  o Using Signal Strength
			  o Using Transmission Failure Count
			  o Using Signatures
			  o Using Sensing Time Measurement)
			• DoS Mitigation (S. Shapsough et al., 2015)
			  o Pushback
			  o Rate limiting
			  o Filtering
			• Identity-based key tree Cryptosystem (J. Hu & A. V. Vasilakos, 2016)
			• Cloud service based hierarchical cryptosystem (J. Hu & A. V. Vasilakos, 2016)
			• Privacy-preserving range query scheme (J. Hu & A. V. Vasilakos, 2016)

continued on following page

Table 4. Continued

	Issues	Solution Techniques
9.	Interruption Eavesdropping Modification Fabrication Message Replay User/Software/ Hardware compromise Internal External Physical Logical attack	• End to End Network Security (Host Identity Protocol (HIP)-based Schemes) (M. Hossain et al., 2017) • End to End Transport Security (Constrained application protocol) (M. Hossain et al., 2017) • Access Control Mechanisms (Role-based Access Control and Capability-based Access Control) (M. Hossain et al., 2017) • Intrusion detection (M. Kowtko, 2011), (Qian, 2013), (M. M. Pour et al., 2017)., (P. Pelargonio & M. Pugliese, 2014), (A. Tyagi et al., 2017) and prevention system (M. Kowtko, 2011), (Qian, 2013), (Z. Drias et al., 2015), (Abdul Razzaq et al., 2013) (A. Razzaq et al., 2013) • Cryptographic counter measures (Z. Drias et al., 2015) • One-time password (OTP) (Deore & Waghmare, 2016) • Firewalls (Qian, 2013), (S. Chatterjee, 2015),, (Deore & Waghmare, 2016) • Anti-virus (Qian, 2013), (M. Radovan & B. Golub, 2017), (Deore & Waghmare, 2016), (P. Sharma et al., 2016) • Log analysis (Deore & Waghmare, 2016) • Distributed Cyber Security Automation Framework for Experiments (DCAFE) (Deore & Waghmare, 2016) • Cisco security framework (Irshad, 2016) • Floodgate security framework (Irshad, 2016) • CoAP: Constrained Application Protocol Frameworks (Irshad, 2016) • OSCAR: Object Security Framework for Internet of Thing (Irshad, 2016)
12.	Radio connection hacking	• Access control (M. Radovan & B. Golub, 2017) • Bidirectional gap (M. Radovan & B. Golub, 2017) • Application white- listing (M. Radovan & B. Golub, 2017) • Asset identification (M. Radovan & B. Golub, 2017) • Anti-virus (Qian, 2013), (M. Radovan & B. Golub, 2017), (Deore & Waghmare, 2016), (P. Sharma et al., 2016) • Assessment (M. Radovan & B. Golub, 2017) • Audit (M. Radovan & B. Golub, 2017) • Continuous monitoring and log analysis (M. Radovan & B. Golub, 2017) • Data forensics (M. Radovan & B. Golub, 2017) • Anomaly detection tools (M. Radovan & B. Golub, 2017)
13.	Exhaustion Tampering Wormhole attack Identity replication attack	• Intrusion Detection (M. Kowtko, 2011), (Qian, 2013), (M. M. Pour et al., 2017)., (P. Pelargonio & M. Pugliese, 2014), (A. Tyagi et al., 2017) • Secure Data aggregation (A. Tyagi et al., 2017) • Secure Group Management (A. Tyagi et al., 2017)
14.	Identity theft Intrusion	• Genomic-based Cryptography (M. Aledhari et al., 2017) • Deterministic Chaos Theory (M. Aledhari et al., 2017) • One-Time Pad Encryption Method (M. Aledhari et al., 2017) • Secure File Transfer Protocol (I. Chiuchisan et al., 2017) • Hyper Text Transfer Protocol Secure (I. Chiuchisan et al., 2017) • Encrypted DB tables (I. Chiuchisan et al., 2017) • Virtual Private Network (I. Chiuchisan et al., 2017) • Personal data stores (PDS) (Chaudhary & Kumar, 2015) • Data mining algorithms (Chaudhary & Kumar, 2015) • Access control, (Chaudhary & Kumar, 2015) • Social graph anonymization (Chaudhary & Kumar, 2015) • DECENT (decentralized online social network architecture) (Chaudhary & Kumar, 2015) • Secure vault (Chaudhary & Kumar, 2015)
15.	Masquerade Message falsification Impersonation	• Encryption techniques (Bowen et al., 2015) • Elliptic Curve Digital Signature Algorithm (ECDSA) (Bowen et al., 2015)
16.	Intrusion (within monitoring devices and networks)	• Intrusion Detection System (Marc Kowtko, 2011),(M. Kowtko, 2011), (Qian, 2013), (M. M. Pour et al., 2017)., (P. Pelargonio & M. Pugliese, 2014), (A. Tyagi et al., 2017) • Object Counting (P. Pelargonio & M. Pugliese, 2014) • Changing scenario Monitoring (P. Pelargonio & M. Pugliese, 2014)
17.	Trojan Horse attack	• Role based access control (RBAC), (M. F. F. Khan & K. Sakamura, 2016) • Discretionary access control (DAC), (M. F. F. Khan & K. Sakamura, 2016) • Mandatory access control (MAC), (M. F. F. Khan & K. Sakamura, 2016) • eTRON&eTNet architecture, (M. F. F. Khan & K. Sakamura, 2016)

continued on following page

Table 4. Continued

		Issues	Solution Techniques
18.		Vandalism Stalking Harassment	• Restricted account settings of social networking site (M. F. F. Khan & K. Sakamura, 2016) • Test (to use one's own data and run queries to discover if any personal data is publicly obtainable). (M. F. F. Khan & K. Sakamura, 2016) • Monitor (continuously monitoring own account's privacy settings) (M. F. F. Khan & K. Sakamura, 2016) • Continuous testing, profiling (S. A. Yadav et al., 2016) • Personal data stores (PDS) (Chaudhary & Kumar, 2015) • Data mining algorithms (Chaudhary & Kumar, 2015) • Access control (Chaudhary & Kumar, 2015) • Social graph anonymization (Chaudhary & Kumar, 2015) • DECENT (decentralized online social network architecture) (Chaudhary & Kumar, 2015) • Secure vault (Chaudhary & Kumar, 2015)
19.		Social scam	• Graph structure-basedmethod (Nandhini & Das, 2016) • Opinion mining method (Nandhini & Das, 2016) • Image processing methods & image verification algorithms (Nandhini & Das, 2016)

20.		Spamming	• Threat detection (T. Mahmood & U. Afzal, 2013)
			  o Network traffic
			  o Web transactions
			  o Network server
			  o Network source
			  o User credentials
			• Corporate Security Analytics Solutions (T. Mahmood & U. Afzal, 2013)
			  o Root Cause Analysis
			  o Pathway analysis
			  o Application discovery
			  o Data leakage discovery
			  o Insider threat analysis
			• Self-Configuring Repeatable Hash Chains (SCRHC) (J. Cashion & Bassiouni, 2011)
			• Rolling Code protocol (J. Cashion & Bassiouni, 2011)

21.		Fraud	• Security Incident and Event Management (B. B. Jayasingh et al., 2016) • Encryption (Rijndeal, Blow fish, Transparent, Application) (B. B. Jayasingh et al., 2016) • API (Application Programming Interface) at application level (B. B. Jayasingh et al., 2016) • Log analysis (Deore & Waghmare, 2016) • VulnerabilityScanners (A. Razzaq et al., 2013) • Intrusion Prevention System (M. Kowtko, 2011), (Qian, 2013),(Z. Drias et al., 2015), (A. Razzaq et al., 2013) • Intrusion Detection System (M. Kowtko, 2011), (Qian, 2013),(M. M. Pour et al., 2017)., (P. Pelargonio & M. Pugliese, 2014), (A. Tyagi et al., 2017) • Network and application Firewall (A. Razzaq et al., 2013)
22.		Session hijacking	• Self-Configuring Repeatable Hash Chains (SCRHC) (J. Cashion & Bassiouni, 2011) • Rolling Code protocol (J. Cashion & Bassiouni, 2011)

Application Areas Under Cybersecurity

Figure 2 shows that research has mainly focused on government and public-sector application areas (17%) followed by the private enterprises and organization sectors (14%). Next were the categories of social network (11%), smart grid (11%), industry applications (11%), business & e-commerce (8%), internet of things (8%) and healthcare applications (8%). The transportation and the rest of the applications (such as wireless sensor network, and industrial control system) made up only 6%.

Figure 2. Analysis of application areas under cyber security in different areas from 2009-2017

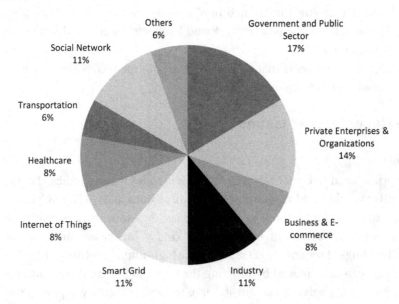

Figure 3. Issues in applications

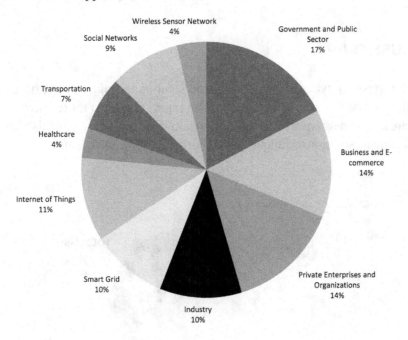

Cyber Security Issues and Solutions

Figure 3 shows that government and public sector had the most threats or issues (17%) whereas wireless sensor network and healthcare were the least prone areas for threats with only 4%.

The Figure 4 represents that there are more solutions (67%) than issues (33%) found or described in the papers reviewed.

Some Recent Cyber Issues

See Table 5.

The prime need in Cyber Security is resolving these rising issues. There is also bigger call to toughen and standardize breach disclosure rules ("Top 7 Cybersecurity Issues of 2018,"). Every small or large business has to incur a greater impact in terms of profit and reputation. Therefore, all kind of businesses must have to offer manifold coatings of security to get ready for the high-impact and unpredicted security threats. This can be achieved by utilizing the emerging technologies and trends in cyber security and applying a useful and apparent cyber security approach with firm security rules and strong antivirus software having all the essential protections in place for securing the networks ("Top 7 Cybersecurity Issues of 2018,").

DISCUSSION

Different types of cyber security issues and solution techniques were observed in this study. Cyber security always remains an interesting topic for research because there are always new types of threats and attacks done by cyber criminals. However, there are many solution techniques for these threats and issues. Unfortunately, there

Figure 4. Total Issues & Solutions provided

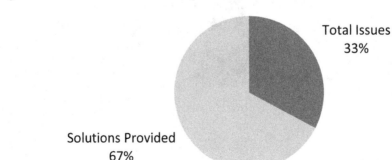

Table 5. Some recent cyber issues

No.	Cyber Issue	Description	References
1.	Cloud is attacked by Ransomware	Ransomware is a malware, which uses some encryption techniques that locks the files on the computer. Hackers use this technique for taking ransoms from the victims to unlock their computer files. Some of these have had a big impact on organizations. Some examples are NotPetya, WannaCry, and Bad Rabbit. In 2017, there were massive attack, known as Ransomware attack, which locked hundreds and thousands of computers. 'FedEx and Maersk' complained millions of dollars in harm because of the 'NotPetya' ransomware attacks in 2017. This year (2018) the main focus is on the businesses of cloud computing. The prevalent cloud players worldwide i.e. Google, Amazon etc. are difficult to hack because of their higher quality digital security techniques set strongly in place. Due to this, smaller businesses, which have lesser resources, are more susceptible to these attacks.	("Top 7 Cybersecurity Issues of 2018,") (Zimba, 2018) (Ali, 2017) (Kotov, 2016)
2.	Hackers use AI Technology Unethically	Artificial Intelligence (AI) technology can be employed to discover, trace, and observe cyber threats, that offer automatic evaluation of computer systems whilst understanding the cyber attacks' patterns, for protecting IT infrastructure by a nimbler cyber defense modification to cyber threats. AI-powered systems can be used to enhance power of human in IT security. However, Corrupt people use it in an immoral way to harm others and own benefits. For instance, hackers perform phishing in more sophisticated way and Viruses are becoming even more clever to circumvent software programs such as 'Sandboxing' intended to identify it.	(Wilkins, 2018) (Grant, 2018) (Zhang, 2018) ("Top 7 Cybersecurity Issues of 2018,")
3.	IoT Device and Augmented Attacks	Through the increased use of the IoT systems, applications and devices such as VR (virtual-reality) devices and wearable devices, probability of cyber-attack threats has also increased. IoT is more likely to be targeted by cyber-attacks because of deficiency in security design settings.	("Top 7 Cybersecurity Issues of 2018,") (Haddadi, 2018) (Darki, 2018)
4.	Targets on Different Infrastructures	Experts in cyber security envisage more attacks on significant infrastructure worldwide. Hackers are aiming at electrical grids, healthcare facilities, transportation systems, and many more attacks by discovering new various and methods to break through the defenses of the older and as well as more vulnerable systems.	(Hilt, 2018) ("Top 7 Cybersecurity Issues of 2018,") (Cornell, 2017)
5.	Mining of Crypto-Currency	Stealing crypto currency such as 'Bitcoin' will persist, as will the computer processing power burglary necessary to mine the crypto currency. If the worth of crypto currency boosts, hackers will definitely make use of enormous numbers of computer networks to aim at those involved in the crypto currencies business.	(Sigler, 2018) (Joy, 2018) (Zimba, 2018) ("Top 7 Cybersecurity Issues of 2018,")

continued on following page

Table 5. Continued

No.	Cyber Issue	Description	References
6.	Escalation of Mobile Threats	Malware specially intended to aim at mobile devices is frequently spread via compromised apps, and cyber criminals that will carry on searching for latest resources and tools to spread malwares i.e on mobile devices worldwide. According to the report of Mobile Security Index 2018 findings show that: • Organizations are giving up mobile security for performance of the business and convenience. Most of the companies have experienced downtime and huge loss of data. • Approximately, all respondents (93%) had same opinion that all mobile devices poses a severe and rising security risk. • In spite of this, many companies were failing to acquire necessary safety measures. Only 39% said they alter all of the default passwords and more than half (51%) did not have any policy of public/open Wi-Fi. • Most of the companies are aware that they are required to take further action. 93% of them agreed that companies should take mobile security more critically.	(Abawajy, 2018) (Roman, 2018) ("Top 7 Cybersecurity Issues of 2018,")

remain advanced types of cyber-attacks, which pose huge threats to organizations. The analysis of the studies showed that the government and public sector is the most cyber-attack prone and the areas least susceptible to cyber-attacks are healthcare and wireless sensor networks. Significantly, the present solutions are costly, so organizations compromise the best solution techniques due to budgetary concerns.

As the analysis showed, over 30% of research was in the government and public sector as well as in private enterprises and organizations. Moreover, the most issues in cyber security were found in the government and public sector with 17%. As such, it could be argued that these sectors have the highest levels of research and issues due to them holding confidential and sensitive information pertaining to the national, public and financial issues. One study (M. Kowtko, 2011) has also found that the government is encountering innumerable cyber-attacks, causing threats on a national level. However, these findings are in contrast to (Kumar et al., 2016) study which discovered that the majority of attacks occurred in industries (28.7%) with government being the second most targeted sector.

In terms of solutions, the studies focused mostly on solutions rather than issues. This focus is justified as businesses need to implement some security measures or guiding principles in order to keep resources (information, devices and networks) safe and protected (A. Razzaq et al., 2013). Therefore, there is a requirement for dynamic solutions that prepare for and adapt to the changing nature of attacks (A. Razzaq et al., 2013). Importantly, the majority of the methods presented in the study were based on signatures that focus on the attack's syntactical representation (A. Razzaq et al., 2013). As such, it is straightforward for an assailant to initiate a hit through small alterations to the signature's syntactical representation. One of the most important challenges is to devise a system, which symbolizes an attack with better concept in order to accommodate for some changes of any particular attack (A. Razzaq et al., 2013). However, the majority of the present solutions utilize primitive (but quite useful) detection mechanisms for static or signature-based attack (A. Razzaq et al., 2013). Therefore, Kowtko (2011) question that if the present solution techniques are insufficient to prevent cyber-attacks what can be done. However, in spite of various solution techniques for the cyber security threats/issues, there are still more complex types of cyber-attacks which can represent some immense threats to organizations.

Security issues (Ekanayake et al., (2018), Munugala et al., (2017), Singh et al., (2018), for example, privacy, secure data transmission, access control, and secure data storage are becoming to be significant obstacles (Zarpelão, 2017) in the cyber space.

As there is an increase in number of cyber-attacks and breaches of security, we must take actions now to address mounting security demands. A country and its financial system depend greatly on the information infrastructure. Therefore, there must be strict protective measures to make sure the information infrastructure is not vulnerable to cyberattacks (M. Kowtko, 2011).

In our research, the second most vulnerable sectors for cyberattacks were business, e-commerce and private organisations. These cyber-attacks can result in great financial crisis to the organisations. The companies can also lose their reputations and customers, if there is a cyberattack which results in a security breach or data breach. Furthermore, other cyberattacks, such as DoS attack and sniffing, can be dangerous for online shopping or e-commerce. As such, there must be adequate security measures for protecting these networks and devices.

Cyber security is becoming a key need today's digital world (Duić et al., 2017). As this study has shown, research has shown the different issues, threats and solutions as well as identified the sectors, which are most vulnerable. On the other hand, this study did not propose any new solution for threats and issues in the cyber world and there is still a need to find such solution techniques, which can fit reasonably into the budget of every organization. Fortunately, researchers are still investigating and experimenting with different security techniques to mitigate the effects of cyber-attacks.

CONCLUSION

In this chapter, a content analysis was performed on 31 peer-reviewed research studies (2009-2017), which discussed the issues and threats as well as the solutions and techniques for cyber security in different application areas. The results of these studies showed that the majority of research has been done in the government and public sector and this is the main target of cyber-attacks. The findings also showed that there are various types of issues and threats in the cyberspaces well as solutions to detect, prevent, and mitigate those threats. Although various security measures have been utilized, attackers are still finding the latest means to hack the systems. Overall, the study shows that the government sector has been targeted more as compared to other sectors, especially compared to healthcare and wireless sensor networks, and therefore cyber security is very crucial national concern. This study can be helpful as a reference for the researchers examining cyber security issues in order to identify the most prominent issues, threats, solutions and sectors.

AUTHOR CONTRIBUTION

G.S. and M.N.H. conceived the study idea and developed the analysis plan. G.S. analyzed the data and wrote the initial paper. M.N.H. helped to prepare the figures and tables and finalizing the manuscript. J.P. completed the final editing of the manuscript. All authors read the manuscript.

REFERENCES

Aledhari, M., Marhoon, A., Hamad, A., & Saeed, F. (2017). *A New Cryptography Algorithm to Protect Cloud-Based Healthcare Services.* Paper presented at the 2017 IEEE/ACM International Conference on Connected Health: Applications, Systems and Engineering Technologies (CHASE).

Aledhari, M., Marhoon, A., Hamad, A., & Saeed, F. (2017). A New Cryptography Algorithm to Protect Cloud-based Healthcare Services. IEEE.

Bowen, Z., Li, W., Deng, P., Gérardy, L., Zhu, Q., & Shankar, N. (2015). *Design and verification for transportation system security.* Paper presented at the 2015 52nd ACM/EDAC/IEEE Design Automation Conference (DAC).

Cashion, J., & Bassiouni, M. (2011). *Protocol for mitigating the risk of hijacking social networking sites.* Paper presented at the 7th International Conference on Collaborative Computing: Networking, Applications and Worksharing (CollaborateCom).

Cashion, J., & Bassiouni, M. (2011). *Protocol for Mitigating the Risk of Hijacking Social Networking Sites.* IEEE.

Chatterjee, S. (2015). *Security and privacy issues in E-Commerce: A proposed guidelines to mitigate the risk.* Paper presented at the 2015 IEEE International Advance Computing Conference (IACC). 10.1109/IADCC.2015.7154737

Chaudhary, M., & Kumar, H. (2015). *Challenges in protecting personnel information in social network space.* Paper presented at the 2015 International Conference on Emerging Trends in Networks and Computer Communications (ETNCC).

Chauhan, P., Singh, N., & Chandra, N. (2013). *Security Breaches in an Organization and Their Countermeasures.* Paper presented at the 2013 5th International Conference and Computational Intelligence and Communication Networks.

Chiuchisan, I., Balan, D., Geman, O., Chiuchisan, I., & Gordin, I. (2017). *A security approach for health care information systems.* Paper presented at the 2017 E-Health and Bioengineering Conference (EHB). Cyber-Physical Systems. Retrieved from https://ptolemy.berkeley.edu/projects/cps/

Deore, U. D., & Waghmare, V. (2016). Cyber Security Automation for Controlling Distributed Data. IEEE.

Drias, Z., Serhrouchni, A., & Vogel, O. (2015). *Analysis of cyber security for industrial control systems.* Paper presented at the 2015 International Conference on Cyber Security of Smart Cities, Industrial Control System and Communications (SSIC).

Duić, I., Cvrtila, V., & Ivanjko, T. (2017). *International cyber security challenges.* Paper presented at the 2017 40th International Convention on Information and Communication Technology, Electronics and Microelectronics (MIPRO).

Ekanayake, B. N. B., Halgamuge, M. N., & Syed, A. (2018). *Review: Security and Privacy Issues of Fog Computing for the Internet of Things (IoT). In Lecture Notes on Data Engineering and Communications Technologies Cognitive Computing for Big Data Systems Over IoT, Frameworks, Tools and Applications* (Vol. 14). Springer.

Fan, X., Fan, K., Wang, Y., & Zhou, R. (2015). *Overview of Cyber-security of Industrial Control System.* Retrieved from https://www.bestcurrentaffairs.com/features-of-smart-cities/

Heikkilä, M., Rättyä, A., Pieskä, S., & Jämsä, J. (2016). *Security challenges in small- and medium-sized manufacturing enterprises.* Paper presented at the 2016 International Symposium on Small-scale Intelligent Manufacturing Systems (SIMS). 10.1109/SIMS.2016.7802895

Hossain, M., Hasan, R., & Skjellum, A. (2017). *Securing the Internet of Things: A Meta-Study of Challenges, Approaches, and Open Problems.* Paper presented at the 2017 IEEE 37th International Conference on Distributed Computing Systems Workshops (ICDCSW).

Hu, J., & Vasilakos, A. V. (2016). Energy Big Data Analytics and Security: Challenges and Opportunities. *IEEE Transactions on Smart Grid, 7*(5), 2423–2436. doi:10.1109/TSG.2016.2563461

Iguer, H., Medromi, H., Sayouti, A., Elhasnaoui, S., & Faris, S. (2014). *The Impact of Cyber Security Issues on Businesses and Governments: A Framework for Implementing a Cyber Security Plan.* Paper presented at the 2014 International Conference on Future Internet of Things and Cloud.

Irshad, M. (2016). A Systematic Review of Information Security Frameworks in the Internet of Things. IEEE.

Jayasingh, B. B., Patra, M. R., & Mahesh, D. B. (2016). *Security issues and challenges of big data analytics and visualization.* Paper presented at the 2016 2nd International Conference on Contemporary Computing and Informatics (IC3I).

Khan, M. F. F., & Sakamura, K. (2016). *A patient-centric approach to delegation of access rights in healthcare information systems.* Paper presented at the 2016 International Conference on Engineering & MIS (ICEMIS).

Khan, Z. C., & Mashiane, T. (2014). An Analysis of Facebook's Graph Search. IEEE.

Kowtko, M. (2011). *Securing our nation and protecting privacy.* Paper presented at the 2011 IEEE Long Island Systems, Applications and Technology Conference.

Kumar, S. R., Yadav, S. A., Sharma, S., & Singh, A. (2016). *Recommendations for effective cyber security execution.* Paper presented at the 2016 International Conference on Innovation and Challenges in Cyber Security (ICICCS-INBUSH). 10.1109/ICICCS.2016.7542327

Ly, K., Sun, W., & Jin, Y. (2016). *Emerging challenges in cyber-physical systems: A balance of performance, correctness, and security.* Paper presented at the 2016 IEEE Conference on Computer Communications Workshops (INFOCOM WKSHPS).

Mahmood, T., & Afzal, U. (2013). *Security Analytics: Big Data Analytics for cybersecurity: A review of trends, techniques and tools.* Paper presented at the 2013 2nd National Conference on Information Assurance (NCIA).

Munugala, S., Brar, G. K., Syed, A., Mohammad, A., & Halgamuge, M. N. (2017). The Much Needed Security and Data Reforms of Cloud Computing in Medical Data Storage. In *Applying Big Data Analytics in Bioinformatics and Medicine.* IGI Global.

Nandhini, D. M., & Das, B. B. (2016). An Assessment And Methodology For Fraud Detection In Online Social. *IEEE Network*, 104–108.

Paladino, S. C., & Fingerman, J. E. (2009). *Cybersecurity Technology Transition: A Practical Approach.* Paper presented at the 2009 Cybersecurity Applications & Technology Conference for Homeland Security.

Pelargonio, P., & Pugliese, M. (2014). *Enhancing security in public transportation services of Roma: The PANDORA system.* Paper presented at the 2014 International Carnahan Conference on Security Technology (ICCST). 10.1109/CCST.2014.6986969

Pour, M. M., Anzalchi, A., & Sarwat, A. (2017). *A review on cyber security issues and mitigation methods in smart grid systems.* Paper presented at the SoutheastCon 2017.

Qian, L. (2013). Study of Information System Security of Government Data Center Based on the Classified Protection. IEEE.

Radovan, M., & Golub, B. (2017). *Trends in IoT security.* Paper presented at the 2017 40th International Convention on Information and Communication Technology, Electronics and Microelectronics (MIPRO).

Razzaq, A., Hur, A., Ahmad, H. F., & Masood, M. (2013). *Cyber security: Threats, reasons, challenges, methodologies and state of the art solutions for industrial applications.* Paper presented at the 2013 IEEE Eleventh International Symposium on Autonomous Decentralized Systems (ISADS). 10.1109/ISADS.2013.6513420

Shapsough, S., Qatan, F., Aburukba, R., Aloul, F., & Ali, A. R. A. (2015). *Smart grid cyber security: Challenges and solutions.* Paper presented at the 2015 International Conference on Smart Grid and Clean Energy Technologies (ICSGCE). 10.1109/ICSGCE.2015.7454291

Sharma, P., Doshi, D., & Prajapati, M. M. (2016). *Cybercrime: Internal security threat.* Paper presented at the 2016 International Conference on ICT in Business Industry & Government (ICTBIG). Smart City. Retrieved from https://www.techopedia.com/definition/31494/smart-city

Singh, M., Halgamuge, M. N., Ekici, G., & Jayasekara, C. S. (2018). *A Review on Security and Privacy Challenges of Big Data. In Lecture Notes on Data Engineering and Communications Technologies Cognitive Computing for Big Data Systems Over IoT, Frameworks, Tools and Applications* (Vol. 14). Springer.

Sood, A. K., & Enbody, R. J. (2013). Targeted Cyberattacks: A Superset of Advanced Persistent Threats. *IEEE Security and Privacy, 11*(1), 54–61. doi:10.1109/MSP.2012.90

Top 7 Cybersecurity Issues of 2018. (n.d.). Retrieved from https://www.cyberreefsolutions.com/top-cybersecurity-concerns-2018/

Tyagi, A., Kushwah, J., & Bhalla, M. (2017). *Threats to security of Wireless Sensor Networks.* Paper presented at the 2017 7th International Conference on Cloud Computing, Data Science & Engineering - Confluence.

Yadav, S. A., Kumar, S. R., Sharma, S., & Singh, A. (2016). *A review of possibilities and solutions of cyber attacks in smart grids.* Paper presented at the 2016 International Conference on Innovation and Challenges in Cyber Security (ICICCS-INBUSH). 10.1109/ICICCS.2016.7542359

Yang, W., & Zhao, Q. (2014). *Cyber security issues of critical components for industrial control system.* Paper presented at the Proceedings of 2014 IEEE Chinese Guidance, Navigation and Control Conference.

Zarpelão, B. B., Miani, R. S., Kawakani, C. T., & Alvarenga, S. C. (2017). A survey of intrusion detection in Internet of Things. *Journal of Network and Computer Applications, 84,* 25–37. doi:10.1016/j.jnca.2017.02.009

Zheng, B., Li, W., Deng, P., Gérard, L., Qi, Z., & Shankary, N. (2015). Design and Verification for Transportation System Security. IEEE.

Chapter 9
Threats and Security Issues in Smart City Devices

Jayapandian N.
Christ University, India

ABSTRACT

The main objective of this chapter is to discuss various security and privacy issues in smart cities. The development of smart cities involves both the private and public sectors. The theoretical background is also discussed in future growth of smart city devices. Thus, the literature survey part discusses different smart devices and their working principle is elaborated. Cyber security and internet security play a major role in smart cities. The primary solution of smart city security issues is to find some encryption methods. The symmetric and asymmetric encryption algorithm is analyzed and given some comparative statement. The final section discusses some possible ways to solve smart city security issues. This chapter showcases the security issues and solutions for smart city devices.

DOI: 10.4018/978-1-5225-7189-6.ch009

INTRODUCTION

The Smart City is the concept of integrate latest technologies and to provide a modern lifestyle for all the citizens. The latest survey shows that, in the year 2030 more than 60% of people moving and living in the urban city. The day to day citizen's lifestyle is growing and moving to the modern technology. The Information and Communication Technology (ICT) plays a vibrant role in smart city concept. The idea of smart city is a combination of traditional and modern technology. It is used to establish ICT in an urban environment. The urban cities are expecting an innovative technology at lesser cost. The final goal of smart city is to provide quality of life in entire citizens. The end of the 20th century smart city concept is evolving. The short duration more innovative technologies developed in smart city. The modern world, people are expecting more sophisticated lifestyle using modern ICT tools. The smart city is to provide social and innovative technology in the existing model. The government service sectors, health and police department is also using these ICT futures (Hernández-Muñoz et al., 2011). The unified information and communication technology is used to optimize and control the urban life. This ICT is also modeling and measuring the sustainable smart city development. The primary objective of this chapter is to discuss about security and privacy issues in smart city and secondary objective is to discuss possible solution to handle this security issues.

The terrorist attack is a major problem in developed countries. Smart city concept is to prevent this type of terrorist attacks. This technology is used to identify the terrorist attackers. The concept of cyber security is to prevent these smart technology devices and provide higher security. In recent days, most of the countries are providing a unique identity number for every citizen. The people are using that unique number or biometric system and access this smart city device. The major problem is once a hacker hacked any one device, they can easy to get user's personal information. The Information Communication Technology is pillar of the smart city. The physical infrastructure and internet connectivity is a major component of smart devices. The entire smart device is connected via internet and sharing information from one smart device to other smart devices. The data transfer speed is also another problem in smart device connectivity. The security concern minimal data handling is not a big task, but they handle higher volumes of data at online cloud storage is really a biggest task. This chapter is discussed on basic structure of smart city. History and development of Information and Communication Technology tools is also discussed. Technology and construction industry is jointly establishing this smart city. The modern world more number of engineering graduates completing the engineering course. The employability is a major task for producing high number of graduates.

This Internet of Things (IoT) technology produces billion and millions of jobs in forthcoming years. The financial growth and economic development of smart city is an important role. The concept of CCTV is old, but smart CCTV is a new one. This part is to elaborate basic technology of smart CCTV technology. The GPS based smart device is hot topic of current technology. The GPS device is used to monitor the vehicle moments and speed.

The smart city is focused on public utility and governance establishment. The modern world industrial development is the most important one; this smart city concept is not a new one. The development of modern industrial products comes to ordinary civilians' usage. The first commercial internet service provider is introduced in the year of 1990s. The intelligent decision mechanism is used for inventing the smart city. The performance of public utility is increased due to this smart city innovation. Smart city is also used to reduce the cost of public infrastructure maintenance. The usage of mobile phone is very high in recent days. Smart mobile phone business is also involved in smart city. The open source android software is used to access and control the smart devices. The major problem with this technology is high-speed wireless internet connection is needed. There is much research is going on public WIFI technology. The reason is public WIFI is required higher bandwidth and higher transmission power. The recent research and development term finally give some solution and provide some new technology named as LTE (Long-Term Evolution). This technology is used to provide high-speed wireless communication for smart mobile (Walravens & Ballon, 2013). Figure 1 is to elaborate evaluation of information technology field.

Figure 1. Evolution of information technology

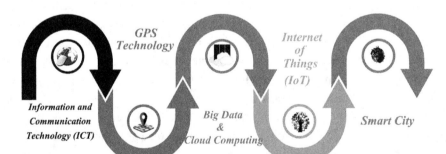

Smart city system is a bridge of the consumer and service provider. That means the consumer get full freedom to give feedback of any public utility. The implementation of this smart city is to provide better QoS (Quality of Service). Advantage of this smart city is to encourage citizens to participate in social activity. The smart city device is a dynamic system because each innovation is a unique concept. The smart city is a combination of science, engineering, innovation and technology. The primary role in this smart device is whenever data needed, they can access it. The information is available at anywhere and anytime. This technology is based on both hardware and software components. The Internet of Things (IoT) is a significant concept of smart city. The modern house is fully automated with the use of IoT technology. IoT is a technology to control and monitor any device through internet. The government is to establish more smart cities under the partnership with private and public sector. The smart city is having many definitions there is no standard definition. A city that is monitor and control infrastructure, including roads, bridges, tunnels, rails, subways, airports, seaports, communications, water, power, even major buildings. Monitor the security aspects, while maximizing services to its citizens (Kitchin, 2014). Most of the government service comes under this internet technology like electricity board, public water service and other bill payments. In future ICT is the core consideration of the smart city establishment. The concept of smart city is too established in the middle of 2008 in United States (Dirks, Gurdgiev & Keeling, 2010).

This chapter is to be discussed on basic of internet security and cyber security. The following security issue is to be discussed in this section, public transportation, security issues in tourist places and security issues in the energy sector. The IoT is a major part of smart city it is directly involved in smart device concepts. The major part of this chapter is internet bandwidth usage in smart devices. The involvement of cloud computing and big data technology in smart city is also being discussed in this chapter. The major part of this chapter is basic working principle of encryption and decryption process. The comparative statement of symmetric and asymmetric encryption is also discussed. The digital signature concept is also involved in smart city. The main part of this chapter is discussed on different encryption algorithm. The following methods is to provide the best solution for handling smart city devices that are firewall method, One Time Password (OTP) system, unique identity number system, Biometric security system, third party security system and finally encryption methods is also used in smart city devices. This chapter is to elaborate the overall security issues and solutions for the smart city devices.

The scope of the smart city is really high. It's indirectly indicating the economic growth of every country. According to the review report, market growth of the smart city is increased around $1.6 trillion by 2020 (Zhao, 2005). The worldwide economic growth of the smart city is $3.48 trillion in the year of 2026. Day to day higher number of peoples is moving from village to urban cities. The people mainly facing two types of problem for this migration, one is pollution and another one is infrastructure. Now a day's crime is the most significant problem in modern cities. The smart city concept is providing a better solution for implementing this latest technology like as CCTV camera and burglar alarm. The main scope of this smart city development is providing a million numbers of job opportunities. The world economic forum is provide statics on engineering graduates; according to that report, Russia produces the highest number of engineering graduates in every year. Russia produces around 4,54,000 engineering graduates in every year. Second position is the United States; it's around 2,37,826 engineering graduates for every year (McCarthy, 2015). The smart city development more enormous volume of money is involved in both private and public sector. In the year 2015, United Kingdom is allocating 40 million dollars for the development of the IoT and smart city. The smart city development is very high, the modern buildings having the sensors and save the electricity power and water resources (Bradshaw, 2010). The development of smart medical field, there is a connection between home to hospital and patient take virtual medical treatments (Dickerson, Gorlin, & Stankovic, 2011).

The CISCO Company is investing 100 million dollars for the development of IoT. The CISCO is selecting some major cities to develop the IoT technology with smart city concept. These cities are Bengaluru, Barcelona, Hamburg, Nice, South Korea, Chicago, San Jose and Songdo. They are establishing Asia's first modern innovation center in India. The primary aim of this research center is to develop the latest IoT technology with a smart city. More than 180 companies join in establishing and developing the innovate products based on IoT. The Indian prime minister recently announced to establish 100 smart cities within the year of 2020 in the investment of 12 billion dollars. The virtual sensor network is for integrating the larger-scale wireless network (Aberer, Hauswirth, & Salehi, 2007). This virtual network aims to connect one virtual smart device to another device.

SMART DEVICES USED IN SMART CITY

Smart CCTV Camera

The CCTV camera is installing in public and private places. The CCTV camera is an essential component of security surveillance. The larger industry, there is no possibility to personal monitoring on the entire campus. This CCTV surveillance system is beneficial for preventing theft and monitor criminal activities. This CCTV surveillance system is installing in most of the cities. This system is conducive for police to track entire criminal and traffic activities. The concept of smart CCTV system is to automatically detecting the objects. The image processing and face detecting mechanism is installed and to identify the criminals. The main advantage of this technology is easy to prevent a terrorist attack. The continuous manual monitoring system is not possible for all aspects. This automated system is handy for monitoring and controlling all criminal activities. The CCTV is a term of closed-circuit television. The input signal received from the camera and stored on server machine. The recent technology of CCTV camera is online storage. There is no need for local servers. The main advantage of this online server system is updating the data in a particular time interval to remote locations. The safety of information is very high compared to traditional methods. The secondary advantage of this virtual cloud system is low cost installation and maintenance.

This face detecting software is installed in single server machine and monitors all the linked CCTV cameras. The single server is maintaining all the data so easy to detect the criminals face. The police department implements this type of technology to identify the criminals. According to the automated imaging association report, the market growth of smart camera has reached around 57 million US dollars. The average growth of this smart camera technology is to increase 15% in every year (Nello, 1983). Decentralization is the primary goal of the smart camera is used to increase the fault tolerance level and improve the surveillance system (Bramberger, Doblander, Maier, Rinner, & Schwabach, 2006). The figure 2 is illustrating different application of smart devices used in real life.

Smart Traffic Management System

Traffic management is a most hectic task for police. The developing country is also facing a lot of problems in this vehicle traffic management. The traffic control and monitoring are very hard. The traffic monitoring system is used to control and manage the traffic. The traditional methods only depend on CCTV cameras. The modern smart systems automatically monitor and suggest the routes. The smart traffic monitoring system is detecting the traffic based on satellite report they change the traffic signal timing. Some service provider gives free road map service. The Google Map is one of the best examples, they suggest shortest route and traffic free routes based on satellite road monitoring report. The implementation cost is very low compared to other traditional systems. The integrated traffic management

Figure 2. Smart devices in different application

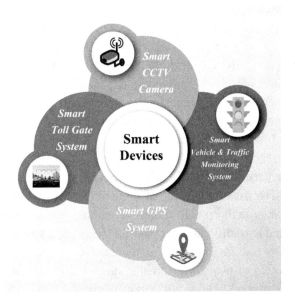

system is used to monitor and view the vehicle traffic information (Jayakrishnan, Mahmassani, & Hu, 1994). The data prediction method is used and analyses the traffic condition. This type of system automatically reduces the vehicle pollution and save the travel time.

Smart Device Using Vehicle Monitoring

Today vehicle monitoring and tracking is the most important one because the vehicle owners always monitor their vehicle using GPS. The GPS is very old technology but additionally added some latest sensors. These sensors used to monitor the vehicle conditions. The modern cars mostly used in these sensors for the entire vehicle control. Example for this is door opening, speed monitoring and all other details monitored on a remote location. This type of technology used for preventing or avoiding the road accidents. Recent survey the developing countries facing more road accident compares to the developed country. The report of world health organization more than 1.25 million people was died because of this road accident. Vehicle tracking system is used for either GPS or GSM technology (Tarapiah, Atalla, & AbuHania, 2013).

GPS Tracking Smart Device

The GPS is a term for Global Positioning System. The purpose of GPS is navigating and tracking the position using satellite. The main advantage of this GPS device is not transferred any unique signal from device to server; this should be automatically tracking the particular device based on the internet or mobile network connection. The GPS system is to give exact location information (Muruganandham, 2010). The GPS system is created by the United States of America, at the time of invention GPS system is only used for military and air forces. The primary aim of this system is to active usage for commercial purpose. The public and private transportation is effectively utilized for this equipment. Most of the countries, this GPS equipment is mandatory for government and private school buses. The purpose of fixing this GPS device is to avoided hijacking and misusage of public vehicles. This GPS system is used for heavy vehicle tracking, like as LPG gas tanker monitoring and other private transportation is monitored continuously. The smart GPS device is some additional advantages. This smart GPS device is to monitors the speed of that particular vehicle. Then this device is helping to child monitoring, and the particular system always tracks the child position.

Smart Technology Used in Toll Gates

The public road transportation is now a day's paid service. The reason for this is an investment of this road establishment is both combined for private and public sector otherwise pure private sector. Once the vehicle crossed that toll gate; they need to pay the road usage charge. In this situation some tool gates, heavy traffic due to manual payment collection. To avoid this problem, they introduce automated payment collection using the technology of IR reader. This IR reader should read the vehicle identity and automatically detect the usage charge from their account (Xiao, Guan, & Zheng, 2008). The implementation cost is low compared to the traditional method.

HANDLING INTERNET OF THINGS, BIG DATA AND CLOUD COMPUTING

Internet of Things

The heart of any smart technology is internet. The entire smart devices should manage and control over this internet. The basic concept of IoT is to establish the connection from one smart device to another device. It's an embedded technology to establish the modern devices like as home appliance, health monitoring and vehicles. The vision and mission of IoT is to provide efficient and secure data service with storage (Gubbi, Buyya, Marusic & Palaniswami, 2013). IoT is still developing architecture. Most of the IoT products are still testing condition. The modern corporate worlds are lead to this type of IoT technology. The implementation point of view, customer expects more security and privacy. They need this IoT technology in some sensitive sectors like as healthcare. The designing principle of IoT is fully depending on internet. The advantage of IoT is to increase efficiency of service and transparency of service. The IoT is not a single technology; it's a combination of some computing technology. Then IoT is involved day to day life products such as mobile, home appliance, vehicles. Then all over the world more than 50,000 devices are connected to this IoT technology.

Internet Usage in Smart Technology

The concept of internet is established connection between entire networks. The information is shared from one network terminal to another network terminal. The internet was developed by Advanced Research Projects Agency (ARPA) in the year of 1969 (Kim, 2005). The early stage internet is not used for commercial purpose. This network is used for communication purpose in US defense department. The development of the network protocol internet is invented. A variety of network applications are created based on this network protocol. The internet is one of the successful applications. Today million and billions of users used in this internet communication. The entire smart devices are connected through this internet medium. The traditional communication is redesigned because of this internet technology. The previous day's text message communication and voice communication use only radio signals. The voices are converted from packets and then send from one terminal to another terminal. The communication cost was gradually reduced compared to previous technology.

Bandwidth Usage in Smart Technology

The bandwidth is generally how much amount of data is to be transferred from one terminal to another terminal at a particular time. The particular smart devices are connected with other smart device via internet. The internet bandwidth is used for transmitting data from one device to another. The developing country internet bandwidth speed is the most important problem. The service providers aren't establishing good infrastructure for communication. The average internet speed of

South Korea is 28.6 Mbps. This is world highest average internet speed. Compare to the United States 65 percentage is faster. The government should take full responsibility to improve the internet speed. The South Korea is one of the fasted developing country in terms of technology and also business. The privacy and security point of view Switzerland is also one of the major technology based country. The worldwide Switzerland passport is one of the best and secure in technology point of view. They provide 21.7 Mbps internet speed. They provide excellent business infrastructure and lifestyle. The king of the world United States only provides 18.7 Mbps internet speed. Compare to other country it's got only 10th position in this category. The internet speed is very slow and they charge high cost compare to other country (Andrew, 2017). The implementation of smart city internet bandwidth is an essential need.

Cloud Computing in Smart City

The general term of cloud computing is sharing of remote resources in terms of software and hardware. The combination of multiple technology cloud computing is framed. The networking, virtualization and grid computing technology are involved in cloud technology. The major services in cloud computing are Infrastructure as a Service (IaaS), Software as a Service (SaaS) and Platform as a Service (PaaS). The smart city technology is directly involved in these three services. The smart city every device need some server storage, the cloud computing is to provide virtual data storage. That service is named as IaaS. The smart device needs some additional software requirement, so they use cloud service provider, which is named as SaaS. The recent days more smart devices enabled with cloud platform. The combination of smart city and cloud computing is providing a variety of additional features that are financial benefit, operational benefit, fast provisioning and pay to use (Hobson & Naccarati, 2011). The new technology is always used in cross platform. The industry or home everywhere people use surveillance camera, all these smart camera need storage space for storing the data. These cameras connected with cloud storage via the internet.

Smart Device in Big Data

The big data is handling larger volume of data. The elements of big data are mainly focused on data volume, variety of data, velocity of data and the value of particular data element. The big data is deals with data storage, data sharing and data analyzing.

The recent days this big data technology is used for predictive analysis. The business point of view, this will be used for analysis and predict the product sales based on the previous sales history. The government and private use more smart devices and these devices generate large volume of data. The normal database technology is not possible to handle this much volume of data. This big data technology is deal with larger data sets. The day to day big data revenue is increased. The global market growth of big data in the year 2017 is almost reached 34 billion U.S dollars. The year 2026 people expect 92 billion U.S dollar for big data revenue. This big data technology is involves many sectors. The most commonly used for this technology is finance, share market, consumer goods and research industry.

SECURITY ISSUES IN SMART CITY

Security

The security is preventing the unauthorized access to any system (Summers, 2004). The information security is blocking many security aspects like as content modification and illegal access. The modern world all the data should be handling in digital format. The handling of digital information is easy at the same time it will face a lot of security threats. The digital world, people expecting more privacy at the same time more freedom. The real factor is no privacy is there in digitalization. For example, the internet search engine is always monitoring our activities. What are all the terms to search, based on that search terms the user received some advertisement.

Internet Security

The internet security is a leaf of computer security system. There are many levels in handling computer security. The following security system is a significant part of managing computer security that are browser security, hardware security and network security. The purpose of internet security is to create a secure protocol and avoid network attackers (Gralla, 1998). The wired communication system information theft is happened only on hacking computer or server system. The wireless communication system is facing many security issues. The information is hacked from the server and also hackers to get information from the communication medium. This wireless technology information is transferred wireless medium through the air in digital signals. The hackers easy to hack the information form this medium. The smart city concept most of the smart devices used in this wireless communication.

Figure 3. Different levels of security issues in smart city

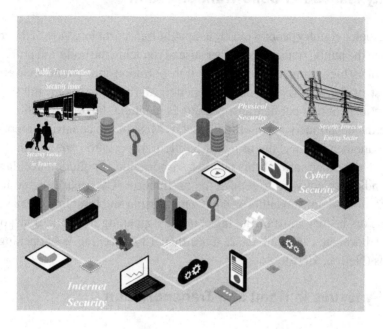

Cyber Security

The cyber security is a technology to prevent and monitor the computer attackers. The primary aim of this cyber security is to control data hacking and damaging. The cyber security has been exploring the major problem in recent days (Choo, 2011). The primary element of this cyber security is network security, information security and application security. The cyber security in a smart city is to prevent the user information from the smart device hackers. The smart city is directly connecting with cyber security technology. The cyber security forum is mainly controlled and maintained by the government body because this is highly sensitive data. To maintain peoples personal data like as biometric information. The market growth is very high compare to the previous years. The developed countries are to given more importance to cyber security issues. The smart city devices are interlinked with some bank transactions, so the people expect more security in this technology. Figure 3 is illustrating different levels of smart city security issues.

Security Issues in Public Transportation

The developing country peoples mostly use public transportation. Most of the countries encourage the public transportation because of reducing the pollution problem. The risk factor is high compared to regular transportation into public transportation. The most of the terrorist attack is happening during this public transportation. The advantage of this public transportation is reducing the transportation cost and increase the economic growth of the country. Most of the countries importing the oil from another country, the usage of public transportation are to reduce the oil imports. The primary aim of this public transportation is pollution avoidance, and the secondary objective is to reduce the oil imports. The security issue in public transportation is some unauthorized person misusing and troubling the passengers. The previous history most of the hijacking is happening in public transportation. The year between 1968 and 1972, more than 130 hijacking is happening in US airplanes (Nelson, 2016).

Security Issues in Road and Transportation

The road safety is an important and government gives more awareness to this issue. The all over the world more than one mission road accident is registered for a particular year. Based on this issue more than 10 million people injured and most of the time they lose out life. Most of the developing country is facing this problem, because developed country maintains much security system to avoid this type of accident (Pearce & Maunder, 2001). They take more safety mechanism to prevent road accident. The TRL research report shows many countries don't properly maintain the public transport vehicles. It's including America and Asian countries; most of the time road accident is occurs in the problem of overloaded. More than 60% road accident is happening laziness of drivers (Jacobs & Aeron-Thomas, 2000). They didn't correctly follow the traffic rules.

Security Issues in Tourism and Public Places

Most of the people spent more time in public places like as hotels, theatres and shopping malls. According to the previous history, the serious of terrorist attacks must happen in public areas only. In the year 2017 more than 1,207 terrorist attacks are happening all over the world. The terrorist mostly chose public places for their attack. This type of incident has mostly happened in public places like as festivals and community halls. The ICT is also one of the significant roles in tourism development. This ICT is to change tourism business and easy to get all the details in a single window (Porter, 2001). Tourism is one of the major revenue generating sectors. Most of the small countries only depend on tourism income. The tourist is expecting higher security then only they choose that country to visit. That's the reason terrorist most concentrate on tourist place for their attacks.

Security Issues in Energy Sector

The worldwide energy is the most crucial factor. It's directly indicated the industrial production and development. China almost number one position in manufacturing sector they provide a continuous energy supply in the industry. The security issue of the energy sector is electricity theft. The developing countries face significant electricity theft problem, both ordinary citizen and industry. The worldwide every year, more than 89 billion dollars will lose for this electricity theft. India makes the highest percentage of loss. Its more than 16 billion dollars are losing this problem. Even some developed country is also facing the same problem; Brazil should lose 10 billion dollars every year for electricity theft (Stephan, 2012). In India, 23 percentages have lost entirely for energy production and distribution (Ahluwalia, 2002).

ENCRYPTION METHODS AND ALGORITHM USED IN SMART CITY

Encryption and Decryption

The basic concept of encryption is changing the data in an unreadable format. There are many techniques used for implementing this encryption. The aim of the encryption and decryption process is to keep high level confidentiality during message communication (Fontaine & Galand, 2009). The original message is accessed only an authorized person. They use some key concept to encode the original message into unreadable format that is named as cipher text. There are two types of encryption is commonly used first one is symmetric encryption, this encryption method is uses the same key for both sender and receiver. The second method is asymmetric encryption; this method is used in different keys for sender and receiver side. In the previous days, this type encryption method is only used for military and government data communication. The customer expects higher security with lower cost. The only solution for the data security is encryption mechanism. The implementation cost is very low compared to other security mechanism. The decryption is the reverse process of encryption method. The cipher text is converted original message for readable format. The concept of this encryption and decryption is, how efficiently used for this technique in smart devices. The smart devices mostly handle sensitive personal data, example user authentication. Here the authentication means personal password, fingerprint recognition and any other user identity (Jayapandian, Rahman, Koushikaa & Radhikadevi, 2016). The hashing is another technique of encryption methods. The data is split into multiple hash function. Each hash value is a unique data element. The small changes are happening in that original message means easy to find it. The reverse process of hash function is very default. This hash method is not pure encryption methods, but it's very secure method to protect the data.

Comparison of Symmetric and Asymmetric Encryption Algorithm

These two encryption method is commonly used. Any encryption algorithm is either symmetric or asymmetric method. The table 1 elaborates and exposes the difference between symmetric and asymmetric method. Based on the comparative statement asymmetric encryption provide higher security. The asymmetric encryption uses the different level of a secret key like as 128,256 and 512 bits (Jayapandian & Rahman, 2017).

Digital Signature

The integrity and authentication of the data are the most important factor. The digital signature is one of the best methods for digital data authentication. In general, digital signature is a mathematical concept to protect the digital document. The digital signature and encryption mechanism is a cyclic process. This is the process of expanding original message bits (Zheng, 1997). The sender and receiver only know a secret key of digital documents. This is the most commonly used crypto methods to protect the financial documents and personal documents. The concept of digital signature is carrying an electronic signature. The digital signature is similar to traditional signature. Its offer more security and solve the digital communication problem. It's providing some additional security mechanism for digital data. This digital signature is mostly used in banking transaction and bank account statement. The people expect more security compared to the traditional method. The digital signature is accepted in legally, depends upon the country. The United State of America is legally accepted in these digital signature methods. The digital signature is a public key crypto system and also it's an asymmetric encryption method. This additional point in the digital signature is providing a secondary security mechanism. The modern electronic mail system supports this digital signature certificate.

Table 1. Comparison of symmetric and asymmetric encryption algorithm

Parameters	Symmetric Encryption	Asymmetric Encryption
Security Level	Moderate Security	High level security
Key Type	Shared Key	Public Key
Size of the Key	128-bit key size is normally used in symmetric encryption	Different size of key length is used 128,256,512
Software & Hardware	Use very simple Algorithm and use low cost Hardware	Use different time complexity algorithm and implement higher configuration hardware
Working Principle	Use same key for encryption and decryption procedure	Encryption and decryption System use different key
Memory Usage	Medium memory usage	Higher memory usage
Computational efficiency	Modest operation used for implementation	Computation operation is very slow, long process to implement complex process

Advanced Encryption Standard Algorithm

The Advanced Encryption Standard (AES) is symmetric key encryption methods. This encryption algorithm is developed by NIST in the year of 2001. This encryption algorithm is used in highly sensitive data maintain both hardware and software. This is very commonly used and powerful algorithm. The fundamental design principle of AES algorithm is two methods one is substitution method, and another one is a permutation method. The AES algorithm is extended version of DES algorithm and compare to DES it should have more mathematical calculation (Bertoni, Breveglieri, Koren, Maistri & Piuri, 2003). The both method provides a higher security mechanism. The commonly used key size of AES algorithm is 128bit, 192bit and 256bit. The key size is used based upon application. The operational structure of AES algorithm is 4 X 4 matrix column. The repetition calculation is based on number of cycles. The 128bit key size is used for 10 cycle repetition. The 192bit key size is used for 12 cycle repetition. Similar to that 256bit key size is used for 14 cycle repetition. The different level of processing is made in this algorithm. The initial level is round key is added. The second level is four steps, the first step is substitution based on a lookup table. The second step is transposition that means shifting the rows from one row to another row. The third step is mixing the column randomly. The fourth step is adding the round key. The third and final level is consists of three phases. The first step is the substitute of the bytes. The second step is shifting the rows. The final step is adding the round key values. The AES algorithm is maintain different steps and it provides a betters security system. This algorithm is mostly used for government sector for protecting the sensitive information. The major advantage of this AES algorithm is provides higher protection compared to the other encryption algorithm. The second advantage is cost; the implementation cost is very low. The third advantage is easy to implement any application on any platform. The main aim of message digest operation is to create the digest in a random manner.

MD5 Encryption Algorithm

The concept of an MD5 algorithm is hash function operation. The value of the hash function is 128bits. This algorithm is used for checksum calculation and data validation. The MD5 algorithm is basically coming from MD4, and the futures are added in this version and increase the security and safety (Xijin & Linxiu, 2003). The input of MD5 algorithm is an arbitrary value, and this will produce a message digest output value in the key length of 128 bits. The most common application of MD5 algorithm is a digital signature. The MD5 algorithm is generally used in buffer concept. The size of buffer length is four words in each 32 bit. That words named as A,B,C and D. The next process of the MD5 algorithm is maintaining the table for the computation purpose. The final stage of the MD5 algorithm is mixing the input words for the auxiliary structure. The four steps commonly used for each level of 16-bit operation.

Blowfish Encryption Algorithm

The blowfish algorithm is a symmetric key encryption method. This is an alternate method of DES and IDEA algorithm. The additional security system of this blowfish algorithm is a variable length of the key is used. Minimum of 32bit and maximum of 448bit key length is used for the encryption process. This algorithm is mainly used for both commercial and non-commercial application. This algorithm is invented in the year of 1993. Implementation cost is very low because it's an open source system. The working method of this algorithm is very easy. This algorithm contains four blocks each named as S-box 1 up to S-box 4. The input of each block includes 8 bit and the subkey array system. The 8bit of input is producing 32bit output after that the XOR operation is performed and combined all the result. The disadvantage of this method is, while the key is changed again the pre-processing method is processed.

Elliptic-Curve Cryptography Algorithm

The basic concept of Elliptic-curve cryptography is an algebraic notation for the curve structure of elliptic. The elliptic curve crypto system is used for various applications; this is based on the factorization of integer values (Lauter, 2004). The implantation point of view, it will take smaller key size compare to other encryption methods. The equation of elliptic curve cryptography is y2=x3+ax+b. The advantage of this algorithm is easy to implement and very small crypto key. The traditional approach is to create a large prime number. But this Elliptic-curve cryptography method generates smaller. The researcher says that higher security level is provided with the key size of 164bit, but expecting the same security level for some other method like 1024 bit key size is needed. The highlight of this method is it consuming very low battery source and computational power. The key generation part Elliptic-curve cryptography should create private and also public key. Based on the receiver public key sender encrypt the original message. The private key is used to decrypt the message from the receiver.

Homomorphic Encryption Algorithm

The operational structure of homomorphic encryption is a computation of the cipher text. The cloud computing platform handles larger data set. This homomorphic encryption produce computational result, it's very difficult to hack the original message. The chaining operation is made without losing the original message. This algorithm is mainly used in some sensitive part like as finance, tax calculation and shipping transactions. This algorithm is also providing better security in the e-Voting system. This algorithm is developed in the year of 1978 (Rivest, Adleman & Dertouzos, 1978). The additional security mechanism is added in after 30 years. The security level is gradually increased and it should be encryption the only one bit at the time of invention. The working model of this algorithm was encrypting the message in already encrypted message (Boneh, Goh, & Nissim, 2005). There are four significant operations in homomorphic encryption. The first step is key generation. The key generation part user will generate the individual secret key. The second step is encryption operation, based on the user secret key the original message is encrypted and forwarded to the server machine. The third step is valuation, that means this step is check the original message is encrypted or not. The fourth and final step is decrypting the original message.

Attribute Based Encryption Algorithm

The Attribute Based Encryption (ABE) algorithm is a concept of the public key encryption method. The original message and the secret key are purely depending upon the attributes. The attributes are pin code or a name of the country. The cipher text is matched with the original attribute value. The primary advantage of this algorithm is collision resistance system. The multiple key values are accessed the original message. There are two types of ABE algorithm, first one is key-policy attribute-based encryption and the second one is Ciphertext-policy attribute-based encryption (Chase, 2007). This is a basic of the log-based encryption method. The log will maintain all the attribute value, and then it executes the encryption process. This algorithm is reducing the key, and then automatically processed.

Probabilistic Encryption Algorithm

The probabilistic encryption algorithm is a randomness algorithm. The same message is randomly encrypted more than one time. The concept of probability algorithm is random either '0' or '1'. The possibility is half of the condition like as coin (Goldwasser & Micali, 1984). This is an asymmetric key and public key crypto system. The asymmetric encryption concept was introduced in the year 1970s. The author Diffie & Hellman is introducing this public key concept (Diffie & Hellman, 1976). The plaintext is YES or NO condition. The probabilistic algorithm uses the concept of arbitraries for the encryption process. This algorithm even prevents the incomplete data. The probabilistic algorithm provides security in $S(m)$. The S means security key and m means data or message.

SOLUTIONS FOR HANDLING SECURITY IN SMART CITY

Firewall Security System

The firewall is the process of protecting network devices from unknown hackers. This system provides better security from unauthorized user. This technology is to block and prevent the computer attacks. The primary module of firewall is policy and authentication. The secondary module of a firewall system is application gateway and packet filtering. There are many types of firewall techniques used. The first one is packet filtering methods; this method is used to filter the unwanted packets from the network. The internet technology is transferring the packets from one terminal to another terminal. The digital data packet, security is the most important one. This method is primarily working on IP protocol technique. The transport layer is working on process to process communication system. This process is converted into packets and sends it from one device to another device. This is also a network traffic security system. The next method is stateful firewall; this method is an active connection track system. The concept of stateful firewall is to allow only for existing connections. The deep packet review firewall is another method. This method is the same concept of intrusion prevention system. The network packet transfer system, deep learning is an essential factor. The application firewall is categorized into two parts network and host-based system. The network firewall is preventing network to network connection and host firewall is preventing host to host connection, it's named as network layer protocol. The last firewall method is a proxy firewall that means to avoid the HTTP or web traffic. This method is validates every data before sending and receives. The implementation point of view is little bit difficulty, because in a particular time it handles single protocol.

The smart city devices are working with backbone of internet technology, the hacker or attackers easy to attack via the internet network. This firewall mechanism is used to prevent and secure the smart device. There are two kinds of firewalls used, hardware and software firewall. The hardware firewall implementation cost is high compared to the software firewall system. The software firewall system no need for a special device, just install the software on that particular device. The compatibility of operating system is only problem. Based on the smart device operating system, separate firewall software is developed and installed.

Mobile Based OTP Security in Smart City

The OTP (One Time Password) security system is used in most of the online service like as banking, finance and trading. This OTP method is used to secure many login processes. The concept of OTP sends a single time password in a particular session. This password will maintain some session timing within that time interval that password is valid. The traditional security method hackers easily find or predict the password. This OTP method is not possible to see the password because every time unique one-time password is generated. That password is sending it to user personal mobile number. There are two methods is used for generating OTP. First one is a time synchronization method; this method is synchronizing the time in between authorized client and server. The second method is using some mathematical algorithm to generate random passwords, that password is dynamically updated in the user database on the server.

The smart city device is also working in online. The device is interconnected with OTP technology based on user mobile identity. The user can use any smart device; they get a one-time password for that particular session. The advantage of this method is only authorized user can access the smart device. The hacker or unauthorized person is not possible to access the smart device. The only drawback of this system is user always keep the mobile and maintain the mobile network. The implementation cost is very low, because no additional hardware component is used. The smart device is configured with software and uses a mobile phone for getting OTP password.

Unique Identity Number System

Most of the developed countries already provided a unique identity number for every individual citizen. This identity number most of the time used for some government purpose like as income tax, government subsidy and other government schemes. The primary aim of this method is linked with the unique identity number and smart city devices. This method easy to find that particular user is authorized or unauthorized. The drawback of this method is only developed country is providing this type of identity number. This smart city concept is also implemented in developed countries because the investment of this smart city establishment is comparatively high. The developing country is not ready to invest this kind of development.

Biometric Identity System for Smart City

The biometric identity system is an old but powerful system. The basic working principle of biometric systems is to find and measure the human identity. The user identity is used for accessing and controlling the smart device. This biometric system uses different types. Most commonly used system is fingerprint recognition. This biometric technology is used in many countries for citizen unique identification. The world largest biometric identification system is implemented in India under the scheme name of Aadhar. The biometric system is easy to inbuilt any hardware. The latest mobile phones this biometric security system is used for mobile phone lock. The body measurement technique is not only implemented in fingerprint technology, it's used in human eye recognition. The recent technology Facebook is used for face recognition, it's intimate to the certain person what are all places used in our face. The government maintains some criminal's records with biometric information to reducing the crime. Figure 4 is an example diagram of biometric control of smart city environment.

Figure 4. Biometric control for smart city

Third Party Security Maintenance

The third party security maintenance is quite regular in now a day. The reason is the small company is not possible to invest large amounts of data security. Solving this problem, the third party service provider develops strong security system gateway, based on that gateway data communication is established. This method is benefited for both service provider and small-scale industry. The user expects better security even that company is small or large. The concept of third party service provider comes from cloud computing. The cloud computing is providing pay for usage, similar to that data security software service is equipped with third-party companies. The only drawback of this third party service is trustworthiness. The customer expects privacy for their data. The private companies misuse the individual personal data. The insurance companies share the customer contact details in some agencies they canvassing their product. This type of data misusing is happening in recent days.

Encryption Mechanism Used in Smart Device

The encryption is very old, but the success rate is very high comparing to other security system. The proposed idea for implementing this encryption method in smart device is each and every data are encrypted before sending to the server. The smart device is always connected with online, the user input data is encrypted using encryption algorithm and then it will send it to the server. There are two ways to implement encryption method one is client-side encryption, and another one is server side encryption. The recent day's both types of encryption are used in a single device. The client-side encryption is used to prevent attackers from outside like hackers. The server-side encryption is used to secure our data from service providers. The year 2014 more than 60% of servers used in virtual server technology like cloud server. The purpose of using virtual server is to reduce the maintenance and implementation cost. The security of their data is still questionable. Sometimes the virtual server maintaining employee or service provider is also misusing the client personal data. The cloud computing encryption tool is the most critical factor without security there is no meaning to use cloud services. The server-side encryption mechanism is keeping the secret key along with our original data. The data is uploaded from client to server on that particular time server is encrypting that data using a secret key received from that particular plain text. The client-side encryption is easy because maintenance of server-side encryption is a little bit difficult in network transmission. Data confidentiality is a major role in online data storage. In our home, there are two locks is to be used for ensure safety. Then digital data is more important in nowadays. Most of the digital data are having all personal and financial information. Then, using two security systems is needed for the smart device communication. The smart device is mostly operated on open source operating system. The client-side encryptions are developed based on that operating system and install it on that particular device. Smart device is encrypting the user information and send to the cloud server via network terminal. If some hackers try to hack the user information means they didn't get the original message. The reason is original message was already encrypted from the client side. That encrypted message is send to the server and again do the same encryptions process once again. The main advantage of this method is service provider is also not getting the user personal information. The previous section many encryption methods discussed and provide a fundamental working principle of encryption algorithms. Figure 5 is elaborating basic working architecture of client and server-side encryption process.

The figure 5 multiple smart devices are interconnected and working with online. The individual security application is to install the smart device, that application is encoding the user input by using data encryption algorithm. The encryption algorithm is selecting based on device to device. The smart mobile use some encryption algorithm and smart biometric device use some other encryption algorithm. Because based on the device algorithm performance is also changed. Once this process is over that encrypted message is sent to a server through cloud infrastructure. Before sending server, one more encryption process is involved that is named as server-side

Figure 5. Architecture diagram of individual encryption method of smart devices

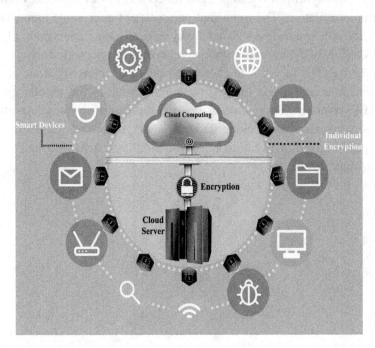

encryption. This encryption process is also taking random encryption algorithm. The advantage of this random encryption algorithm is a hacker doesn't predict the algorithm and secret key. The proposed method provides both server and client side security. The execution time of increased because double encryption process is happened.

SUMMARY

The information security and privacy is a major problem of this smart world. This chapter is deals with the concept of smart city and security fears of smart devices. The day to day life smart device is unavoidable. Data security is a secondary part of this chapter. The user data is stored in online cloud server. The customers are expecting higher data security in lower cost. To solve this data security problem the encryption algorithm is used. The important part of the chapter is to use both client and server side encryption algorithm. The client-side encryption algorithm is to enhance the security level up to reaching of the cloud network. The server-side encryption algorithm is providing security for cloud networks to the server system.

This encryption algorithm is categorized based on various secret key length and level of implementation. The important parameter of encryption algorithm is the compatibility of a smart device. That's why here dynamic encryption algorithms used. The concept of dynamic encryption is choosing an encryption algorithm based on smart device. The server-side encryption is not selecting dynamic encryption because dynamic encryption method takes more time to the regular encryption system. The server-side encryption uses some standard encryption algorithm and stores the smart device data in cloud. This chapter is analysis various security issues and encryption algorithm. In future idea is too inbuilt the light weighted encryption method in smart device.

REFERENCES

Aberer, K., Hauswirth, M., & Salehi, A. (2007, May). Infrastructure for data processing in large-scale interconnected sensor networks. In *Mobile Data Management, 2007 International Conference on* (pp. 198-205). IEEE. 10.1109/MDM.2007.36

Ahluwalia, M. S. (2002). Economic reforms in India since 1991: Has gradualism worked? *The Journal of Economic Perspectives*, *16*(3), 67–88. doi:10.1257/089533002760278721 PMID:15179979

Andrew, H. (2017). *10 Countries with the Fastest Internet Speed in the World.* Retrieved from http://nomadcapitalist.com/2013/12/01/top-5-countries-fastest-internet-speeds-world/

Bertoni, G., Breveglieri, L., Koren, I., Maistri, P., & Piuri, V. (2003). Error analysis and detection procedures for a hardware implementation of the advanced encryption standard. *IEEE Transactions on Computers*, *52*(4), 492–505. doi:10.1109/TC.2003.1190590

Boneh, D., Goh, E. J., & Nissim, K. (2005). Evaluating 2-DNF formulas on ciphertexts. In *Proceedings of the theory of cryptography conference* (pp. 325-341). Springer. 10.1007/978-3-540-30576-7_18

Bradshaw, V. (2010). *The building environment: Active and passive control systems.* John Wiley & Sons.

Bramberger, M., Doblander, A., Maier, A., Rinner, B., & Schwabach, H. (2006). Distributed embedded smart cameras for surveillance applications. *Computer*, *39*(2), 68-75.

Chase, M. (2007, February). Multi-authority attribute based encryption. In *Theory of Cryptography Conference* (pp. 515-534). Springer. 10.1007/978-3-540-70936-7_28

Cho, J.-H., Swami, A., & Chen, R. (2011). A survey on trust management for mobile ad hoc networks. *IEEE Communications Surveys and Tutorials*, *13*(4), 562–583. doi:10.1109/SURV.2011.092110.00088

Choo, K. K. R. (2011). The cyber threat landscape: Challenges and future research directions. *Computers & Security*, *30*(8), 719–731. doi:10.1016/j.cose.2011.08.004

Dickerson, R. F., Gorlin, E. I., & Stankovic, J. A. (2011, October). Empath: a continuous remote emotional health monitoring system for depressive illness. In *Proceedings of the 2nd Conference on Wireless Health* (p. 5). ACM. 10.1145/2077546.2077552

Diffie, W., & Hellman, M. E. (1976). New Directions in Cryptography. *IEEE Transactions on Information Theory*, *22*(6), 644–654. doi:10.1109/TIT.1976.1055638

Dirks, S., Gurdgiev, C., & Keeling, M. (2010). *Smarter cities for smarter growth: How cities can optimize their systems for the talent-based economy*. IBM Institute for Business Value. Retrieved from https://ssrn.com/abstract=2001907

Fontaine, C., & Galand, F. (2007). A survey of homomorphic encryption for nonspecialists. *EURASIP Journal on Information Security*, *1*(1), 41–50.

Gralla, P. (1998). *How the Internet works*. Macmillan Computer Publishing.

Gubbi, J., Buyya, R., Marusic, S., & Palaniswami, M. (2013). Internet of Things (IoT): A vision, architectural elements, and future directions. *Future Generation Computer Systems*, *29*(7), 1645–1660. doi:10.1016/j.future.2013.01.010

Hernández-Muñoz, J. M., Vercher, J. B., Muñoz, L., Galache, J. A., Presser, M., Gómez, L. A. H., & Pettersson, J. (2011). Smart cities at the forefront of the future internet. In *The future internet assembly* (pp. 447–462). Berlin: Springer. doi:10.1007/978-3-642-20898-0_32

Hobson, S., & Naccarati, F. (2011). *IBM Smarter City Solutions on Cloud White paper*. IBM Global Services.

Jacobs, G., & Aeron-Thomas, A. (2005). Africa Road Safety Review. Final Report. US Department of Transportation/Federal Highway Administration.

Jayapandian, N., & Rahman, A. M. Z. (2017). Secure and efficient online data storage and sharing over cloud environment using probabilistic with homomorphic encryption. *Cluster Computing*, *20*(2), 1561–1573. doi:10.100710586-017-0809-4

Jayapandian, N., Rahman, A. M. Z., Koushikaa, M., & Radhikadevi, S. (2016, February). A novel approach to enhance multi level security system using encryption with fingerprint in cloud. In *Futuristic Trends in Research and Innovation for Social Welfare (Startup Conclave), World Conference on* (pp. 1-5). IEEE. 10.1109/STARTUP.2016.7583903

Kim, B. K. (2005). *Internationalizing the Internet: the Co-evolution of Influence and Technology*. Edward Elgar Publishing. doi:10.4337/9781845426750

Kitchin, R. (2014). The real-time city? Big data and smart urbanism. *GeoJournal*, *79*(1), 1–14. doi:10.100710708-013-9516-8

Lauter, K. (2004). The advantages of elliptic curve cryptography for wireless security. *IEEE Wireless Communications, 11*(1), 62–67. doi:10.1109/MWC.2004.1269719

McCarthy, N. (2015). The Countries with the Most Engineering Graduates. *Forbes. Business (Atlanta, Ga.), 8*(1), 33–38.

Muruganandham, P. R. M. (2010). Real time web based vehicle tracking using GPS. *World Academy of Science, Engineering and Technology, 61*(1), 91–96.

Nello, Z. (2004). *Technologies Impacting Machine Vision – Vision Engines.* Retrieved from https://www.visiononline.org/vision-resources-details.cfm/vision-resources/Technologies-Impacting-Machine-Vision-Vision-Engines-By-Nello-Zuech-President-Vision-Systems-International-Consultancy/content_id/1243

Nelson, L. (2016). *The US once had more than 130 hijackings in 4 years. Here's why they finally stopped.* Retrieved from https://www.vox.com/2016/3/29/11326472/hijacking-airplanes-egyptair

Pearce, T., & Maunder, D. A. C. (2000). *Public Transport Safety in Four Emerging Nations.* Washington, DC: Transport Research Laboratory.

Porter, M. E. (2001). Strategy and the Internet. *Harvard Business Review, 79*(3), 63–78. PMID:11246925

Rivest, R. L., Adleman, L., & Dertouzos, M. L. (1978). On data banks and privacy homomorphisms. *Foundations of Secure Computation, 4*(11), 169-180.

Shafi, G., & Micali, S. (1984). Probabilistic encryption. *Journal of Computer and System Sciences, 28*(2), 270–299. doi:10.1016/0022-0000(84)90070-9

Stephan, N. (2012). *Smart Meters Help Brazil Zap Electricity Theft. Brazilian utilities battle theft and deadbeats with smart meters.* Retrieved from https://www.bloomberg.com/news/ articles/2012-03-08/smart-meters-help-brazil-zap-electricity-theft

Summers, G. (2004). *Data and databases.* Nelson Australia Pty Limited.

Tarapiah, S., Atalla, S., & AbuHania, R. (2013). Smart on-board transportation management system using gps/gsm/gprs technologies to reduce traffic violation in developing countries. *International Journal of Digital Information and Wireless Communications, 3*(4), 430–439.

Walravens, N., & Ballon, P. (2013). Platform business models for smart cities: From control and value to governance and public value. *IEEE Communications Magazine*, *51*(6), 72–79. doi:10.1109/MCOM.2013.6525598

Xiao, Z. H., Guan, Z. Q., & Zheng, Z. H. (2008). The research and development of the highway's electronic toll collection system. In *Proceedings of Knowledge Discovery and Data Mining* (pp. 359–362). University of Adelaide. doi:10.1109/WKDD.2008.100

Xijin, W., & Linxiu, F. (2003). The application research of MD5 encryption algorithm in DCT digital Watermarking. *Physics Procedia*, *25*(1), 1264–1269.

Zhao, F. (2005). *Maximize business profits through e-partnerships*. IGI Global.

Zheng, Y. (1997). Digital signcryption or how to achieve cost (signature & encryption)≪cost (signature)+cost (encryption). In *Annual International Cryptology Conference* (pp. 165-179). Springer. 10.1007/BFb0052234

ADDITIONAL READING

Hargreaves, T., Wilson, C., & Hauxwell-Baldwin, R. (2018). Learning to live in a smart home. *Building Research and Information*, *46*(1), 127–139. doi:10.1080/09 613218.2017.1286882

Herrero, S. T., Nicholls, L., & Strengers, Y. (2018). Smart home technologies in everyday life: Do they address key energy challenges in households? *Current Opinion in Environmental Sustainability*, *31*(1), 65–70. doi:10.1016/j.cosust.2017.12.001

Hubaux, J. P., Capkun, S., & Luo, J. (2004). The security and privacy of smart vehicles. *IEEE Security and Privacy*, *2*(3), 49–55. doi:10.1109/MSP.2004.26

Palanca, J., del Val, E., Garcia-Fornes, A., Billhardt, H., Corchado, J. M., & Julián, V. (2018). Designing a goal-oriented smart-home environment. *Information Systems Frontiers*, *20*(1), 125–142. doi:10.100710796-016-9670-x

Rai, S., Chukwuma, P., & Cozart, R. (2016). *Security and Auditing of Smart Devices: Managing Proliferation of Confidential Data on Corporate and BYOD Devices*. London, New York: CRC Press. doi:10.1201/9781315369372

Stimmel, C. L. (2015). *Building smart cities: analytics, ICT, and design thinking*. London, New York: CRC Press. doi:10.1201/b18827

Xu, J., Zhu, W. T., & Feng, D. G. (2009). An improved smart card based password authentication scheme with provable security. *Computer Standards & Interfaces*, *31*(4), 723–728. doi:10.1016/j.csi.2008.09.006

KEY TERMS AND DEFINITIONS

Cyber Security: Cyber security is term as provides an internet data security for system to system communication.

Digital Data: The term of digital data is a binary format of information. The computer is converted into some machine-readable digital format.

ICT: The information and communication technology is a development of software and hardware of computer technology.

OTP: The term of OTP is one-time password. This method purely depends on mobile and email systems. The one-time password is generated from server and sends it to the user mobile or email.

Smart Device: The smart device is generally wireless electronic device. The connection is established from one device to another device in wireless medium.

Smart Encryption: Smart encryption is to provide higher security in smart device. This method is implemented in individual smart device.

Smart Security: The smart security provides information privacy and authentication of smart device.

Glossary

Agent: An agent is a program installed at the host to do the following tasks; event filtering, event aggregation, normalization of the aggregated events, and sending the aggregated results regarding the events to the data analyzer of a centralized intrusion detection program for the further inspection and decision making.

Aggregation: A process of gathering information together from different sources.

Anomaly: An abnormal behavior of a system or user that is deviating from the normal set of usage. An anomaly can be an indication of a malfunction, an error or more importantly an intrusion.

Anonymization: The process of hiding a person's true identity.

Bits Rotation: The process of shifting or rearranging bits by employing the same pattern (number of positions and direction) over the whole set of bits.

Building Area Network (BAN): A network that combines several local area networks (LANs) together to cover an entire building.

Cyber Security: Cyber security is term as provides an internet data security for system to system communication.

Denial of Service (DoS): A class of attack in which targeted network or system disconnects and quits from the intended mode of operation. It is one of the most dangerous attacks capable of taking down any type of network that is of any size.

Digital Data: The term of digital data is a binary format of information. The computer is converted into some machine-readable digital format.

Event: An expected or unexpected happening related to the systems that are in operation. An example of events maybe; arrival of connection commands, the request for the permission to some certain files, the request for escalation in the permissions, etc.

Firewall: A software or hardware that is designed to block the unwanted or unauthorized network traffic between computer networks or hosts. Firewalls are mostly considered as a part of IPS that constitutes the first line of defense in the provisioning of the information security services.

Hashing: The process of transforming a set of characters into a smaller fixed-length key or parameter that uniquely represents the original set of characters.

Home Area Network (HAN): A network limited within a user's home that connects all appliances.

ICT: The information and communication technology is a development of software and hardware of computer technology.

Industrial Area Network (IAN): A network that covers an entire industrial environment and connects its machines to control and monitoring systems.

Intrusion: An event where unauthorized users, generally referred to as hackers, gather information or access rights that he/she is normally not allowed to.

Log File: A file that keeps records of events that happen in an operating system or in other software. Logging is the function of keeping a log in a specific place. Following an intrusion event, log files help information security officers to reveal what went wrong and which damages happened during the intruding activity.

Masking Technique: A process of concealing authentic data with arbitrary an input.

OTP: The term of OTP is one-time password. This method purely depends on mobile and email systems. The one-time password is generated from server and sends it to the user mobile or email.

Pseudonyms: A set of character used to hide person's true identity.

Secret Key: This is a parameter utilized for encryption and decryption of messages in a secret-key or symmetric encryption technique.

Security Risk Assessment: Identifying the vulnerabilities of a system along with the possible worst-case scenarios as well as their probabilities and the evaluation of total property losses in case of such events. This activity generally performed during the establishment of security services for a network or computer system as a part of the provisioning of information security services.

Sensors: These are responsible for collecting evidence regarding the events. It is one of the most critical components of the IDS. The input for a sensor may be any component of a network that could generate useful information regarding intrusions. Some kind of input for a sensor may be composed of the log files, network packets, traces of the system calls, etc. Sensors also send the intrusion-related data to the data analyzer of the IDS 5 .

Shamir's Secret Sharing: The process of dividing a secret message into a number of segments before sending it and to reconstruct the original message a certain number (threshold) of the segments is required.

Smart Device: The smart device is generally wireless electronic device. The connection is established from one device to another device in wireless medium.

Smart Encryption: Smart encryption is to provide higher security in smart device. This method is implemented in individual smart device.

Smart Meter: An internet enabled device that records and send energy consumption information of buildings to the utility companies for accurate billing and monitoring.

Smart Security: The smart security provides information privacy and authentication of smart device.

Snort: It is a lightweight and dedicated host-based IDS, which is composed of the four components: packet decoder, the detection engine, logger, alert generator. Snort employs an easy and flexible rule definition language that creates rules used by the intrusion detection engine.

Time Perturbation: The process of obfuscating the real time interval of recording an information by introducing random time delay.

Compilation of References

Abass, Xiao, Mandayam, & Gajic. (2017). Evolutionary game theoretic analysis of advanced persistent threats against cloud storage. *IEEE Access*, *5*, 8482-8491.

Abdallah, A. R., & Shen, X. S. (2017). A Lightweight Lattice-based Security and Privacy- Preserving Scheme for Smart Grid Customer side Networks. *IEEE Transactions on Smart Grid*, *8*(3), 1064–1074. doi:10.1109/TSG.2015.2463742

Abdallah, A., & Shen, X. S. (2018). A Lightweight Lattice-Based Homomorphic Privacy-Preserving Data Aggregation Scheme for Smart Grid. *IEEE Transactions on Smart Grid*, *9*(1), 396–405. doi:10.1109/TSG.2016.2553647

Aberer, K., Hauswirth, M., & Salehi, A. (2007, May). Infrastructure for data processing in large-scale interconnected sensor networks. In *Mobile Data Management, 2007 International Conference on* (pp. 198-205). IEEE. 10.1109/MDM.2007.36

Afify, M., Abuabed, A. S. A., & Alsbou, N. (2018). Smart engine speed control system with ECU system interface. *IEEE International Instrumentation and Measurement Technology Conference (I2MTC)*, 1-6. 10.1109/I2MTC.2018.8409871

Ahluwalia, M. S. (2002). Economic reforms in India since 1991: Has gradualism worked? *The Journal of Economic Perspectives*, *16*(3), 67–88. doi:10.1257/089533002760278721 PMID:15179979

Airehrour, D., Gutierrez, J., & Ray, S. K. (2017). A trust-aware RPL routing protocol to detect blackhole and selective forwarding attacks. *Australian Journal of Telecommunications and the Digital Economy*, *5*(1), 50. doi:10.18080/ajtde.v5n1.88

Akhunzada, A., Gani, A., Anuar, N. B., Abdelaziz, A., Khan, M. K., Hayat, A., & Khan, S. U. (2016). Secure and dependable software defined networks. *Journal of Network and Computer Applications*, *61*, 199–221. doi:10.1016/j.jnca.2015.11.012

Al F., Dalloro, L., Ludwig, H., Claus, J., Frohlich, R., & Butun I. (2012). *Networking elements as a patch distribution platform for distributed automation and control domains*. Patent App. PCT/US2012/043,084.

Al Nuaimi, E., Al Neyadi, H., Mohamed, N., & Al-Jaroodi, J. (2015). Applications of big data to smart cities. *Journal of Internet Services and Applications, 6*(1), 1–15. doi:10.118613174-015-0041-5

Al-Anbagi, I., & Mouftah, H. T. (2016). WAVE 4 V2G: Wireless access in vehicular environments for vehicle-to-grid applications. *Vehicular Communications, 3*, 31–42. doi:10.1016/j.vehcom.2015.12.002

Albanese, J., & Sonnenreich, W. (2004). *Network Security Illustrated*. McGraw-Hill.

Aledhari, M., Marhoon, A., Hamad, A., & Saeed, F. (2017). A New Cryptography Algorithm to Protect Cloud-based Healthcare Services. IEEE.

Aledhari, M., Marhoon, A., Hamad, A., & Saeed, F. (2017). *A New Cryptography Algorithm to Protect Cloud-Based Healthcare Services*. Paper presented at the 2017 IEEE/ACM International Conference on Connected Health: Applications, Systems and Engineering Technologies (CHASE).

Alharbi, K., & Lin, X. (2012). LPDA: A Lightweight Privacy-preserving Data Aggregation Scheme for Smart Grid. *2012 International Conference on Wireless Communications and Signal Processing (WCSP)*. 10.1109/WCSP.2012.6542936

Alshawish, R. A., Alfagih, S. A. M., & Musbah, M. S. (2016). Big data applications in smart cities. *International Conference on Engineering & MIS (ICEMIS)*, 1-7.

Alsheikh, M. A., Lin, S., Niyato, D., & Tan, H. P. (2018). Machine learning in wireless sensor networks: Algorithms, strategies, and applications. IEEE Commun. Surveys Tuts., 16(4), 1996-2018.

Al-Sultan, S., Al-Doori, M. M., Al-Bayatti, A. H., & Zedan, H. (2013). A comprehensive survey on vehicular Ad Hoc network. *Journal of Network and Computer Applications, 37*(1), 380–392. doi:10.1016/j.jnca.2013.02.036

Aly, A. (2015). *Network flow problems with secure multiparty computation*. Universt´e catholique de Louvain.

Amin, Sherratt, Giri, Islam, & Khan. (2017). A software agent enabled biometric security algorithm for secure _le access in consumer storage devices. *IEEE Trans. Consum. Electron., 63*(1), 53-61.

Aminanto, Choi, Tanuwidjaja, Yoo, & Kim. (2018). Deep abstraction and weighted feature selection forWi-Fi impersonation detection. *IEEE Trans. Inf. Forensics Security, 13*(3), 621-636.

Andrew, H. (2017). *10 Countries with the Fastest Internet Speed in the World.* Retrieved from http://nomadcapitalist.com/2013/12/01/top-5-countries-fastest-internet-speeds-world/

Angelakis, V., Tragos, E., Pöhls, H. C., Kapovits, A., & Bassi, A. (2017). *Designing, Developing, and Facilitating Smart Cities: Urban Design to IoT.* Springer. doi:10.1007/978-3-319-44924-1

Awadalla, M. H. A. (2017). Design of a smart traffic information system. *International Conference on Intelligent Computing, Instrumentation and Control Technologies (ICICICT),* 757-762.

Baldini, G., Kounelis, I., Fovino, I. N., & Neisse, R. (2013). A framework for privacy protection and usage control of personal data in a smart city scenario. In *Critical Information Infrastructures Security (LNCS 8328)* (pp. 212–217). Cham, Switzerland: Springer. doi:10.1007/978-3-319-03964-0_20

Banafa, A. (2016). *The Internet of Everything (IoE).* Retrieved from https://www.bbvaopenmind.com/en/the-internet-of-everything-ioe/

Bao, H., & Chen, L. (2016). A lightweight privacy-preserving scheme with data integrity for smart grid communications. *Concurrency and Computation: Practice and Experience, 28,* 1094–1110. doi:10.1002/cpe

Bao, F., & Chen, I.-R. (2012). Dynamic trust management for internet of things applications. *Proceedings of the 2012 international workshop on Self-aware internet of things.* 10.1145/2378023.2378025

Batty. (2012). Smart cities of the future. *Eur. Phys. J. Special Topics, 214*(1), 481-518.

Bello, A., Liu, W., Bai, Q., & Narayanan, A. (2015a). *Exploring the Role of Structural Similarity in Securing Smart Metering Infrastructure.* Paper presented at the Data Science and Data Intensive Systems (DSDIS), 2015 IEEE International Conference on. 10.1109/DSDIS.2015.95

Bello, A., Liu, W., Bai, Q., & Narayanan, A. (2015b). Revealing the Role of Topological Transitivity in Efficient Trust and Reputation System in Smart Metering Network. *Proceedings of the 2015 IEEE International Conference on Data Science and Data Intensive Systems (DSDIS).* 10.1109/DSDIS.2015.114

Ben-Or, M., Goldwasser, S., & Wigderson, A. (1988). Completeness theorems for non-cryptographic fault-tolerant distributed computation. In *STOC* (pp. 1–10). Chicago: ACM. doi:10.1145/62212.62213

Berthelot, A., Tamke, A., Dang, T., & Breuel, G. (2012). A novel approach for the probabilistic computation of Time-To-Collision. *Intelligent Vehicles Symposium*. 10.1109/IVS.2012.6232221

Bertoni, G., Breveglieri, L., Koren, I., Maistri, P., & Piuri, V. (2003). Error analysis and detection procedures for a hardware implementation of the advanced encryption standard. *IEEE Transactions on Computers, 52*(4), 492–505. doi:10.1109/TC.2003.1190590

Bhuiyan, T. (2013). *Online Survey on Trust and Interest Similarity. In Trust for Intelligent Recommendation* (pp. 53–61). Springer. doi:10.1007/978-1-4614-6895-0_4

Biggio, Fumera, Russu, Didaci, & Roli. (2015). Adversarial biometric recognition: A review on biometric system security from the adversarial machine-learning perspective. *IEEE Signal Process. Mag., 32*(5), 31-41.

Blaze, M., Feigenbaum, J., & Lacy, J. (1996). Decentralized trust management. In *Proceedings 1996 IEEE Symposium on Security and Privacy*. IEEE. 10.1109/SECPRI.1996.502679

Boel, R., Marinica, N., Moradzadeh, M., & Sutarto, H. (2013). Some paradigms for coordinating feedback control with applications to urban traffic control and smart grids. *2013 3rd International Conference on Instrumentation Control and Automation (ICA)*, 1-6. 10.1109/ICA.2013.6734036

Bogatinoska, D. C., Malekian, R., Trengoska, J., & Nyako, W. A. (2016). Advanced sensing and internet of things in smart cities. *2016 39th International Convention on Information and Communication Technology, Electronics and Microelectronics (MIPRO)*, 632-637.

Boneh, D., Goh, E. J., & Nissim, K. (2005). Evaluating 2-DNF formulas on ciphertexts. In *Proceedings of the theory of cryptography conference* (pp. 325-341). Springer. 10.1007/978-3-540-30576-7_18

Boreiko, O., & Teslyuk, V. (2017). Information model of the control system for passenger traffic registration of public transport in the "smart" city. *12th International Scientific and Technical Conference on Computer Sciences and Information Technologies (CSIT)*, 113-116. 10.1109/STC-CSIT.2017.8098749

Borgia, E. (2014). The Internet of Things vision: Key features, applications and open issues. *Computer Communications*, *54*, 1–31. doi:10.1016/j.comcom.2014.09.008

Bowen, Z., Li, W., Deng, P., Gérardy, L., Zhu, Q., & Shankar, N. (2015). *Design and verification for transportation system security.* Paper presented at the 2015 52nd ACM/EDAC/IEEE Design Automation Conference (DAC).

Bradshaw, V. (2010). *The building environment: Active and passive control systems.* John Wiley & Sons.

Bramberger, M., Doblander, A., Maier, A., Rinner, B., & Schwabach, H. (2006). Distributed embedded smart cameras for surveillance applications. *Computer*, *39*(2), 68-75.

Burange, A. W., & Misalkar, H. D. (2015). Review of Internet of Things in development of smart cities with data management & privacy. *Proc. Int. Conf. Adv. Comput. Eng. Appl. (ICACEA)*, 189–195. 10.1109/ICACEA.2015.7164693

Butun, I. (2013). *Prevention and Detection of Intrusions in Wireless Sensor Networks* (Ph.D. Dissertation). University of South Florida.

Butun, I., Morgera, S. D., & Sankar, R. (2014). A survey of intrusion detection systems in wireless sensor networks. *IEEE Communications Surveys and Tutorials*, *16*(1), 266–282. doi:10.1109/SURV.2013.050113.00191

Camacho, F., Cárdenas, C., & Muñoz, D. (2018). Emerging technologies and research challenges for intelligent transportation systems: 5G, HetNets, and SDN. *International Journal on Interactive Design and Manufacturing*, *12*(1), 327–335. doi:10.100712008-017-0391-2

Camenisch, J., & Lysyanskaya, A. (2003). A signature scheme with efficient protocols. In *3rd Int. Conf. Security Commun. Netw.* (pp. 268–289). Academic Press. 10.1007/3-540-36413-7_20

Camenisch, J., Hohenberger, S., & Pedersen, M. Ø. (2012). Batch verification of short signatures. *Journal of Cryptology*, *25*(4), 723–747. doi:10.100700145-011-9108-z

Canetti, R. (2000). Security and composition of multiparty cryptographic protocols. *Journal of Cryptology*, *13*(1), 143–202. doi:10.1007001459910006

Cao, L. (2016). Two intersections traffic signal control method based on ADHDP. *IEEE International Conference on Vehicular Electronics and Safety (ICVES)*, 1-5. 10.1109/ICVES.2016.7548166

Car 2 Car Consortium. (n.d.). Retrieved from https://www.car-2-car.org

Cardenas, A. A., Amin, S., & Sastry, S. (2008, July). Research Challenges for the Security of Control Systems. HotSec.

Cardenas, A., Amin, S., Sinopoli, B., Giani, A., Perrig, A., & Sastry, S. (2009, July). Challenges for securing cyber physical systems. In *Workshop on future directions in cyber-physical systems security* (*Vol. 5*). Academic Press.

Carlsen, L. H. (2014). *The location of privacy—A case study of Copenhagen connecting's smart city* (M.S. thesis). Dept. Commun. Bus. Inf. Technol., Roskilde Univ., Roskilde, Denmark.

Cashion, J., & Bassiouni, M. (2011). *Protocol for Mitigating the Risk of Hijacking Social Networking Sites*. IEEE.

Cashion, J., & Bassiouni, M. (2011). *Protocol for mitigating the risk of hijacking social networking sites*. Paper presented at the 7th International Conference on Collaborative Computing: Networking, Applications and Worksharing (CollaborateCom).

Cebe, M., Erdin, E., Akkaya, K., Aksu, H., & Uluagac, S. (2018). *Block4Forensic: An Integrated Lightweight Blockchain Framework for Forensics Applications of Connected Vehicles*. arXiv preprint arXiv:1802.00561

Centea, D., Singh, I., & Elbestawi, M. (2018). Framework for the Development of a Cyber-Physical Systems Learning Centre. In *Online Engineering & Internet of Things* (pp. 919–930). Cham: Springer. doi:10.1007/978-3-319-64352-6_86

Chahal, Harit, Mishra, Sangaiah, & Zheng. (2017). A Survey on software-defined networking in vehicular ad hoc networks: Challenges, applications and use cases. *Sustainable Cities and Society, 35*, 830-840. doi:10.1016/j.scs.2017.07.007

Chang, K.-D., & Chen, J.-L. (2012). A survey of trust management in WSNs, internet of things and future internet. *Transactions on Internet and Information Systems (Seoul), 6*(1).

Chase, M. (2007, February). Multi-authority attribute based encryption. In *Theory of Cryptography Conference* (pp. 515-534). Springer. 10.1007/978-3-540-70936-7_28

Chatterjee, S. (2015). *Security and privacy issues in E-Commerce: A proposed guidelines to mitigate the risk*. Paper presented at the 2015 IEEE International Advance Computing Conference (IACC). 10.1109/IADCC.2015.7154737

Chatterjee, S., Kar, A. K., & Gupta, M. P. (2017). Critical success factors to establish 5G network in smart cities: Inputs for security and privacy. *Journal of Global Information Management, 25*(2), 15–37. doi:10.4018/JGIM.2017040102

Chaudhary, M., & Kumar, H. (2015). *Challenges in protecting personnel information in social network space.* Paper presented at the 2015 International Conference on Emerging Trends in Networks and Computer Communications (ETNCC).

Chauhan, P., Singh, N., & Chandra, N. (2013). *Security Breaches in an Organization and Their Countermeasures.* Paper presented at the 2013 5th International Conference and Computational Intelligence and Communication Networks.

Chaum, D. (1982). Blind signatures for untraceable payments. In Advances in Cryptology - Crypto '82 (pp. 199–203). Springer-Verlag.

Chaum, D., Cŕepeau, C., & Damgard, I. (1988). Multiparty unconditionally secure protocols. In *STOC. ACM* (pp. 11–19). ACM.

Chen, R., Bao, F., & Guo, J. (2016). Trust-based service management for social internet of things systems. *IEEE Transactions on Dependable and Secure Computing, 13*(6), 684–696. doi:10.1109/TDSC.2015.2420552

Chen, S., Irissappane, A. A., & Zhang, J. (2017). POMDP-Based Decision Making for Fast Event Handling in VANETs. *The Thirty-Second AAAI Conference on Artificial Intelligence.*

Chih-Ju, C., Sheng-Hao, S., Kuo-Hsiung, T., & To-Cheng, L. (2015). A novel SCADA system design and application for intelligent traffic control. *The 27th Chinese Control and Decision Conference,* 726-730. 10.1109/CCDC.2015.7162015

Chiti, F., Fantacci, R., Giuli, D., Paganelli, F., & Rigazzi, G. (2017). *Communications protocol design for 5G vehicular networks. In 5G Mobile Communications* (pp. 625–649). Cham: Springer.

Chiuchisan, I., Balan, D., Geman, O., Chiuchisan, I., & Gordin, I. (2017). *A security approach for health care information systems.* Paper presented at the 2017 E-Health and Bioengineering Conference (EHB). Cyber-Physical Systems. Retrieved from https://ptolemy.berkeley.edu/projects/cps/

Cho, J.-H., Swami, A., & Chen, I.-R. (2011). A Survey on Trust Management for Mobile Ad Hoc Networks. *IEEE Communications Surveys and Tutorials, 13*(4), 562–583. doi:10.1109/SURV.2011.092110.00088

Choo, K. K. R. (2011). The cyber threat landscape: Challenges and future research directions. *Computers & Security*, *30*(8), 719–731. doi:10.1016/j.cose.2011.08.004

Christidis & Devetsikiotis. (2016). Blockchains and smart contracts for the Internet of Things. *IEEE Access*, *4*, 2292-2303.

Chze, P. L. R., & Leong, K. S. (2014). *A secure multi-hop routing for IoT communication*. Paper presented at the Internet of Things (WF-IoT), 2014 IEEE World Forum on. 10.1109/WF-IoT.2014.6803204

Cikhardtová, K., Bělinová, Z., Tichý, T., & Růžička, J. (2016). Evaluation of traffic control impact on smart cities environment. *2016 Smart Cities Symposium Prague (SCSP)*, 1-4. 10.1109/SCSP.2016.7501011

Cimmino, A., Pecorella, T., Fantacci, R., Granelli, F., Rahman, T. F., Sacchi, C., ... Harsh, P. (2014). The role of small cell technology in future smart city applications. *Trans. Emerg. Telecommun. Technol. Special Issue Smart Cities*, *25*(1), 11–20. doi:10.1002/ett.2766

Clarke, R. (n.d.). *What's Privacy?* Retrieved from www.rogerclarke.com/DV/Privacy.html

Cohen, R., Zhang, J., Finnson, J., Tran, T., & Minhas, U. F. (2014). A trust-based framework for vehicular travel with non-binary reports and its validation via an extensive simulation testbed. Journal of Trust Management. doi:10.118640493-014-0010-0

Coleman, J. S. (2000). *Social capital in the creation of human capital. In Knowledge and social capital* (pp. 17–41). Elsevier. doi:10.1016/B978-0-7506-7222-1.50005-2

Cybersecurity for smart city architecture. (2015). National Institute of Standard and Technology (NIST). Retrieved from http://nist.gov/cps/cybersec_smartcities.cfm

Dai, H., Jia, Z., & Qin, Z. (2009). Trust Evaluation and Dynamic Routing Decision Based on Fuzzy Theory for MANETs. *JSW*, *4*(10), 1091–1101. doi:10.4304/jsw.4.10.1091-1101

Damásio, A. (1994). *Descartes' Error: emotion, reason, and the human brain*. New York: G. P. Putnam's Sons.

Dashtinezhad, S., Nadeem, T., Dorohonceanu, B., Borcea, C., Kang, P., & Iftode, L. (2004). TrafficView: a driver assistant device for traffic monitoring based on car-to-car communication. *IEEE 59th Vehicular Technology Conference.* 10.1109/VETECS.2004.1391464

deMeer, J., & Rennoch, A. (2011). The ETSI TVRA Security-Measurement Methodology by means of TTCN-3 Notation. *10th TTCN-3 User Conference.*

Deng, S., Zhou, A. H., Yue, D., Hu, B., & Zhu, L. P. (2017). Distributed intrusion detection based on hybrid gene expression programming and cloud computing in a cyber physical power system. *IET Control Theory & Applications, 11*(11), 1822–1829. doi:10.1049/iet-cta.2016.1401

Deore, U. D., & Waghmare, V. (2016). Cyber Security Automation for Controlling Distributed Data. IEEE.

Department of Economic and Social Affairs. (2014). *World Urbanization Prospects: The 2014 Revision, Highlights.* New York: United Nations Population Division.

Desmedt, Y. G. (1994). Threshold cryptography. *European Transactions on Telecommunications, 5*(4), 449–458. doi:10.1002/ett.4460050407

Dhungana, D., Engelbrecht, G., Parreira, J. X., Schuster, A., & Valerio, D. (2015). Aspern smart ICT: Data analytics and privacy challenges in a smart city. *Proc. IEEE 2nd World Forum Internet Things (WF IoT),* 447–452. 10.1109/WF-IoT.2015.7389096

Diao, F., Zhang, F., & Cheng, X. (2015). A Privacy-Preserving Smart Metering Scheme Using Linkable Anonymous Credential. *IEEE Transactions on Smart Grid, 6*(1), 461–467. doi:10.1109/TSG.2014.2358225

Dickerson, R. F., Gorlin, E. I., & Stankovic, J. A. (2011, October). Empath: a continuous remote emotional health monitoring system for depressive illness. In *Proceedings of the 2nd Conference on Wireless Health* (p. 5). ACM. 10.1145/2077546.2077552

Diffie, W., & Hellman, M. E. (1976). New Directions in Cryptography. *IEEE Transactions on Information Theory, 22*(6), 644–654. doi:10.1109/TIT.1976.1055638

Ding, D., Han, Q. L., Xiang, Y., Ge, X., & Zhang, X. M. (2018). A survey on security control and attack detection for industrial cyber-physical systems. *Neurocomputing, 275,* 1674–1683. doi:10.1016/j.neucom.2017.10.009

Dirks, S., Gurdgiev, C., & Keeling, M. (2010). *Smarter cities for smarter growth: How cities can optimize their systems for the talent-based economy.* IBM Institute for Business Value. Retrieved from https://ssrn.com/abstract=2001907

Do, C. T., Tran, N. H., Hong, C., Kamhoua, C. A., Kwiat, K. A., Blasch, E., ... Iyengar, S. S. (2017). Game theory for cyber security and privacy. *ACM Computing Surveys*, *50*(2), 30. doi:10.1145/3057268

Dorri, A., Kanhere, S. S., Jurdak, R., & Gauravaram, P. (2017). Blockchain for IoT security and privacy: The case study of a smart home. *Proc. IEEE Int. Conf. Pervasive Comput. Commun. Workshops (PerCom Workshops)*, 618-623. 10.1109/PERCOMW.2017.7917634

Dorri, A., Kanhere, S. S., Jurdak, R., & Gauravaram, P. (2017, March). Blockchain for IoT security and privacy: The case study of a smart home. In *Pervasive Computing and Communications Workshops (PerCom Workshops), 2017 IEEE International Conference on* (pp. 618-623). IEEE.

Drias, Z., Serhrouchni, A., & Vogel, O. (2015). *Analysis of cyber security for industrial control systems.* Paper presented at the 2015 International Conference on Cyber Security of Smart Cities, Industrial Control System and Communications (SSIC).

Dubey, Akshdeep, & Rane. (2017). Implementation of an intelligent traffic control system and real time traffic statistics broadcasting. *International conference of Electronics, Communication and Aerospace Technology (ICECA)*, 33-37. 10.1109/ICECA.2017.8212827

Ducq, Y., Agostinho, C., Chen, D., Zacharewicz, G., & Goncalves, R. (n.d.). Generic methodology for service engineering based on service modelling and model transformation. In Manufacturing Service Ecosystem. Achievements of the European 7th FP FoF-ICT Project MSEE: Manufacturing SErvice Ecosystem (Grant No. 284860). Academic Press.

Duić, I., Cvrtila, V., & Ivanjko, T. (2017). *International cyber security challenges.* Paper presented at the 2017 40th International Convention on Information and Communication Technology, Electronics and Microelectronics (MIPRO).

Efthymiou, C., & Kalogridis, G. (2010). Smart Grid Privacy via Anonymization of Smart Metering Data. In *2010 First IEEE International Conference on Smart Grid Communications* (pp. 238–243). IEEE. 10.1109/SMARTGRID.2010.5622050

Ekanayake, B. N. B., Halgamuge, M. N., & Syed, A. (2018). *Review: Security and Privacy Issues of Fog Computing for the Internet of Things (IoT). In Lecture Notes on Data Engineering and Communications Technologies Cognitive Computing for Big Data Systems Over IoT, Frameworks, Tools and Applications* (Vol. 14). Springer.

Electronic Privacy Information Center. (2011). *The Smart Grid and Privacy*. Retrieved from https://epic.org/privacy/smartgrid/smartgrid.html

ElGamal, T. (1985). A public key cryptosystem and a signature scheme based on discrete logarithms. *IEEE Transactions on Information Theory, 31*(4), 469–472. doi:10.1109/TIT.1985.1057074

Elleuch, H., & Rouis, J. (2017). Devising a smart control and urban management network based on determination of the obstruction points of road traffic in Sfax. *International Conference on Smart, Monitored and Controlled Cities (SM2C)*, 111-116. 10.1109/SM2C.2017.8071830

Engle, S. L. (2001). *Structural holes and Simmelian ties: Exploring social capital, task interdependence, and individual effectiveness*. Academic Press.

European Commission. (2006). *European smart grids technology platform: vision and strategy for Europe's electricity*. Retrieved from http://www.ec.europa.eu/

Eze, E. C., Zhang, S. J., Liu, E. J., & Eze, J. C. (2016). Advances in vehicular ad-hoc networks (VANETs): Challenges and road-map for future development. *International Journal of Automation and Computing, 13*(1), 1–18. doi:10.100711633-015-0913-y

Fan, X., Fan, K., Wang, Y., & Zhou, R. (2015). *Overview of Cyber-security of Industrial Control System*. Retrieved from https://www.bestcurrentaffairs.com/features-of-smart-cities/

Ferrara, A. L., Green, M., Hohenberger, S., & Pedersen, M. Ø. (2009). Practical Short Signature Batch Verification. In Proc.2009 CT-RSA (pp. 309–324). Academic Press. doi:10.1007/978-3-642-00862-7_21

Field, S. A. (2013). *Tagging obtained content for white and black listing*. U.S. Patent No. 8,544,086. Washington, DC: U.S. Patent and Trademark Office.

Fitzgerald, E., & Landfeldt, B. (2012). A System for Coupled Road Traffic Utility Maximisation and Risk Management Using VANET. *15th International IEEE Conference on Intelligent Transportation Systems*. 10.1109/ITSC.2012.6338630

Fitzgerald, E., & Landfeldt, B. (2013). On Road Network Utility Based on Risk-Aware Link Choice. *International IEEE Conference on Intelligent Transportation Systems*. 10.1109/ITSC.2013.6728361

Fitzgerald, E., & Landfeldt, B. (2015). Increasing Road Traffic Throughput through Dynamic Traffic Accident Risk Mitigation. *Journal of Transportation Technologies*, *5*(5), 223–239. doi:10.4236/jtts.2015.54021

Fog Computing and the Internet of Things: Extend the Cloud to Where the Things Are. (2015). Available: http://www.cisco.com/c/dam/en_us/solutions/trends/iot/docs/computingoverview.pdf

Fontaine, C., & Galand, F. (2007). A survey of homomorphic encryption for nonspecialists. *EURASIP Journal on Information Security*, *1*(1), 41–50.

Fu, Y., Jia, S., & Hao, J. (2015). A scalable cloud for the Internet of Things in smart cities. *Journal of Computers*, *26*(3), 63–75.

GazetteS. (2018). Retrieved 2018, from http://saudigazette.com.sa/article/524118/SAUDI-ARABIA/One-accident-every-minute-20-deaths-daily-on-Saudi-roads

Geronimo, D., Lopez, A. M., Sappa, A. D., & Graf, T. (2010). Survey of pedestrian detection for advanced driver assistance systems. *IEEE Transactions on Pattern Analysis and Machine Intelligence*, *32*(7), 1239–1258. doi:10.1109/TPAMI.2009.122 PMID:20489227

Gindele, T., Brechtel, S., & Dillmann, R. (2010). A Probabilistic Model for Estimating Driver Behaviors and Vehicle Trajectories in Traffic Environments. *Annual Conference on Intelligent Transportation Systems*. 10.1109/ITSC.2010.5625262

Glaser, S., Vanholme, B., Mammar, S., Gruyer, D., & Nouveliere, L. (2010). Maneuver-Based Trajectory Planning for Highly Autonomous Vehicles on Real Road with Traffic and Driver Interaction. *IEEE Transactions on Intelligent Transportation Systems*, *11*(3), 589–606. doi:10.1109/TITS.2010.2046037

Goldwasser, Micali, & Rackoff. (1989). The knowledge complexity of interactive proof systems. *SIAM J. Comput.*, *18*(1), 186-208.

Gomez, C., & Paradells, J. (2010). Wireless home automation networks: A survey of architectures and technologies. *IEEE Communications Magazine*, *48*(6), 92–101. doi:10.1109/MCOM.2010.5473869

Goodman, E. P. (2015). *Rapporteur. In The Atomic Age of Data: Policies for the Internet of Things* (p. 5). The Aspen Institute.

Gope, P., Sikdar, B., & Member, S. (2018). An Efficient Data Aggregation Scheme for Privacy-Friendly Dynamic Pricing-based Billing and Demand-Response Management in Smart Grids. *IEEE Internet of Things Journal*. doi:10.1109/JIOT.2018.2846299

Goudar, R. H., & Megha, H. N. (2017). Next generation intelligent traffic management system and analysis for smart cities. *International Conference On Smart Technologies For Smart Nation (SmartTechCon)*, 999-1003. 10.1109/SmartTechCon.2017.8358521

Govindan, K., & Mohapatra, P. (2012). Trust computations and trust dynamics in mobile adhoc networks: A survey. *IEEE Communications Surveys and Tutorials*, *14*(2), 279–298. doi:10.1109/SURV.2011.042711.00083

Gralla, P. (1998). *How the Internet works*. Macmillan Computer Publishing.

Greveler, U., Glösekötterz, P., Justusy, B., & Loehr, D. (2012). Multimedia content identification through smart meter power usage profiles. *Proc. Int. Conf. Inf. Knowl. Eng. (IKE)*, 383–390.

Gubbi, J., Buyya, R., Marusic, S., & Palaniswami, M. (2013). Internet of Things (IoT): A vision, architectural elements, and future directions. *Future Generation Computer Systems*, *29*(7), 1645–1660. doi:10.1016/j.future.2013.01.010

Guo, J. (2018). *Trust-based Service Management of Internet of Things Systems and Its Applications*. Virginia Tech.

Hainalkar, G. N., & Vanjale, M. S. (2017). Smart parking system with pre & post reservation, billing and traffic app. *International Conference on Intelligent Computing and Control Systems*, 500-505.

Hajy, S., Zargar, M., & Yaghmaee, M. H. (2013). Privacy Preserving via Group Signature in Smart Grid. In *1st Electric Industry Automation Congress, EIAC 2013* (pp. 1–5). Academic Press.

Haller, P., & Genge, B. (2017). Using sensitivity analysis and cross-association for the design of intrusion detection systems in industrial cyber-physical systems. *IEEE Access: Practical Innovations, Open Solutions*, *5*, 9336–9347. doi:10.1109/ACCESS.2017.2703906

Handfield, R. B., & Bechtel, C. (2002). The role of trust and relationship structure in improving supply chain responsiveness. *Industrial Marketing Management*, *31*(4), 367–382. doi:10.1016/S0019-8501(01)00169-9

Hart, G. (1992). Nonintrusive appliance load monitoring. *IEEE*, *80*(12), 1870–1891.

Hasrouny, H., Bassil, C., Samhat, A. E., & Laouiti, A. (2017). Security risk analysis of a trust model for secure group leader-based communication in VANET. In *Vehicular Ad-Hoc Networks for Smart Cities* (pp. 71–83). Singapore: Springer. doi:10.1007/978-981-10-3503-6_6

He, D., Kumar, N., Zeadally, S., Vinel, A., & Yang, L. T. (2017). Efficient and Privacy-Preserving Data Aggregation Scheme for Smart Grid Against Internal Adversaries. *IEEE Transactions on Smart Grid*, 8(5), 2411–2419. doi:10.1109/TSG.2017.2720159

Heikkilä, M., Rättyä, A., Pieskä, S., & Jämsä, J. (2016). *Security challenges in small- and medium-sized manufacturing enterprises.* Paper presented at the 2016 International Symposium on Small-scale Intelligent Manufacturing Systems (SIMS). 10.1109/SIMS.2016.7802895

Henningsen, S., Dietzel, S., & Scheuermann, B. (2018). Misbehavior Detection in Industrial Wireless Networks: Challenges and Directions. *Mobile Networks and Applications*, 1–7.

Hernández-Muñoz, J. M., Vercher, J. B., Muñoz, L., Galache, J. A., Presser, M., Gómez, L. A. H., & Pettersson, J. (2011). Smart cities at the forefront of the future internet. In *The future internet assembly* (pp. 447–462). Berlin: Springer. doi:10.1007/978-3-642-20898-0_32

Hillenbrand, J., Spieker, A. M., & Kroschel, K. (2006). A Multilevel Collision Mitigation Approach—Its Situation Assessment, Decision Making, and Performance Tradeoffs. *IEEE Transactions on Intelligent Transportation Systems*, 7(4), 528–540. doi:10.1109/TITS.2006.883115

Hobson, S., & Naccarati, F. (2011). *IBM Smarter City Solutions on Cloud White paper.* IBM Global Services.

Hoepman, J.-H. (2014). Privacy design strategies. In *ICT Systems Security and Privacy Protection (SEC)* (pp. 446–459). Heidelberg, Germany: Springer. doi:10.1007/978-3-642-55415-5_38

Hoffman, K., Zage, D., & Nita-Rotaru, C. (2009). A survey of attack and defense techniques for reputation systems. *ACM Computing Surveys*, 42(1), 1–31. doi:10.1145/1592451.1592452

Ho, J.-W., Wright, M., & Das, S. K. (2012). ZoneTrust: Fast zone-based node compromise detection and revocation in wireless sensor networks using sequential hypothesis testing. *IEEE Transactions on Dependable and Secure Computing*, 9(4), 494–511. doi:10.1109/TDSC.2011.65

Hossain, M., Hasan, R., & Skjellum, A. (2017). *Securing the Internet of Things: A Meta-Study of Challenges, Approaches, and Open Problems.* Paper presented at the 2017 IEEE 37th International Conference on Distributed Computing Systems Workshops (ICDCSW).

Huang, Z., Ruj, S., Cavenaghi, M., & Nayak, A. (2011). Limitations of Trust Management Schemes in VANET and Countermeasures. IEEE 22nd International Symposium on Personal, Indoor and Mobile Radio Communications. doi:10.1109/PIMRC.2011.6139695

Hui, L. C. K., & Li, V. O. K. (2011). Credential-based Privacy-preserving Power Request Scheme for Smart Grid Network. In *2011 IEEE Global Telecommunications Conference - GLOBECOM 2011*. Kathmandu, Nepal: IEEE.

Hu, J., & Vasilakos, A. V. (2016). Energy Big Data Analytics and Security: Challenges and Opportunities. *IEEE Transactions on Smart Grid*, 7(5), 2423–2436. doi:10.1109/TSG.2016.2563461

Hurst, W., Shone, N., El Rhalibi, A., Happe, A., Kotze, B., & Duncan, B. (2017). Advancing the micro-CI testbed for IoT cyber-security research and education. *Proc. CLOUD Comput.*, 139.

Hwang, J., Lee, S., Chung, B., Cho, H., & Nyang, D. H. (2013). Group signatures with controllable linkability for dynamic membership. *Information Sciences*, 222, 761–778. doi:10.1016/j.ins.2012.07.065

Ianuale, N., Schiavon, D., & Capobianco, E. (2016). Smart Cities, Big Data, and Communities: Reasoning From the Viewpoint of Attractors. IEEE Access, 4, 41-47. doi:10.1109/ACCESS.2015.2500733

Iguer, H., Medromi, H., Sayouti, A., Elhasnaoui, S., & Faris, S. (2014). *The Impact of Cyber Security Issues on Businesses and Governments: A Framework for Implementing a Cyber Security Plan.* Paper presented at the 2014 International Conference on Future Internet of Things and Cloud.

Irshad, M. (2016). A Systematic Review of Information Security Frameworks in the Internet of Things. IEEE.

Islam, Razzaque, Hassan, Nagy, & Song. (2017). Mobile Cloud-Based Big Healthcare Data Processing in Smart Cities. *IEEE Access*.

Jabbar & Najim. (2016). Using fully homomorphic encryption to secure cloud computing. *Internet Things Cloud Comput.*, *4*(2), 13-18.

Jacobs, G., & Aeron-Thomas, A. (2005). Africa Road Safety Review. Final Report. US Department of Transportation/Federal Highway Administration.

Jan, M. A., Khan, F., Alam, M., & Usman, M. (2017). A payload-based mutual authentication scheme for Internet of Things. *Future Generation Computer Systems*.

Jayapandian, N., Rahman, A. M. Z., Koushikaa, M., & Radhikadevi, S. (2016, February). A novel approach to enhance multi level security system using encryption with fingerprint in cloud. In *Futuristic Trends in Research and Innovation for Social Welfare (Startup Conclave), World Conference on* (pp. 1-5). IEEE. 10.1109/STARTUP.2016.7583903

Jayapandian, N., & Rahman, A. M. Z. (2017). Secure and efficient online data storage and sharing over cloud environment using probabilistic with homomorphic encryption. *Cluster Computing*, *20*(2), 1561–1573. doi:10.100710586-017-0809-4

Jayasingh, B. B., Patra, M. R., & Mahesh, D. B. (2016). *Security issues and challenges of big data analytics and visualization.* Paper presented at the 2016 2nd International Conference on Contemporary Computing and Informatics (IC3I).

Jimenez, J. M., Romero Martínez, J. O., Rego, A., Dilendra, A., & Lloret, J. (2015). Study of multimedia delivery over software defined networks. In Network Protocols and Algorithms (Vol. 7, No. 4, pp. 37-62). Macrothink Institute.

Jin, Gubbi, Marusic, & Palaniswami. (n.d.). An Information Framework for Creating a Smart City Through Internet of Things. In *IoT Architecture to Enable Intercommunication Through REST API and UPnP Using IP, ZigBee and Arduino*. Academic Press.

Jing, Vasilakos, Wan, Lu, & Qiu. (2014). Security of the Internet of Things: Perspectives and challenges. *Wireless Netw.*, *20*(8), 2481-2501.

Kamvar, S. D., Schlosser, M. T., & Garcia-Molina, H. (2003). The eigentrust algorithm for reputation management in p2p networks. *Proceedings of the 12th international conference on World Wide Web*.

Kang, Y.-S., Park, I.-H., Rhee, J., & Lee, Y.-H. (2016). MongoDB-Based Repository Design for IoT-Generated RFID/Sensor Big Data. *IEEE Sensors Journal, 16*(2), 485–497. doi:10.1109/JSEN.2015.2483499

Kanungo, A., Sharma, A., & Singla, C. (2014). *Smart traffic lights switching and traffic density calculation using video processing.* Chandigarh: Recent Advances in Engineering and Computational Sciences. doi:10.1109/RAECS.2014.6799542

Kaplan, S., & Garrick, B. J. (1981). *On The Quantitative Definition of Risk.* Academic Press.

Katsaros, K., & Dianati, M. (2017). *A conceptual 5G vehicular networking architecture. In 5G Mobile Communications* (pp. 595–623). Cham: Springer.

Khan, M. F. F., & Sakamura, K. (2016). *A patient-centric approach to delegation of access rights in healthcare information systems.* Paper presented at the 2016 International Conference on Engineering & MIS (ICEMIS).

Khan, Z. C., & Mashiane, T. (2014). An Analysis of Facebook's Graph Search. IEEE.

Khatoun, R., & Zeadally, S. (2017). Cybersecurity and privacy solutions in smart cities. *IEEE Communications Magazine, 55*(3), 51–59. doi:10.1109/MCOM.2017.1600297CM

Kim, K. D., & Kumar, P. R. (2012). Cyber–physical systems: A perspective at the centennial. *Proceedings of the IEEE, 100*, 1287-1308.

Kim, Ko, & Kim. (2017). Quality of private information (QoPI) model for effective representation and prediction of privacy controls in mobile computing. *Comput. Secur., 66*, 1-19.

Kim, B. K. (2005). *Internationalizing the Internet: the Co-evolution of Influence and Technology.* Edward Elgar Publishing. doi:10.4337/9781845426750

Kim, J. H. (2017). A survey of IoT security: Risks, requirements, trends, and key technologies. *Journal of Industrial Integration and Management, 2*(02), 1750008. doi:10.1142/S2424862217500087

Kitchin. (2016). *Getting smarter about smart cities: Improving data privacy and data security.* Dept. Taoiseach, Data Protection Unit, Dublin, Ireland, Tech. Rep.

Kitchin, R. (2014). The real-time city? Big data and smart urbanism. *GeoJournal, 79*(1), 1–14. doi:10.100710708-013-9516-8

Kowtko, M. (2011). *Securing our nation and protecting privacy.* Paper presented at the 2011 IEEE Long Island Systems, Applications and Technology Conference.

Kraijak, S., & Tuwanut, P. (2015). *A survey on IoT architectures, protocols, applications, security, privacy, real-world implementation and future trends.* Academic Press.

Kumar, S. R., Yadav, S. A., Sharma, S., & Singh, A. (2016). *Recommendations for effective cyber security execution.* Paper presented at the 2016 International Conference on Innovation and Challenges in Cyber Security (ICICCS-INBUSH). 10.1109/ICICCS.2016.7542327

Kumar, K. K., Durai, S., Vadivel, M. T., & Kumar, K. A. (2017). Smart traffic system using raspberry pi by applying dynamic color changer algorithm. *IEEE International Conference on Smart Technologies and Management for Computing, Communication, Controls, Energy and Materials (ICSTM)*, 146-150. 10.1109/ICSTM.2017.8089141

Kumar, S., & Dutta, K. (2016). Intrusion detection in mobile ad hoc networks: Techniques, systems, and future challenges. *Security and Communication Networks*, *9*(14), 2484–2556. doi:10.1002ec.1484

La, Quek, Lee, Jin, & Zhu. (2016). Deceptive attack and defense game in honeypot-enabled networks for the Internet of Things. *IEEE Internet Things J.*, *3*(6), 1025-1035.

Langner, R. (2011). Stuxnet: Dissecting a cyberwarfare weapon. *IEEE Security and Privacy*, *9*(3), 49–51. doi:10.1109/MSP.2011.67

Lauter, K. (2004). The advantages of elliptic curve cryptography for wireless security. *IEEE Wireless Communications*, *11*(1), 62–67. doi:10.1109/MWC.2004.1269719

Lee, W.-H., & Lee, R. B. (2015). Multi-sensor authentication to improve smartphone security. *Proc. Int. Conf. Inf. Syst. Secur. Privacy (ICISSP)*, 1-11.

Lee, H., & Jung, S. (2017). Line-up formation control of intelligent traffic cones. *17th International Conference on Control, Automation and Systems (ICCAS)*, 616-618. 10.23919/ICCAS.2017.8204303

Lei, A., Cruickshank, H., Cao, Y., Asuquo, P., Ogah, C. P. A., & Sun, Z. (2017). Blockchain-based dynamic key management for heterogeneous intelligent transportation systems. *IEEE Internet of Things Journal*, *4*(6), 1832–1843. doi:10.1109/JIOT.2017.2740569

Li, F., Luo, B., & Liu, P. (2010). Secure Information Aggregation for Smart Grids Using Homomorphic Encryption. In *2010 First IEEE International Conference on Smart Grid Communications* (pp. 327–332). IEEE. 10.1109/SMARTGRID.2010.5622064

Li, H., Lin, X., Yang, H., Liang, X., Lu, R., & Shen, X. (2014). EPPDR: An efficient privacy-preserving demand response scheme with adaptive key evolution in smart grid. *IEEE Transactions on Parallel and Distributed Systems*, 25(8), 2053–2064. doi:10.1109/TPDS.2013.124

Lin, B., Guo, W., Xiong, N., Chen, G., Vasilakos, A. V., & Zhang, H. (2016). A Pretreatment Workflow Scheduling Approach for Big Data Applications in Multicloud Environments. *IEEE eTransactions on Network and Service Management*, 13(3), 581–594. doi:10.1109/TNSM.2016.2554143

Liu, X., Liu, K., Guo, L., Li, X., & Fang, Y. (2013). A game-theoretic approach for achieving k-anonymity in location based services. *Proc. IEEE INFOCOM*, 2985-2993. 10.1109/INFCOM.2013.6567110

Liu, S., Liu, P. X., & Saddik, A. E. (2015). Modeling and Stability Analysis of Automatic Generation Control Over Cognitive Radio Networks in Smart Grids. *IEEE Transactions on Systems, Man, and Cybernetics. Systems*, 45(2), 223–234. doi:10.1109/TSMC.2014.2351372

Liu, X., Cao, J., Yang, Y., & Jiang, S. (2018). CPS-Based Smart Warehouse for Industry 4.0: A Survey of the Underlying Technologies. *Computers*, 7(1), 13. doi:10.3390/computers7010013

Liu, X., Xiong, N., Zhang, N., Liu, A., Shen, H., & Huang, C. (2018). A trust with abstract information verified routing scheme for cyber-physical network. *IEEE Access: Practical Innovations, Open Solutions*, 6, 3882–3898. doi:10.1109/ACCESS.2018.2799681

Liu, Y., Guo, W., Fan, C., Chang, L., & Cheng, C. (2018, February). A Practical Privacy-Preserving Data Aggregation (3PDA) Scheme for Smart Grid. *IEEE Transactions on Industrial Informatics*, 1–1.

Lize, G., Jingpei, W., & Bin, S. (2014). Trust management mechanism for Internet of Things. *China Communications*, 11(2), 148–156. doi:10.1109/CC.2014.6821746

LondonM. U. (2015). *TRIG Project*. Retrieved from http://www.vanet.mdx.ac.uk/research/trig-project/

Luo, Zhang, Yang, Liu, Chang, & Ning. (2016). A kernel machine-based secure data sensing and fusion scheme in wireless sensor networks for the cyber-physical systems. *Future Gener. Comput. Syst.*, *61*, 85-96.

Lu, R., Liang, X., Member, S., & Li, X. (2012). EPPA: An Efficient and Privacy-Preserving Aggregation Scheme for Secure Smart Grid Communications. *IEEE Transactions on Parallel and Distributed Systems*, *23*(9), 1621–1632. doi:10.1109/TPDS.2012.86

Lv, Z. (2016). Managing Big City Information Based on WebVRGIS. IEEE Access, 4, 407-415.

Ly, K., Sun, W., & Jin, Y. (2016). *Emerging challenges in cyber-physical systems: A balance of performance, correctness, and security.* Paper presented at the 2016 IEEE Conference on Computer Communications Workshops (INFOCOM WKSHPS).

Ma, H., Ding, H., Yang, Y., Mi, Z., Yang, J. Y., & Xiong, Z. (2016). Bayes-based ARP attack detection algorithm for cloud centers. *Tsinghua Science and Technology*, *21*(1), 17–28. doi:10.1109/TST.2016.7399280

Mahmood, T., & Afzal, U. (2013). *Security Analytics: Big Data Analytics for cybersecurity: A review of trends, techniques and tools.* Paper presented at the 2013 2nd National Conference on Information Assurance (NCIA).

Mahmood, Z., Ning, H., & Ghafoor, A. (2016). Lightweight two-level session key management for end user authentication in Internet of Things. *Proc. IEEE Int. Conf. Internet Things (iThings) IEEE Green Comput. Com-mun. (GreenCom) IEEE Cyber, Phys. Social Comput. (CPSCom) IEEE Smart Data (SmartData)*, 323-327. 10.1109/iThings-GreenCom-CPSCom-SmartData.2016.78

Mamata, R. (2018). A Methodical Analysis of Application of Emerging Ubiquitous Computing Technology With Fog Computing and IoT in Diversified Fields and Challenges of Cloud Computing. *International Journal of Information Communication Technologies and Human Development, 10*(2).

Mamata, R. B. P. (2018). Communication Improvement and Traffic Control Based on V2I in Smart City Framework. *International Journal of Vehicular Telematics and Infotainment Systems, 2*(1).

Manadhata, P. K., & Wing, J. M. (2011). An Attack Surface Metric. *IEEE Transactions on Software Engineering*, *37*(3), 371–386. doi:10.1109/TSE.2010.60

Marsh, S., & Dibben, M. R. (2005). The Role of Trust in Information Science and Technology. *Annual Review of Information Science & Technology*. doi:10.1002/aris.1440370111

Martínez-Ballesté, A., Pérez-Martínez, P. A., & Solanas, A. (2013, June). The pursuit of citizens' privacy: A privacy-aware smart city is possible. *IEEE Communications Magazine*, *51*(6), 136–141. doi:10.1109/MCOM.2013.6525606

Mayer, R. C., Davis, J. H., & Schoorman, F. D. (1995). An integrative model of organizational trust. *Academy of Management Review*, *20*(3), 709–734. doi:10.5465/amr.1995.9508080335

McCall, J. C., & Trivedi, M. M. (2007). Driver Behavior and Situation Aware Brake Assistance for Intelligent Vehicles. *Proceedings of the IEEE*. 10.1109/JPROC.2006.888388

McCarthy, N. (2015). The Countries with the Most Engineering Graduates. *Forbes. Business (Atlanta, Ga.)*, *8*(1), 33–38.

Mehmood, Y., Ahmad, F., Yaqoob, I., Adnane, A., Imran, M., & Guizani, S. J. I. C. M. (2017). *Internet-of-things-based smart cities: Recent advances and challenges*. Academic Press.

Melchor, C. A., Castagnos, G., & Gaborit, P. (2008). Lattice-based homomorphic encryption of vector. In *2008 IEEE International Symposium on Information Theory* (pp. 1858–1862). Toronto: IEEE. 10.1109/ISIT.2008.4595310

Menouar, H., Guvenc, I., Akkaya, K., Uluagac, A. S., Kadri, A., & Tuncer, A. (2017). UAV-enabled intelligent transportation systems for the smart city: Applications and challenges. *IEEE Communications Magazine*, *55*(3), 22–28. doi:10.1109/MCOM.2017.1600238CM

Mishra, S., & Lohani, V. (2009). Adaptive Traffic Monitoring and Controlling System (ATMC). *International Conference on Advanced Computer Control*, 74-78. 10.1109/ICACC.2009.125

Misra, S., Maheswaran, M., & Hashmi, S. (2017). *Security Challenges an Approaches in Internet of Things*. Springer. doi:10.1007/978-3-319-44230-3

Mitchel, S., Villa, N., & Stewart-Weeks, M. (2013). *The Internet of Everything for Cities: Connecting People, Process, Data, and Things to Improve the 'Livability' of Cities and Communities*. Point of View, Cisco.

Mitchell, R., & Chen, R. (2013). Effect of intrusion detection and response on reliability of cyber physical systems. *IEEE Transactions on Reliability, 62*(1), 199–210. doi:10.1109/TR.2013.2240891

Moghadam, M. H., & Mozayani, N. (2011). A street lighting control system based on holonic structures and traffic system. *3rd International Conference on Computer Research and Development*, 92-96.

Moraru, L., Leone, P., Nikoletseas, S., & Rolim, J. D. (2007). Near optimal geographic routing with obstacle avoidance in wireless sensor networks by fast-converging trust-based algorithms. *Proceedings of the 3rd ACM workshop on QoS and security for wireless and mobile networks*. 10.1145/1298239.1298246

Moreno, M. V., Terroso-Saenz, F., Gonzalez-Vidal, A., Valdes-Vela, M., Skarmeta, A. F., Zamora, M. A., & Chang, V. (2017). Applicability of Big Data Techniques to Smart Cities Deployments. *IEEE Transactions on Industrial Informatics, 13*(2), 800–809. doi:10.1109/TII.2016.2605581

Mo, Y., Kim, T. H. J., Brancik, K., Dickinson, D., Lee, H., Perrig, A., & Sinopoli, B. (2012). Cyber–physical security of a smart grid infrastructure. *Proceedings of the IEEE, 100*(1), 195–209. doi:10.1109/JPROC.2011.2161428

Mozzaquatro, B. A., Jardim-Goncalves, R., & Agostinho, C. (2015, October). Towards a reference ontology for security in the internet of things. In *Measurements & Networking (M&N), 2015 IEEE International Workshop on* (pp. 1-6). IEEE. 10.1109/IWMN.2015.7322984

Mui, L., Mohtashemi, M., & Halberstadt, A. (2002). A computational model of trust and reputation. *Proceedings of the 35th Hawaii International Conference on System Science (HICSS)*. 10.1109/HICSS.2002.994181

Munugala, S., Brar, G. K., Syed, A., Mohammad, A., & Halgamuge, M. N. (2017). The Much Needed Security and Data Reforms of Cloud Computing in Medical Data Storage. In *Applying Big Data Analytics in Bioinformatics and Medicine*. IGI Global.

Muruganandham, P. R. M. (2010). Real time web based vehicle tracking using GPS. *World Academy of Science, Engineering and Technology, 61*(1), 91–96.

Mustafa, M. A., Cleemput, S., Aly, A., Abidin, A., & Leuven, K. U. (2017). An MPC-based Protocol for Secure and Privacy-Preserving Smart Metering. In 2017 IEEE PES Innovative Smart Grid Technologies Conference Europe (ISGT-Europe) (pp. 1–6). IEEE. doi:10.1109/ISGTEurope.2017.8260202

Namal, S., Ahmad, I., Saud, S., Jokinen, M., & Gurtov, A. (2016). Implementation of OpenFlow based cognitive radio network architecture: SDN&R. *Wireless Networks*, 22(2), 663–677. doi:10.100711276-015-0973-5

Nandhini, D. M., & Das, B. B. (2016). An Assessment And Methodology For Fraud Detection In Online Social. *IEEE Network*, 104–108.

Naranjo, J. E., Gonzalez, C., Garcia, R., & Pedro, T. (2008). Lane-Change Fuzzy Control in Autonomous Vehicles for the Overtaking Maneuver. *IEEE Transactions on Intelligent Transportation Systems*, 9(3), 438–450. doi:10.1109/TITS.2008.922880

National Institute of Standards and Technology. (2013). *National Institute of Standards and Technology (2013): NIST framework and roadmap for smart grid interoperability standards, Release 2.0. smart grid interoperability panel (SGIP).* Retrieved from http:// j.mp/1rs1tKs http://collaborate.nist.gov/twiki- sggrid/pub/ SmartGrid/IKBFramework/ NIST_Framework_Release_2-0_corr.pdf

Nawa, K., Chandrasiri, N. P., Yanagihara, T., & Oguchi, K. (2014). Cyber physical system for vehicle application. *Transactions of the Institute of Measurement and Control*, 36(7), 898–905. doi:10.1177/0142331213510018

Nello, Z. (2004). *Technologies Impacting Machine Vision – Vision Engines*. Retrieved from https://www.visiononline.org/vision-resources-details.cfm/vision-resources/ Technologies-Impacting-Machine-Vision-Vision-Engines-By-Nello-Zuech-President-Vision-Systems-International-Consultancy/content_id/1243

Nelson, L. (2016). *The US once had more than 130 hijackings in 4 years. Here's why they finally stopped.* Retrieved from https://www.vox.com/2016/3/29/11326472/ hijacking-airplanes-egyptair

Nest Inc. (2018). *Nest Learning Thermostat, programs itself, and then pays for itself.* Retrieved from https://nest.com/thermostats/nest-learning-thermostat/overview/

Niehaus, A., & Stengel, R. (1991). An expert system for automated highway driving. *IEEE Control Systems Magazine*, 11(3), 53–61. doi:10.1109/37.75579

Niehaus, A., & Stengel, R. F. (1994). Probability-Based Decision Making for Automated Highway Driving. *IEEE Transactions on Vehicular Technology*, 43(3), 626–634. doi:10.1109/25.312814

NIST 800-122. (2010). *Guide to Protecting the Confidentiality of Personally Identifiable Information (PII)*. National Institute of Standards and Technology.

Noh, S., & An, K. (2017). Risk Assessment for Automatic Lane Change Maneuvers on Highways. *International Conference on Robotics and Automation (ICRA)*. 10.1109/ICRA.2017.7989031

Noh, S., An, K., & Han, W. (2015). High-Level Data Fusion based Probabilistic Situation Assessment for Highly Automated Driving. *18th International Conference on Intelligent Transportation Systems*. 10.1109/ITSC.2015.259

North-east group llc. (2017). *Western Europe Smart Grid : Market Forecast.* Retrieved from http://www.northeast-group.com/

O'Donovan, P., Gallagher, C., Bruton, K., & O'Sullivan, D. T. (2018). A fog computing industrial cyber-physical system for embedded low-latency machine learning Industry 4.0 applications. *Manufacturing Letters, 15*, 139–142. doi:10.1016/j.mfglet.2018.01.005

Olejnik, K., Dacosta, I., Machado, J. S., Huguenin, K., Khan, M. E., & Hubaux, J.-P. (2017). SmarPer: Context-aware and automatic run time permissions for mobile devices. *Proc. 38th IEEE Symp. Secur. Privacy (SP)*, 1058-1076.

Paladino, S. C., & Fingerman, J. E. (2009). *Cybersecurity Technology Transition: A Practical Approach.* Paper presented at the 2009 Cybersecurity Applications & Technology Conference for Homeland Security.

Pasqualetti, F., Dörfler, F., & Bullo, F. (2011, December). Cyber-physical attacks in power networks: Models, fundamental limitations and monitor design. In *Decision and Control and European Control Conference (CDC-ECC), 2011 50th IEEE Conference on* (pp. 2195-2201). IEEE. 10.1109/CDC.2011.6160641

Patel, S., Park, H., Bonato, P., Chan, L., & Rodgers, M. (2012). A review of wearable sensors and systems with application in rehabilitation. *Journal of Neuroengineering and Rehabilitation, 9*(1), 21. doi:10.1186/1743-0003-9-21 PMID:22520559

Pathan, A. S. K. (Ed.). (2014). *The state of the art in intrusion prevention and detection.* CRC Press. doi:10.1201/b16390

Pathan, A.-S. K. (2016). *Security of self-organizing networks: MANET, WSN, WMN, VANET.* CRC Press.

Pearce, T., & Maunder, D. A. C. (2000). *Public Transport Safety in Four Emerging Nations.* Washington, DC: Transport Research Laboratory.

Pelargonio, P., & Pugliese, M. (2014). *Enhancing security in public transportation services of Roma: The PANDORA system.* Paper presented at the 2014 International Carnahan Conference on Security Technology (ICCST). 10.1109/CCST.2014.6986969

Piro, G., Cianci, I., Grieco, L. A., Boggia, G., & Camarda, P. (2014). Information centric services in smart cities. *Journal of Systems and Software, 88,* 169–188. doi:10.1016/j.jss.2013.10.029

Pohls, Angelakis, Suppan, Fischer, Oikonomou, Tragos, ... Mouroutis. (2014). Rerum: Building a reliable iot upon privacyand security-enabled smart objects. In Wireless Communications and Networking Conference Workshops (WCNCW) (pp. 122–127). IEEE.

Poovendran, R. A. D. H. A. (2010). Cyber–physical systems: Close encounters between two parallel worlds. *Proceedings of the IEEE, 98*(8), 1363–1366. doi:10.1109/JPROC.2010.2050377

Portela, C. M. (2013). A flexible, privacy enhanced and secured ICT architecture for a smart grid project with active consumers in the city of Zwolle—NL. *Proc. 22nd Int. Conf. Electricity Distrib. (CIRED),* 1–4. 10.1049/cp.2013.0628

Porter, M. E. (2001). Strategy and the Internet. *Harvard Business Review, 79*(3), 63–78. PMID:11246925

Pour, M. M., Anzalchi, A., & Sarwat, A. (2017). *A review on cyber security issues and mitigation methods in smart grid systems.* Paper presented at the SoutheastCon 2017.

Qian, L. (2013). Study of Information System Security of Government Data Center Based on the Classified Protection. IEEE.

Qu, H., Shang, P., Lin, X. J., & Sun, L. (2015). Cryptanalysis of A Privacy-Preserving Smart Metering Scheme Using Linkable Anonymous Credential. *IACR Cryptology EPrint Archive, 2015,* 1066.

Radovan, M., & Golub, B. (2017). *Trends in IoT security.* Paper presented at the 2017 40th International Convention on Information and Communication Technology, Electronics and Microelectronics (MIPRO).

Rath & Oreku. (2018). Security Issues in Mobile Devices and Mobile Adhoc Networks. In Mobile Technologies and Socio-Economic Development in Emerging Nations. IGI Global. doi:10.4018/978-1-5225-4029-8.ch009

Rath, M. (2018). An Exhaustive Study and Analysis of Assorted Application and Challenges in Fog Computing and Emerging Ubiquitous Computing Technology. *International Journal of Applied Evolutionary Computation, 9*(2), 17-32. Retrieved from www.igi-global.com/ijaec

Rath, M., & Panda, M. R. (2017). MAQ system development in mobile ad-hoc networks using mobile agents. *IEEE 2nd International Conference on Contemporary Computing and Informatics (IC3I),* 794-798.

Rath, M., & Pattanayak, B. K. (2014). A methodical survey on real time applications in MANETS: Focussing On Key Issues. *International Conference on, High Performance Computing and Applications (IEEE ICHPCA),* 1-5. 10.1109/ICHPCA.2014.7045301

Rath, M., & Pattanayak, B. K. (2018). Monitoring of QoS in MANET Based Real Time Applications. In Information and Communication Technology for Intelligent Systems (vol. 2). Springer. doi:10.1007/978-3-319-63645-0_64

Rath, M., Pati, B., & Pattanayak, B. K. (2016). Inter-Layer Communication Based QoS Platform for Real Time Multimedia Applications in MANET. Wireless Communications, Signal Processing and Networking (IEEE WiSPNET), 613-617. doi:10.1109/WiSPNET.2016.7566203

Rath. (2018). Effective Routing in Mobile Ad-hoc Networks With Power and End-to-End Delay Optimization: Well Matched With Modern Digital IoT Technology Attacks and Control in MANET. In *Advances in Data Communications and Networking for Digital Business Transformation.* IGI Global. Doi:10.4018/978-1-5225-5323-6.ch007

Rath, M. (2017). Resource provision and QoS support with added security for client side applications in cloud computing. *International Journal of Information Technology, 9*(3), 1–8.

Rath, M., & Pati, B. (2017). *Load balanced routing scheme for MANETs with power and delay optimization. International Journal of Communication Network and Distributed Systems, 19.*

Rath, M., Pati, B., & Pattanayak, B. K. (2017). Cross layer based QoS platform for multimedia transmission in MANET. *11th International Conference on Intelligent Systems and Control (ISCO),* 402-407. 10.1109/ISCO.2017.7856026

Rath, M., & Pattanayak, B. (2017). MAQ: A Mobile Agent Based QoS Platform for MANETs. *International Journal of Business Data Communications and Networking, IGI Global, 13*(1), 1–8. doi:10.4018/IJBDCN.2017010101

Rath, M., & Pattanayak, B. K. (2018). SCICS: A Soft Computing Based Intelligent Communication System in VANET. Smart Secure Systems – IoT and Analytics Perspective. *Communications in Computer and Information Science, 808*, 255–261. doi:10.1007/978-981-10-7635-0_19

Rath, M., Pattanayak, B. K., & Pati, B. (2017). *Energetic Routing Protocol Design for Real-time Transmission in Mobile Ad hoc Network. In Computing and Network Sustainability, Lecture Notes in Networks and Systems* (Vol. 12). Singapore: Springer.

Rath, M., Swain, J., Pati, B., & Pattanayak, B. K. (2018). *Attacks and Control in MANET. In Handbook of Research on Network Forensics and Analysis Techniques* (pp. 19–37). IGI Global. doi:10.4018/978-1-5225-4100-4.ch002

Raut, S. B., & Malik, L. G. (2014, December). Survey on vehicle collision prediction in VANET. In *Computational Intelligence and Computing Research (ICCIC), 2014 IEEE International Conference on* (pp. 1-5). IEEE. 10.1109/ICCIC.2014.7238552

Razzaq, A., Hur, A., Ahmad, H. F., & Masood, M. (2013). *Cyber security: Threats, reasons, challenges, methodologies and state of the art solutions for industrial applications.* Paper presented at the 2013 IEEE Eleventh International Symposium on Autonomous Decentralized Systems (ISADS). 10.1109/ISADS.2013.6513420

Razzaq, Anwar, Ahmad, Latif, & Munir. (2014). Ontology for attack detection: An intelligent approach to Web application security. *Comput. Secur., 45*, 124-146.

Reichardt, D. M. (2008). Approaching Driver Models Which Integrate Models Of Emotion And Risk. *IEEE Intelligent Vehicles Symposium.* 10.1109/IVS.2008.4621284

Rivest, R. L., Adleman, L., & Dertouzos, M. L. (1978). On data banks and privacy homomorphisms. *Foundations of Secure Computation, 4*(11), 169-180.

Rizwan, P., Suresh, K., & Babu, M. R. (2016). Real-time smart traffic management system for smart cities by using Internet of Things and big data. *International Conference on Emerging Technological Trends (ICETT),* 1-7. 10.1109/ICETT.2016.7873660

Road traffic safety. (n.d.). In *Wikipedia.* Retrieved from https://en.wikipedia.org/wiki/Road_traffic_safety

Růžička, J., Šilar, J., Bělinová, Z., & Langr, M. (2018). Methods of traffic surveys in cities for comparison of traffic control systems — A case study. *Smart City Symposium Prague (SCSP),* 1-6. 10.1109/SCSP.2018.8402666

Samaila, M. G., Neto, M., Fernandes, D. A., Freire, M. M., & Inácio, P. R. (2017). Security challenges of the Internet of Things. In *Beyond the Internet of Things* (pp. 53–82). Cham: Springer. doi:10.1007/978-3-319-50758-3_3

Samian, N., & Seah, W. K. (2017). Trust-based Scheme for Cheating and Collusion Detection in Wireless Multihop Networks. *Proceedings of the 14th EAI International Conference on Mobile and Ubiquitous Systems: Computing, Networking and Services (MobiQuitous)*. 10.1145/3144457.3144486

Sarkar, S., & Datta, R. (2012). *A trust based protocol for energy-efficient routing in self-organized MANETs.* Paper presented at the India Conference (INDICON), 2012 Annual IEEE. 10.1109/INDCON.2012.6420778

Schleicher, J. M., Vögler, M., Dustdar, S., & Inzinger, C. (2016). Application Architecture for the Internet of Cities: Blueprints for Future Smart City Applications. *IEEE Internet Computing*, *20*(6), 68–75. doi:10.1109/MIC.2016.130

Schneider, J., Wilde, A., & Naab, K. (2008). Probabilistic Approach for Modeling and Identifying Driving Situations. *IEEE Intelligent Vehicles Symposium.* 10.1109/IVS.2008.4621145

Science, C. (2015). Massachusetts Institute of Technology (MIT). Retrieved from http://cities.media.mit.edu/

Scott, M. (2011). *Miracl Library*. Retrieved from http://www.shamus.ie

Sedjelmaci, Senouci, & Taleb. (2017). An accurate security game for low-resource IoT devices. *IEEE Trans. Veh. Technol.*, *66*(10), 9381-9393.

Sen, J. (2010). *Reputation-and trust-based systems for wireless self-organizing networks*. Aurbach Publications, CRC Press.

Sfar, A. R., Natalizio, E., Challal, Y., & Chtourou, Z. (2018). A roadmap for security challenges in the Internet of Things. *Digital Communications and Networks*, *4*(2), 118–137. doi:10.1016/j.dcan.2017.04.003

Shafi, G., & Micali, S. (1984). Probabilistic encryption. *Journal of Computer and System Sciences*, *28*(2), 270–299. doi:10.1016/0022-0000(84)90070-9

Shahzad, G., Yang, H., Ahmad, A. W., & Lee, C. (2016). Energy-Efficient Intelligent Street Lighting System Using Traffic-Adaptive Control. *IEEE Sensors Journal*, *16*(13), 5397–5405. doi:10.1109/JSEN.2016.2557345

Shaikh, R. A. (2016). Fuzzy Risk-based Decision Method for Vehicular Ad Hoc Networks. *International Journal of Advanced Computer Science and Applications*, 7(9), 54–62. doi:10.14569/IJACSA.2016.070908

Shamshirband, Patel, Anuar, Kiah, & Abraham. (2014). Cooperative game theoretic approach using fuzzy Q-learning for detecting and preventing intrusions in wireless sensor networks. *Eng. Appl. Artif. Intell.*, *32*, 228-241.

Shapsough, S., Qatan, F., Aburukba, R., Aloul, F., & Ali, A. R. A. (2015). *Smart grid cyber security: Challenges and solutions.* Paper presented at the 2015 International Conference on Smart Grid and Clean Energy Technologies (ICSGCE). 10.1109/ICSGCE.2015.7454291

Sharma, Chen, & Park. (2017). A software de_ned fog node based distributed blockchain cloud architecture for IoT. *IEEE Access*, *6*, 115-124.

Sharma, P., Doshi, D., & Prajapati, M. M. (2016). *Cybercrime: Internal security threat.* Paper presented at the 2016 International Conference on ICT in Business Industry & Government (ICTBIG). Smart City. Retrieved from https://www.techopedia.com/definition/31494/smart-city

Shukla, S., Balachandran, K., & Sumitha, V. S. (2016). A framework for smart transportation using Big Data. *2016 International Conference on ICT in Business Industry & Government (ICTBIG)*, 1-3. 10.1109/ICTBIG.2016.7892720

Sicari, S., Rizzardi, A., Grieco, L. A., & Coen-Porisini, A. (2015). Security, privacy and trust in Internet of Things: The road ahead. *Computer Networks*, *76*, 146–164. doi:10.1016/j.comnet.2014.11.008

Siddique, K., Akhtar, Z., Yoon, E. J., Jeong, Y. S., Dasgupta, D., & Kim, Y. (2016). Apache Hama: An Emerging Bulk Synchronous Parallel Computing Framework for Big Data Applications. IEEE Access, 4, 8879-8887. doi:10.1109/ACCESS.2016.2631549

Simmel, G. (2011). *Georg Simmel on individuality and social forms.* University of Chicago Press.

Şimşek, M. U., Okay, F. Y., Mert, D., & Özdemir, S. (2018). TPS3 : A privacy preserving data collection protocol for smart grids. *Information Security Journal: A Global Perspective, 27*(2), 102–118.

Singh, D., Vishnu, C., & Mohan, C. K. (2016). Visual Big Data Analytics for Traffic Monitoring in Smart City. *15th IEEE International Conference on Machine Learning and Applications (ICMLA)*, 886-891. 10.1109/ICMLA.2016.0159

Singh, M., Halgamuge, M. N., Ekici, G., & Jayasekara, C. S. (2018). *A Review on Security and Privacy Challenges of Big Data. In Lecture Notes on Data Engineering and Communications Technologies Cognitive Computing for Big Data Systems Over IoT, Frameworks, Tools and Applications* (Vol. 14). Springer.

Sobh, T. S. (2006). Wired and wireless intrusion detection system: Classifications, good characteristics and state-of-the-art. *Elsevier J. Computer Standards & Interfaces, 28*(6), 670–694. doi:10.1016/j.csi.2005.07.002

Solanas, A., Patsakis, C., Conti, M., Vlachos, I., Ramos, V., Falcone, F., ... Martinez-Balleste, A. (2014, August). Smart health: A context-aware health paradigm within smart cities. *IEEE Communications Magazine, 52*(8), 74–81. doi:10.1109/MCOM.2014.6871673

Soliman, M., Abiodun, T., Hamouda, T., Zhou, J., & Lung, C. (2013). Smart Home: Integrating Internet of Things with Web Services and Cloud Computing. *IEEE 5th International Conference on Cloud Computing Technology and Science.*

Song, G., & Wu, L. (2012). Smart city in perspective of innovation 2.0. City Management, 9, 53-60.

Song, H. M., Kim, H. R., & Kim, H. K. (2016, January). Intrusion detection system based on the analysis of time intervals of CAN messages for in-vehicle network. In *2016 international conference on information networking (ICOIN)* (pp. 63-68). IEEE.

Son, M., & Jung, H. (2017). Development and Construction of Security-Enhanced LTE traffic signal system in Korea. *2nd International Conference on Computer and Communication Systems (ICCCS),* 91-95. 10.1109/CCOMS.2017.8075274

Sood, A. K., & Enbody, R. J. (2013). Targeted Cyberattacks: A Superset of Advanced Persistent Threats. *IEEE Security and Privacy, 11*(1), 54–61. doi:10.1109/MSP.2012.90

Sridhar, S., Hahn, A., & Govindarasu, M. (2012). Cyber–physical system security for the electric power grid. *Proceedings of the IEEE, 100*(1), 210–224. doi:10.1109/JPROC.2011.2165269

Srinivasan, A., Li, F., & Wu, J. (2008). *A novel CDS-based reputation monitoring system for wireless sensor networks.* Paper presented at the Distributed computing systems workshops, 2008. ICDCS'08. 28th international conference on. 10.1109/ICDCS.Workshops.2008.17

Stallings, W., & Brown, L. (2015). *Computer security: principles and practice* (4th ed.). Pearson Education.

Stamp, J., Dillinger, J., Young, W., & DePoy, J. (2003). *Common vulnerabilities in critical infrastructure control systems. SAND2003-1772C*. Sandia National Laboratories.

Stephan, N. (2012). *Smart Meters Help Brazil Zap Electricity Theft. Brazilian utilities battle theft and deadbeats with smart meters*. Retrieved from https://www.bloomberg.com/news/ articles/2012-03-08/smart-meters-help-brazil-zap-electricity-theft

Stockholm City Executive Office. (2010). *Living in Stockholm Should Be e-asy*. Available: http://international.stockholm.se/globalassets/ovriga-bilder-och-filer/e-tjanster_broschyr-16-sid_4.pdf

Stoneburner, G., Goguen, A. Y., & Feringa, A. (2002). *Risk Management Guide for Information Technology Systems*. National Institute of Standards and Technology. doi:10.6028/NIST.SP.800-30

Sudrich, S., Borges, J., & Beigl, M. (2017, August). Graph-based anomaly detection for smart cities: A survey. In *2017 IEEE SmartWorld, Ubiquitous Intelligence & Computing, Advanced & Trusted Computed, Scalable Computing & Communications, Cloud & Big Data Computing, Internet of People and Smart City Innovation (SmartWorld/SCALCOM/UIC/ATC/CBDCom/IOP/SCI)*. IEEE.

Summers, G. (2004). *Data and databases*. Nelson Australia Pty Limited.

Sun, Y., Song, H., Jara, A. J., & Bie, R. (2016). Internet of Things and Big Data Analytics for Smart and Connected Communities. IEEE Access, 4, 766-773. doi:10.1109/ACCESS.2016.2529723

Sun, W., Yu, F. R., Tang, T., & You, S. (2017). A Cognitive Control Method for Cost-Efficient CBTC Systems With Smart Grids. *IEEE Transactions on Intelligent Transportation Systems*, *18*(3), 568–582. doi:10.1109/TITS.2016.2586938

Suo, H., Wan, J., Zou, C., & Liu, J. (2012). Security in the internet of things: a review. In *Computer Science and Electronics Engineering (ICCSEE), 2012 International Conference on* (vol. *3*, pp. 648–651). Academic Press. 10.1109/ICCSEE.2012.373

Tahmid, T., & Hossain, E. (2017). Density based smart traffic control system using canny edge detection algorithm for congregating traffic information. *3rd International Conference on Electrical Information and Communication Technology (EICT)*, 1-5. 10.1109/EICT.2017.8275131

Tajeddine, A., Kayssi, A., Chehab, A., Elhajj, I., & Itani, W. (2015). CENTERA: A centralized trust-based efficient routing protocol with authentication for wireless sensor networks. *Sensors (Basel)*, *15*(2), 3299–3333. doi:10.3390150203299 PMID:25648712

Talpur, Bhuiyan, & Wang. (2014). Shared-node IoT network architecture with ubiquitous homomorphic encryption for healthcare monitoring. *Int. J. Embedded Syst.*, *7*(1), 43-54.

Tang, Chen, Hefferman, Pei, Tao, He, & Yang. (2017).Incorporating Intelligence in Fog Computing for Big Data Analysis in Smart Cities. *IEEE Transactions on Industrial Informatics*.

Tangade, Sun, & Manvi. (2013). A Survey on Attacks, Security and Trust Management Solutions in VANETs. In *2013 Fourth International Conference on Computing, Communications and Networking Technologies (ICCCNT)*. IEEE. 10.1109/ICCCNT.2013.6726668

Tao, Zuo, Liu, Castiglione, & Palmieri. (2016). Multi-layer cloud architectural model and ontology-based security service framework for IoT-based smart homes. *Future Gener. Comput. Syst.*, *78*, 1040-1051.

Tarapiah, S., Atalla, S., & AbuHania, R. (2013). Smart on-board transportation management system using gps/gsm/gprs technologies to reduce traffic violation in developing countries. *International Journal of Digital Information and Wireless Communications*, *3*(4), 430–439.

Tavakolifard, M., & Almeroth, K. C. (2012). Social computing: An intersection of recommender systems, trust/reputation systems, and social networks. *IEEE Network*, *26*(4), 53–58. doi:10.1109/MNET.2012.6246753

Tawalbeh, L. A., Mehmood, R., Benkhlifa, E., & Song, H. (2016). Mobile Cloud Computing Model and Big Data Analysis for Healthcare Applications. IEEE Access, *4*, 6171-6180. doi:10.1109/ACCESS.2016.2613278

Ten, C. W., Liu, C. C., & Manimaran, G. (2008). Vulnerability assessment of cybersecurity for SCADA systems. *IEEE Transactions on Power Systems*, *23*(4), 1836–1846. doi:10.1109/TPWRS.2008.2002298

Tewari, A., & Gupta, B. B. (2017). Cryptanalysis of a novel ultra-lightweight mutual authentication protocol for IoT devices using RFID tags. *The Journal of Supercomputing*, *73*(3), 1085–1102. doi:10.100711227-016-1849-x

Thayananthan, V., Abdulkader, O., Jambi, K., & Bamahdi, A. M. (2017, June). Analysis of Cybersecurity based on Li-Fi in green data storage environments. In *Cyber Security and Cloud Computing (CSCloud), 2017 IEEE 4th International Conference on* (pp. 327-332). IEEE. 10.1109/CSCloud.2017.32

Thayananthan, V., & Albeshri, A. (2015). Big data security issues based on quantum cryptography and privacy with authentication for mobile data center. *Procedia Computer Science, 50*, 149–156. doi:10.1016/j.procs.2015.04.077

Thayananthan, V., Alzahrani, A., & Qureshi, M. S. (2015). Efficient techniques of key management and quantum cryptography in RFID networks. *Security and Communication Networks, 8*(4), 589–597. doi:10.1002ec.1005

Thayananthan, V., & Shaikh, R. A. (2016). Contextual Risk-based Decision Modeling for Vehicular Networks. *International Journal of Computer Network and Information Security, 8*(9), 1–9. doi:10.5815/ijcnis.2016.09.01

Thenmozhi, T., & Somasundaram, R. (2016). Towards modelling a trusted and secured centralised reputation system for VANET's. *Proceedings of the International Conference on Soft Computing Systems.* 10.1007/978-81-322-2674-1_64

Thoma, C., Cui, T., & Franchetti, F. (2012). Secure Multiparty Computation Based Privacy Preserving Smart Metering System. In *44th North American Power Symposium (NAPS)* (pp. 1–6). Academic Press. 10.1109/NAPS.2012.6336415

Top 7 Cybersecurity Issues of 2018. (n.d.). Retrieved from https://www.cyberreefsolutions.com/top-cybersecurity-concerns-2018/

Tsai, Lai, Chiang, & Yang. (2014). Data mining for Internet of Things: A survey. *IEEE Commun. Surveys Tuts., 16*(1), 77-97.

Tyagi, A., Kushwah, J., & Bhalla, M. (2017). *Threats to security of Wireless Sensor Networks.* Paper presented at the 2017 7th International Conference on Cloud Computing, Data Science & Engineering - Confluence.

Ullah, S., Khan, E., Ullah, S., & Ali, W. (2017). A Light-Weight Secret Key-Based Privacy Preserving Technique for Home Area Networks in Smart Grid. In *2017 13th International Conference on Natural Computation, Fuzzy Systems and Knowledge Discovery (ICNC-FSKD)* (pp. 895–899). IEEE. 10.1109/FSKD.2017.8393395

Uludag, S., Zeadally, S., & Badra, M. (2015). *Techniques, Taxonomy, and Challenges of Privacy Protection in the Smart Grid: Privacy in Digital, Networked World.* Springer. doi:10.1007/978-3-319-08470-1_15

US Department of Energy. (2009). *Smart grid system report*. Retrieved from http://www.doe.energy.gov/

Usman, A. B., & Gutierrez, J. J. A. o. O. R. (2018). *DATM: a dynamic attribute trust model for efficient collaborative routing*. Academic Press.

Usman, A. B. (2018). *Trust-based protocols for secure collaborative routing in wireless mobile networks*. Auckland University of Technology.

Usman, A. B., & Gutierrez, J. (2016). A Reliability-Based Trust Model for Efficient Collaborative Routing in Wireless Networks. *Proceedings of the 11th International Conference on Queueing Theory and Network Applications*. 10.1145/3016032.3016057

Usman, A. B., & Gutierrez, J. (2018). Toward Trust Based Protocols in a Pervasive and Mobile Computing: A Survey. *Ad Hoc Networks*.

Vasserman, E. Y., & Hopper, N. (2013). Vampire attacks: Draining life from wireless ad hoc sensor networks. *IEEE Transactions on Mobile Computing, 12*(2), 318–332. doi:10.1109/TMC.2011.274

Vlacic, A. F. (n.d.). Multiple Criteria-Based Real-Time Decision Making by Autonomous City Vehicles. Institute of Integrated and Intelligent Systems. *Griffith University*.

Walravens. (2012). Mobile business and the smart city: Developing a business model framework to include public design parameters for mobile city services. *J. Theor. Appl. Electron. Commerce Res., 7*(3), 121-135.

Walravens, N., & Ballon, P. (2013). Platform business models for smart cities: From control and value to governance and public value. *IEEE Communications Magazine, 51*(6), 72–79. doi:10.1109/MCOM.2013.6525598

Wang, Wan, Guo, Cheung, & Yuen. (2017). Inference-based similarity search in randomized Montgomery domains for privacy preserving biometric identification. *IEEE Trans. Pattern Anal. Mach. Intell., 40*(7), 1611-1624.

Wang, L., Törngren, M., & Onori, M. (2015). Current status and advancement of cyber-physical systems in manufacturing. *Journal of Manufacturing Systems, 37*, 517–527. doi:10.1016/j.jmsy.2015.04.008

Wang, X.-F., And, Y. M., & Chen, R.-M. (2016). An efficient privacy-preserving aggregation and billing protocol for smart grid. *Security and Communication Networks, 9*(17), 4536–4547. doi:10.1002ec.1645

Wang, Y., Chen, Q., Kang, C., & Xia, Q. (2016). Clustering of Electricity Consumption Behavior Dynamics Toward Big Data Applications. *IEEE Transactions on Smart Grid*, 7(5), 2437–2447. doi:10.1109/TSG.2016.2548565

Wang, Y., & Vassileva, J. (2007). Toward trust and reputation based web service selection: A survey. *International Transactions on Systems Science and Applications*, 3(2), 118–132.

Wei, Y.-C., & Chen, Y.-M. (2014). Adaptive decision making for improving trust establishment in VANET. *The 16th Asia-Pacific Network Operations and Management Symposium*. 10.1109/APNOMS.2014.6996523

Wikipedia. (2018). Retrieved November Saturday, 2018, from https://en.wikipedia.org/wiki/Risk

Wu, J., Ota, K., Dong, M., & Li, C. (2016). A Hierarchical Security Framework for Defending Against Sophisticated Attacks on Wireless Sensor Networks in Smart Cities. *IEEE Access: Practical Innovations, Open Solutions*, 4(4), 416–424. doi:10.1109/ACCESS.2016.2517321

Wu, X., He, J., Zhang, X., & Xu, F. (2009). A Distributed Decision-Making Mechanism for Wireless P2P Networks. *Journal of Communications and Networks (Seoul)*, 11(4), 359–367. doi:10.1109/JCN.2009.6391349

Xiao, Li, Han, Liu, & Zhuang. (2016). PHY-layer spoofing detection with reinforcement learning in wireless networks. *IEEE Trans. Veh. Technol.*, 65(12), 10037-10047.

Xiao, Z. H., Guan, Z. Q., & Zheng, Z. H. (2008). The research and development of the highway's electronic toll collection system. In *Proceedings of Knowledge Discovery and Data Mining* (pp. 359–362). University of Adelaide. doi:10.1109/WKDD.2008.100

Xijin, W., & Linxiu, F. (2003). The application research of MD5 encryption algorithm in DCT digital Watermarking. *Physics Procedia*, 25(1), 1264–1269.

Xing, Hu, Yu, Cheng, & Zhang. (2017). Mutual privacy preserving *k*-means clustering in social participatory sensing. *IEEE Trans. Ind. Informat.*, 13(4), 2066-2076.

Xu, Cao, Ren, Li, & Feng. (2017). Network security situation awareness based on semantic ontology and user-defined rules for Internet of Things. *IEEE Access*, 5, 21046-21056.

Yadav, S. A., Kumar, S. R., Sharma, S., & Singh, A. (2016). *A review of possibilities and solutions of cyber attacks in smart grids*. Paper presented at the 2016 International Conference on Innovation and Challenges in Cyber Security (ICICCS-INBUSH). 10.1109/ICICCS.2016.7542359

Yan, Z. (2007). *Trust Management for Mobile Computing Platforms*. Academic Press.

Yan, Z., & Holtmanns, S. (2008). Trust modeling and management: from social trust to digital trust. IGI Global.

Yang, W., & Zhao, Q. (2014). *Cyber security issues of critical components for industrial control system*. Paper presented at the Proceedings of 2014 IEEE Chinese Guidance, Navigation and Control Conference.

Yan, Z., Zhang, P., & Vasilakos, A. V. (2014). A survey on trust management for Internet of Things. *Journal of Network and Computer Applications*, *42*, 120–134. doi:10.1016/j.jnca.2014.01.014

Yu. (2016). Big privacy: Challenges and opportunities of privacy study in the age of big data. *IEEE Access*, *4*, 2751-2763.

Yuan, W. (2015). A Smart Work Performance Measurement System for Police Officers. IEEE Access, 3, 1755-1764. doi:10.1109/ACCESS.2015.2481927

Yu, H., Shen, Z., Miao, C., Leung, C., & Niyato, D. (2010). A survey of trust and reputation management systems in wireless communications. *Proceedings of the IEEE*, *98*(10), 1755–1772. doi:10.1109/JPROC.2010.2059690

Zahariadis, T., Leligou, H. C., Trakadas, P., & Voliotis, S. (2010). Trust management in wireless sensor networks. *Transactions on Emerging Telecommunications Technologies*, *21*(4), 386–395.

Zaid, A. A., Suhweil, Y., & Yaman, M. A. (2017). Smart controlling for traffic light time. *IEEE Jordan Conference on Applied Electrical Engineering and Computing Technologies (AEECT)*, 1-5.

Zarpelão, B. B., Miani, R. S., Kawakani, C. T., & Alvarenga, S. C. (2017). A survey of intrusion detection in Internet of Things. *Journal of Network and Computer Applications*, *84*, 25–37. doi:10.1016/j.jnca.2017.02.009

Zeadally, S., Hunt, R., Chen, Y.-S., Irwin, A., & Hassan, A. (2010). *Vehicular ad hoc networks (VANETS): status, results, and challenges*. Springer Science.

Zeadally, S., Pathan, A.-S. K., Alcaraz, C., & Badra, M. (2013). Towards Privacy Protection in Smart Grid. *Wireless Personal Communications*, *73*(1), 23–50. doi:10.100711277-012-0939-1

Zhan, X. S. (2013). Promoting the construction of information security system. Information Security and Communications Privacy, 5, 9-12.

Zhang, Ni, Yang, Liang, Ren, & Shen. (2017). Security and privacy in smart city applications: Challenges and solutions. *IEEE Communication Magazine, 55*(1), 122-129.

Zhan, G., Shi, W., & Deng, J. (2012). Design and implementation of TARF: A trust-aware routing framework for WSNs. *IEEE Transactions on Dependable and Secure Computing*, *9*(2), 184–197. doi:10.1109/TDSC.2011.58

Zhang, J. (2011). A Survey on Trust Management for VANETs. *International Conference on Advanced Information Networking and Applications*. 10.1109/AINA.2011.86

Zhang, J. (2012). *Trust Management for VANETs: Challenges, Desired Properties and Future Directions.* Singapore: IGI Global. doi:10.4018/jdst.2012010104

Zhang, L. H., Zhang, G. H., Yu, L., Zhang, J., & Bai, Y. C. (2004). Intrusion detection using rough set classification. *Journal of Zhejiang University. Science A*, *5*(9), 1076–1086. doi:10.1631/jzus.2004.1076 PMID:15323002

Zhao, F. (2005). *Maximize business profits through e-partnerships.* IGI Global.

Zheng, B., Li, W., Deng, P., Gérard, L., Qi, Z., & Shankary, N. (2015). Design and Verification for Transportation System Security. IEEE.

Zheng, Y. (1997). Digital signcryption or how to achieve cost (signature & encryption)≪cost(signature)+cost(encryption). In *Annual International Cryptology Conference* (pp. 165-179). Springer. 10.1007/BFb0052234

ZhiJie & RuiBing. (2016). Intelligent Traffic Control System Based Single Chip Microcomputer. *International Conference on Intelligent Transportation, Big Data & Smart City (ICITBS)*, 577-579.

Zhou, L., Su, C., Chiu, W., & Yeh, K.-H. (n.d.). You think, therefore you are: Transparent authentication system with brainwave-oriented bio-features for IoT networks. *IEEE Trans. Emerg. Topics Comput.* Available: https://ieeexplore.ieee.org/abstract/document/8057810/

Zhou, X., Zou, Z., Song, R., Wang, Y., & Yu, Z. (2016). *Cooperative caching strategies for mobile peer-to-peer networks: A survey. In Information Science and Applications (ICISA) 2016* (pp. 279–287). Springer. doi:10.1007/978-981-10-0557-2_28

Zhu, Q., & Başar, T. (2011, December). Robust and resilient control design for cyber-physical systems with an application to power systems. In *Decision and Control and European Control Conference (CDC-ECC), 2011 50th IEEE Conference on* (pp. 4066-4071). IEEE. 10.1109/CDC.2011.6161031

Zhu, F., Li, Z., Chen, S., & Xiong, G. (2016). Parallel Transportation Management and Control System and Its Applications in Building Smart Cities. *IEEE Transactions on Intelligent Transportation Systems, 17*(6), 1576–1585. doi:10.1109/TITS.2015.2506156

Zimmer, C., Bhat, B., Mueller, F., & Mohan, S. (2010, April). Time-based intrusion detection in cyber-physical systems. In *Proceedings of the 1st ACM/IEEE International Conference on Cyber-Physical Systems* (pp. 109-118). ACM. 10.1145/1795194.1795210

Related References

To continue our tradition of advancing information science and technology research, we have compiled a list of recommended IGI Global readings. These references will provide additional information and guidance to further enrich your knowledge and assist you with your own research and future publications.

Aasi, P., Rusu, L., & Vieru, D. (2017). The Role of Culture in IT Governance Five Focus Areas: A Literature Review. *International Journal of IT/Business Alignment and Governance, 8*(2), 42-61. doi:10.4018/IJITBAG.2017070103

Abdrabo, A. A. (2018). Egypt's Knowledge-Based Development: Opportunities, Challenges, and Future Possibilities. In A. Alraouf (Ed.), *Knowledge-Based Urban Development in the Middle East* (pp. 80–101). Hershey, PA: IGI Global. doi:10.4018/978-1-5225-3734-2.ch005

Abu Doush, I., & Alhami, I. (2018). Evaluating the Accessibility of Computer Laboratories, Libraries, and Websites in Jordanian Universities and Colleges. *International Journal of Information Systems and Social Change, 9*(2), 44–60. doi:10.4018/IJISSC.2018040104

Adeboye, A. (2016). Perceived Use and Acceptance of Cloud Enterprise Resource Planning (ERP) Implementation in the Manufacturing Industries. *International Journal of Strategic Information Technology and Applications, 7*(3), 24–40. doi:10.4018/IJSITA.2016070102

Adegbore, A. M., Quadri, M. O., & Oyewo, O. R. (2018). A Theoretical Approach to the Adoption of Electronic Resource Management Systems (ERMS) in Nigerian University Libraries. In A. Tella & T. Kwanya (Eds.), *Handbook of Research on Managing Intellectual Property in Digital Libraries* (pp. 292–311). Hershey, PA: IGI Global. doi:10.4018/978-1-5225-3093-0.ch015

Adhikari, M., & Roy, D. (2016). Green Computing. In G. Deka, G. Siddesh, K. Srinivasa, & L. Patnaik (Eds.), *Emerging Research Surrounding Power Consumption and Performance Issues in Utility Computing* (pp. 84–108). Hershey, PA: IGI Global. doi:10.4018/978-1-4666-8853-7.ch005

Afolabi, O. A. (2018). Myths and Challenges of Building an Effective Digital Library in Developing Nations: An African Perspective. In A. Tella & T. Kwanya (Eds.), *Handbook of Research on Managing Intellectual Property in Digital Libraries* (pp. 51–79). Hershey, PA: IGI Global. doi:10.4018/978-1-5225-3093-0.ch004

Agarwal, R., Singh, A., & Sen, S. (2016). Role of Molecular Docking in Computer-Aided Drug Design and Development. In S. Dastmalchi, M. Hamzeh-Mivehroud, & B. Sokouti (Eds.), *Applied Case Studies and Solutions in Molecular Docking-Based Drug Design* (pp. 1–28). Hershey, PA: IGI Global. doi:10.4018/978-1-5225-0362-0.ch001

Ali, O., & Soar, J. (2016). Technology Innovation Adoption Theories. In L. Al-Hakim, X. Wu, A. Koronios, & Y. Shou (Eds.), *Handbook of Research on Driving Competitive Advantage through Sustainable, Lean, and Disruptive Innovation* (pp. 1–38). Hershey, PA: IGI Global. doi:10.4018/978-1-5225-0135-0.ch001

Alsharo, M. (2017). Attitudes Towards Cloud Computing Adoption in Emerging Economies. *International Journal of Cloud Applications and Computing*, 7(3), 44–58. doi:10.4018/IJCAC.2017070102

Amer, T. S., & Johnson, T. L. (2016). Information Technology Progress Indicators: Temporal Expectancy, User Preference, and the Perception of Process Duration. *International Journal of Technology and Human Interaction*, 12(4), 1–14. doi:10.4018/IJTHI.2016100101

Amer, T. S., & Johnson, T. L. (2017). Information Technology Progress Indicators: Research Employing Psychological Frameworks. In A. Mesquita (Ed.), *Research Paradigms and Contemporary Perspectives on Human-Technology Interaction* (pp. 168–186). Hershey, PA: IGI Global. doi:10.4018/978-1-5225-1868-6.ch008

Related References

Anchugam, C. V., & Thangadurai, K. (2016). Introduction to Network Security. In D. G., M. Singh, & M. Jayanthi (Eds.), Network Security Attacks and Countermeasures (pp. 1-48). Hershey, PA: IGI Global. doi:10.4018/978-1-4666-8761-5.ch001

Anchugam, C. V., & Thangadurai, K. (2016). Classification of Network Attacks and Countermeasures of Different Attacks. In D. G., M. Singh, & M. Jayanthi (Eds.), Network Security Attacks and Countermeasures (pp. 115-156). Hershey, PA: IGI Global. doi:10.4018/978-1-4666-8761-5.ch004

Anohah, E. (2016). Pedagogy and Design of Online Learning Environment in Computer Science Education for High Schools. *International Journal of Online Pedagogy and Course Design*, 6(3), 39–51. doi:10.4018/IJOPCD.2016070104

Anohah, E. (2017). Paradigm and Architecture of Computing Augmented Learning Management System for Computer Science Education. *International Journal of Online Pedagogy and Course Design*, 7(2), 60–70. doi:10.4018/IJOPCD.2017040105

Anohah, E., & Suhonen, J. (2017). Trends of Mobile Learning in Computing Education from 2006 to 2014: A Systematic Review of Research Publications. *International Journal of Mobile and Blended Learning*, 9(1), 16–33. doi:10.4018/IJMBL.2017010102

Assis-Hassid, S., Heart, T., Reychav, I., & Pliskin, J. S. (2016). Modelling Factors Affecting Patient-Doctor-Computer Communication in Primary Care. *International Journal of Reliable and Quality E-Healthcare*, 5(1), 1–17. doi:10.4018/IJRQEH.2016010101

Bailey, E. K. (2017). Applying Learning Theories to Computer Technology Supported Instruction. In M. Grassetti & S. Brookby (Eds.), *Advancing Next-Generation Teacher Education through Digital Tools and Applications* (pp. 61–81). Hershey, PA: IGI Global. doi:10.4018/978-1-5225-0965-3.ch004

Balasubramanian, K. (2016). Attacks on Online Banking and Commerce. In K. Balasubramanian, K. Mala, & M. Rajakani (Eds.), *Cryptographic Solutions for Secure Online Banking and Commerce* (pp. 1–19). Hershey, PA: IGI Global. doi:10.4018/978-1-5225-0273-9.ch001

Baldwin, S., Opoku-Agyemang, K., & Roy, D. (2016). Games People Play: A Trilateral Collaboration Researching Computer Gaming across Cultures. In K. Valentine & L. Jensen (Eds.), *Examining the Evolution of Gaming and Its Impact on Social, Cultural, and Political Perspectives* (pp. 364–376). Hershey, PA: IGI Global. doi:10.4018/978-1-5225-0261-6.ch017

Banerjee, S., Sing, T. Y., Chowdhury, A. R., & Anwar, H. (2018). Let's Go Green: Towards a Taxonomy of Green Computing Enablers for Business Sustainability. In M. Khosrow-Pour (Ed.), *Green Computing Strategies for Competitive Advantage and Business Sustainability* (pp. 89–109). Hershey, PA: IGI Global. doi:10.4018/978-1-5225-5017-4.ch005

Basham, R. (2018). Information Science and Technology in Crisis Response and Management. In M. Khosrow-Pour, D.B.A. (Ed.), Encyclopedia of Information Science and Technology, Fourth Edition (pp. 1407-1418). Hershey, PA: IGI Global. doi:10.4018/978-1-5225-2255-3.ch121

Batyashe, T., & Iyamu, T. (2018). Architectural Framework for the Implementation of Information Technology Governance in Organisations. In M. Khosrow-Pour, D.B.A. (Ed.), Encyclopedia of Information Science and Technology, Fourth Edition (pp. 810-819). Hershey, PA: IGI Global. doi:10.4018/978-1-5225-2255-3.ch070

Bekleyen, N., & Çelik, S. (2017). Attitudes of Adult EFL Learners towards Preparing for a Language Test via CALL. In D. Tafazoli & M. Romero (Eds.), *Multiculturalism and Technology-Enhanced Language Learning* (pp. 214–229). Hershey, PA: IGI Global. doi:10.4018/978-1-5225-1882-2.ch013

Bennett, A., Eglash, R., Lachney, M., & Babbitt, W. (2016). Design Agency: Diversifying Computer Science at the Intersections of Creativity and Culture. In M. Raisinghani (Ed.), *Revolutionizing Education through Web-Based Instruction* (pp. 35–56). Hershey, PA: IGI Global. doi:10.4018/978-1-4666-9932-8.ch003

Bergeron, F., Croteau, A., Uwizeyemungu, S., & Raymond, L. (2017). A Framework for Research on Information Technology Governance in SMEs. In S. De Haes & W. Van Grembergen (Eds.), *Strategic IT Governance and Alignment in Business Settings* (pp. 53–81). Hershey, PA: IGI Global. doi:10.4018/978-1-5225-0861-8.ch003

Bhatt, G. D., Wang, Z., & Rodger, J. A. (2017). Information Systems Capabilities and Their Effects on Competitive Advantages: A Study of Chinese Companies. *Information Resources Management Journal, 30*(3), 41–57. doi:10.4018/IRMJ.2017070103

Related References

Bogdanoski, M., Stoilkovski, M., & Risteski, A. (2016). Novel First Responder Digital Forensics Tool as a Support to Law Enforcement. In M. Hadji-Janev & M. Bogdanoski (Eds.), *Handbook of Research on Civil Society and National Security in the Era of Cyber Warfare* (pp. 352–376). Hershey, PA: IGI Global. doi:10.4018/978-1-4666-8793-6.ch016

Boontarig, W., Papasratorn, B., & Chutimaskul, W. (2016). The Unified Model for Acceptance and Use of Health Information on Online Social Networks: Evidence from Thailand. *International Journal of E-Health and Medical Communications*, 7(1), 31–47. doi:10.4018/IJEHMC.2016010102

Brown, S., & Yuan, X. (2016). Techniques for Retaining Computer Science Students at Historical Black Colleges and Universities. In C. Prince & R. Ford (Eds.), *Setting a New Agenda for Student Engagement and Retention in Historically Black Colleges and Universities* (pp. 251–268). Hershey, PA: IGI Global. doi:10.4018/978-1-5225-0308-8.ch014

Burcoff, A., & Shamir, L. (2017). Computer Analysis of Pablo Picasso's Artistic Style. *International Journal of Art, Culture and Design Technologies*, 6(1), 1–18. doi:10.4018/IJACDT.2017010101

Byker, E. J. (2017). I Play I Learn: Introducing Technological Play Theory. In C. Martin & D. Polly (Eds.), *Handbook of Research on Teacher Education and Professional Development* (pp. 297–306). Hershey, PA: IGI Global. doi:10.4018/978-1-5225-1067-3.ch016

Calongne, C. M., Stricker, A. G., Truman, B., & Arenas, F. J. (2017). Cognitive Apprenticeship and Computer Science Education in Cyberspace: Reimagining the Past. In A. Stricker, C. Calongne, B. Truman, & F. Arenas (Eds.), *Integrating an Awareness of Selfhood and Society into Virtual Learning* (pp. 180–197). Hershey, PA: IGI Global. doi:10.4018/978-1-5225-2182-2.ch013

Carlton, E. L., Holsinger, J. W. Jr, & Anunobi, N. (2016). Physician Engagement with Health Information Technology: Implications for Practice and Professionalism. *International Journal of Computers in Clinical Practice*, 1(2), 51–73. doi:10.4018/IJCCP.2016070103

Carneiro, A. D. (2017). Defending Information Networks in Cyberspace: Some Notes on Security Needs. In M. Dawson, D. Kisku, P. Gupta, J. Sing, & W. Li (Eds.), Developing Next-Generation Countermeasures for Homeland Security Threat Prevention (pp. 354-375). Hershey, PA: IGI Global. doi:10.4018/978-1-5225-0703-1.ch016

Cavalcanti, J. C. (2016). The New "ABC" of ICTs (Analytics + Big Data + Cloud Computing): A Complex Trade-Off between IT and CT Costs. In J. Martins & A. Molnar (Eds.), *Handbook of Research on Innovations in Information Retrieval, Analysis, and Management* (pp. 152–186). Hershey, PA: IGI Global. doi:10.4018/978-1-4666-8833-9.ch006

Chase, J. P., & Yan, Z. (2017). Affect in Statistics Cognition. In *Assessing and Measuring Statistics Cognition in Higher Education Online Environments: Emerging Research and Opportunities* (pp. 144–187). Hershey, PA: IGI Global. doi:10.4018/978-1-5225-2420-5.ch005

Chen, C. (2016). Effective Learning Strategies for the 21st Century: Implications for the E-Learning. In M. Anderson & C. Gavan (Eds.), *Developing Effective Educational Experiences through Learning Analytics* (pp. 143–169). Hershey, PA: IGI Global. doi:10.4018/978-1-4666-9983-0.ch006

Chen, E. T. (2016). Examining the Influence of Information Technology on Modern Health Care. In P. Manolitzas, E. Grigoroudis, N. Matsatsinis, & D. Yannacopoulos (Eds.), *Effective Methods for Modern Healthcare Service Quality and Evaluation* (pp. 110–136). Hershey, PA: IGI Global. doi:10.4018/978-1-4666-9961-8.ch006

Cimermanova, I. (2017). Computer-Assisted Learning in Slovakia. In D. Tafazoli & M. Romero (Eds.), *Multiculturalism and Technology-Enhanced Language Learning* (pp. 252–270). Hershey, PA: IGI Global. doi:10.4018/978-1-5225-1882-2.ch015

Cipolla-Ficarra, F. V., & Cipolla-Ficarra, M. (2018). Computer Animation for Ingenious Revival. In F. Cipolla-Ficarra, M. Ficarra, M. Cipolla-Ficarra, A. Quiroga, J. Alma, & J. Carré (Eds.), *Technology-Enhanced Human Interaction in Modern Society* (pp. 159–181). Hershey, PA: IGI Global. doi:10.4018/978-1-5225-3437-2.ch008

Cockrell, S., Damron, T. S., Melton, A. M., & Smith, A. D. (2018). Offshoring IT. In M. Khosrow-Pour, D.B.A. (Ed.), Encyclopedia of Information Science and Technology, Fourth Edition (pp. 5476-5489). Hershey, PA: IGI Global. doi:10.4018/978-1-5225-2255-3.ch476

Coffey, J. W. (2018). Logic and Proof in Computer Science: Categories and Limits of Proof Techniques. In J. Horne (Ed.), *Philosophical Perceptions on Logic and Order* (pp. 218–240). Hershey, PA: IGI Global. doi:10.4018/978-1-5225-2443-4.ch007

Dale, M. (2017). Re-Thinking the Challenges of Enterprise Architecture Implementation. In M. Tavana (Ed.), *Enterprise Information Systems and the Digitalization of Business Functions* (pp. 205–221). Hershey, PA: IGI Global. doi:10.4018/978-1-5225-2382-6.ch009

Das, A., Dasgupta, R., & Bagchi, A. (2016). Overview of Cellular Computing-Basic Principles and Applications. In J. Mandal, S. Mukhopadhyay, & T. Pal (Eds.), *Handbook of Research on Natural Computing for Optimization Problems* (pp. 637–662). Hershey, PA: IGI Global. doi:10.4018/978-1-5225-0058-2.ch026

De Maere, K., De Haes, S., & von Kutzschenbach, M. (2017). CIO Perspectives on Organizational Learning within the Context of IT Governance. *International Journal of IT/Business Alignment and Governance, 8*(1), 32-47. doi:10.4018/IJITBAG.2017010103

Demir, K., Çaka, C., Yaman, N. D., İslamoğlu, H., & Kuzu, A. (2018). Examining the Current Definitions of Computational Thinking. In H. Ozcinar, G. Wong, & H. Ozturk (Eds.), *Teaching Computational Thinking in Primary Education* (pp. 36–64). Hershey, PA: IGI Global. doi:10.4018/978-1-5225-3200-2.ch003

Deng, X., Hung, Y., & Lin, C. D. (2017). Design and Analysis of Computer Experiments. In S. Saha, A. Mandal, A. Narasimhamurthy, S. V, & S. Sangam (Eds.), Handbook of Research on Applied Cybernetics and Systems Science (pp. 264-279). Hershey, PA: IGI Global. doi:10.4018/978-1-5225-2498-4.ch013

Denner, J., Martinez, J., & Thiry, H. (2017). Strategies for Engaging Hispanic/Latino Youth in the US in Computer Science. In Y. Rankin & J. Thomas (Eds.), *Moving Students of Color from Consumers to Producers of Technology* (pp. 24–48). Hershey, PA: IGI Global. doi:10.4018/978-1-5225-2005-4.ch002

Devi, A. (2017). Cyber Crime and Cyber Security: A Quick Glance. In R. Kumar, P. Pattnaik, & P. Pandey (Eds.), *Detecting and Mitigating Robotic Cyber Security Risks* (pp. 160–171). Hershey, PA: IGI Global. doi:10.4018/978-1-5225-2154-9.ch011

Dores, A. R., Barbosa, F., Guerreiro, S., Almeida, I., & Carvalho, I. P. (2016). Computer-Based Neuropsychological Rehabilitation: Virtual Reality and Serious Games. In M. Cruz-Cunha, I. Miranda, R. Martinho, & R. Rijo (Eds.), *Encyclopedia of E-Health and Telemedicine* (pp. 473–485). Hershey, PA: IGI Global. doi:10.4018/978-1-4666-9978-6.ch037

Doshi, N., & Schaefer, G. (2016). Computer-Aided Analysis of Nailfold Capillaroscopy Images. In D. Fotiadis (Ed.), *Handbook of Research on Trends in the Diagnosis and Treatment of Chronic Conditions* (pp. 146–158). Hershey, PA: IGI Global. doi:10.4018/978-1-4666-8828-5.ch007

Doyle, D. J., & Fahy, P. J. (2018). Interactivity in Distance Education and Computer-Aided Learning, With Medical Education Examples. In M. Khosrow-Pour, D.B.A. (Ed.), Encyclopedia of Information Science and Technology, Fourth Edition (pp. 5829-5840). Hershey, PA: IGI Global. doi:10.4018/978-1-5225-2255-3.ch507

Elias, N. I., & Walker, T. W. (2017). Factors that Contribute to Continued Use of E-Training among Healthcare Professionals. In F. Topor (Ed.), *Handbook of Research on Individualism and Identity in the Globalized Digital Age* (pp. 403–429). Hershey, PA: IGI Global. doi:10.4018/978-1-5225-0522-8.ch018

Eloy, S., Dias, M. S., Lopes, P. F., & Vilar, E. (2016). Digital Technologies in Architecture and Engineering: Exploring an Engaged Interaction within Curricula. In D. Fonseca & E. Redondo (Eds.), *Handbook of Research on Applied E-Learning in Engineering and Architecture Education* (pp. 368–402). Hershey, PA: IGI Global. doi:10.4018/978-1-4666-8803-2.ch017

Estrela, V. V., Magalhães, H. A., & Saotome, O. (2016). Total Variation Applications in Computer Vision. In N. Kamila (Ed.), *Handbook of Research on Emerging Perspectives in Intelligent Pattern Recognition, Analysis, and Image Processing* (pp. 41–64). Hershey, PA: IGI Global. doi:10.4018/978-1-4666-8654-0.ch002

Filipovic, N., Radovic, M., Nikolic, D. D., Saveljic, I., Milosevic, Z., Exarchos, T. P., ... Parodi, O. (2016). Computer Predictive Model for Plaque Formation and Progression in the Artery. In D. Fotiadis (Ed.), *Handbook of Research on Trends in the Diagnosis and Treatment of Chronic Conditions* (pp. 279–300). Hershey, PA: IGI Global. doi:10.4018/978-1-4666-8828-5.ch013

Fisher, R. L. (2018). Computer-Assisted Indian Matrimonial Services. In M. Khosrow-Pour, D.B.A. (Ed.), Encyclopedia of Information Science and Technology, Fourth Edition (pp. 4136-4145). Hershey, PA: IGI Global. doi:10.4018/978-1-5225-2255-3.ch358

Fleenor, H. G., & Hodhod, R. (2016). Assessment of Learning and Technology: Computer Science Education. In V. Wang (Ed.), *Handbook of Research on Learning Outcomes and Opportunities in the Digital Age* (pp. 51–78). Hershey, PA: IGI Global. doi:10.4018/978-1-4666-9577-1.ch003

García-Valcárcel, A., & Mena, J. (2016). Information Technology as a Way To Support Collaborative Learning: What In-Service Teachers Think, Know and Do. *Journal of Information Technology Research, 9*(1), 1–17. doi:10.4018/JITR.2016010101

Gardner-McCune, C., & Jimenez, Y. (2017). Historical App Developers: Integrating CS into K-12 through Cross-Disciplinary Projects. In Y. Rankin & J. Thomas (Eds.), *Moving Students of Color from Consumers to Producers of Technology* (pp. 85–112). Hershey, PA: IGI Global. doi:10.4018/978-1-5225-2005-4.ch005

Garvey, G. P. (2016). Exploring Perception, Cognition, and Neural Pathways of Stereo Vision and the Split–Brain Human Computer Interface. In A. Ursyn (Ed.), *Knowledge Visualization and Visual Literacy in Science Education* (pp. 28–76). Hershey, PA: IGI Global. doi:10.4018/978-1-5225-0480-1.ch002

Ghafele, R., & Gibert, B. (2018). Open Growth: The Economic Impact of Open Source Software in the USA. In M. Khosrow-Pour (Ed.), *Optimizing Contemporary Application and Processes in Open Source Software* (pp. 164–197). Hershey, PA: IGI Global. doi:10.4018/978-1-5225-5314-4.ch007

Ghobakhloo, M., & Azar, A. (2018). Information Technology Resources, the Organizational Capability of Lean-Agile Manufacturing, and Business Performance. *Information Resources Management Journal, 31*(2), 47–74. doi:10.4018/IRMJ.2018040103

Gianni, M., & Gotzamani, K. (2016). Integrated Management Systems and Information Management Systems: Common Threads. In P. Papajorgji, F. Pinet, A. Guimarães, & J. Papathanasiou (Eds.), *Automated Enterprise Systems for Maximizing Business Performance* (pp. 195–214). Hershey, PA: IGI Global. doi:10.4018/978-1-4666-8841-4.ch011

Gikandi, J. W. (2017). Computer-Supported Collaborative Learning and Assessment: A Strategy for Developing Online Learning Communities in Continuing Education. In J. Keengwe & G. Onchwari (Eds.), *Handbook of Research on Learner-Centered Pedagogy in Teacher Education and Professional Development* (pp. 309–333). Hershey, PA: IGI Global. doi:10.4018/978-1-5225-0892-2.ch017

Gokhale, A. A., & Machina, K. F. (2017). Development of a Scale to Measure Attitudes toward Information Technology. In L. Tomei (Ed.), *Exploring the New Era of Technology-Infused Education* (pp. 49–64). Hershey, PA: IGI Global. doi:10.4018/978-1-5225-1709-2.ch004

Grace, A., O'Donoghue, J., Mahony, C., Heffernan, T., Molony, D., & Carroll, T. (2016). Computerized Decision Support Systems for Multimorbidity Care: An Urgent Call for Research and Development. In M. Cruz-Cunha, I. Miranda, R. Martinho, & R. Rijo (Eds.), *Encyclopedia of E-Health and Telemedicine* (pp. 486–494). Hershey, PA: IGI Global. doi:10.4018/978-1-4666-9978-6.ch038

Gupta, A., & Singh, O. (2016). Computer Aided Modeling and Finite Element Analysis of Human Elbow. *International Journal of Biomedical and Clinical Engineering*, *5*(1), 31–38. doi:10.4018/IJBCE.2016010104

H., S. K. (2016). Classification of Cybercrimes and Punishments under the Information Technology Act, 2000. In S. Geetha, & A. Phamila (Eds.), *Combating Security Breaches and Criminal Activity in the Digital Sphere* (pp. 57-66). Hershey, PA: IGI Global. doi:10.4018/978-1-5225-0193-0.ch004

Hafeez-Baig, A., Gururajan, R., & Wickramasinghe, N. (2017). Readiness as a Novel Construct of Readiness Acceptance Model (RAM) for the Wireless Handheld Technology. In N. Wickramasinghe (Ed.), *Handbook of Research on Healthcare Administration and Management* (pp. 578–595). Hershey, PA: IGI Global. doi:10.4018/978-1-5225-0920-2.ch035

Hanafizadeh, P., Ghandchi, S., & Asgarimehr, M. (2017). Impact of Information Technology on Lifestyle: A Literature Review and Classification. *International Journal of Virtual Communities and Social Networking*, *9*(2), 1–23. doi:10.4018/IJVCSN.2017040101

Harlow, D. B., Dwyer, H., Hansen, A. K., Hill, C., Iveland, A., Leak, A. E., & Franklin, D. M. (2016). Computer Programming in Elementary and Middle School: Connections across Content. In M. Urban & D. Falvo (Eds.), *Improving K-12 STEM Education Outcomes through Technological Integration* (pp. 337–361). Hershey, PA: IGI Global. doi:10.4018/978-1-4666-9616-7.ch015

Related References

Haseski, H. İ., Ilic, U., & Tuğtekin, U. (2018). Computational Thinking in Educational Digital Games: An Assessment Tool Proposal. In H. Ozcinar, G. Wong, & H. Ozturk (Eds.), *Teaching Computational Thinking in Primary Education* (pp. 256–287). Hershey, PA: IGI Global. doi:10.4018/978-1-5225-3200-2.ch013

Hee, W. J., Jalleh, G., Lai, H., & Lin, C. (2017). E-Commerce and IT Projects: Evaluation and Management Issues in Australian and Taiwanese Hospitals. *International Journal of Public Health Management and Ethics, 2*(1), 69–90. doi:10.4018/IJPHME.2017010104

Hernandez, A. A. (2017). Green Information Technology Usage: Awareness and Practices of Philippine IT Professionals. *International Journal of Enterprise Information Systems, 13*(4), 90–103. doi:10.4018/IJEIS.2017100106

Hernandez, A. A., & Ona, S. E. (2016). Green IT Adoption: Lessons from the Philippines Business Process Outsourcing Industry. *International Journal of Social Ecology and Sustainable Development, 7*(1), 1–34. doi:10.4018/IJSESD.2016010101

Hernandez, M. A., Marin, E. C., Garcia-Rodriguez, J., Azorin-Lopez, J., & Cazorla, M. (2017). Automatic Learning Improves Human-Robot Interaction in Productive Environments: A Review. *International Journal of Computer Vision and Image Processing, 7*(3), 65–75. doi:10.4018/IJCVIP.2017070106

Horne-Popp, L. M., Tessone, E. B., & Welker, J. (2018). If You Build It, They Will Come: Creating a Library Statistics Dashboard for Decision-Making. In L. Costello & M. Powers (Eds.), *Developing In-House Digital Tools in Library Spaces* (pp. 177–203). Hershey, PA: IGI Global. doi:10.4018/978-1-5225-2676-6.ch009

Hossan, C. G., & Ryan, J. C. (2016). Factors Affecting e-Government Technology Adoption Behaviour in a Voluntary Environment. *International Journal of Electronic Government Research, 12*(1), 24–49. doi:10.4018/IJEGR.2016010102

Hu, H., Hu, P. J., & Al-Gahtani, S. S. (2017). User Acceptance of Computer Technology at Work in Arabian Culture: A Model Comparison Approach. In M. Khosrow-Pour (Ed.), *Handbook of Research on Technology Adoption, Social Policy, and Global Integration* (pp. 205–228). Hershey, PA: IGI Global. doi:10.4018/978-1-5225-2668-1.ch011

Huie, C. P. (2016). Perceptions of Business Intelligence Professionals about Factors Related to Business Intelligence input in Decision Making. *International Journal of Business Analytics, 3*(3), 1–24. doi:10.4018/IJBAN.2016070101

Hung, S., Huang, W., Yen, D. C., Chang, S., & Lu, C. (2016). Effect of Information Service Competence and Contextual Factors on the Effectiveness of Strategic Information Systems Planning in Hospitals. *Journal of Global Information Management*, 24(1), 14–36. doi:10.4018/JGIM.2016010102

Ifinedo, P. (2017). Using an Extended Theory of Planned Behavior to Study Nurses' Adoption of Healthcare Information Systems in Nova Scotia. *International Journal of Technology Diffusion*, 8(1), 1–17. doi:10.4018/IJTD.2017010101

Ilie, V., & Sneha, S. (2018). A Three Country Study for Understanding Physicians' Engagement With Electronic Information Resources Pre and Post System Implementation. *Journal of Global Information Management*, 26(2), 48–73. doi:10.4018/JGIM.2018040103

Inoue-Smith, Y. (2017). Perceived Ease in Using Technology Predicts Teacher Candidates' Preferences for Online Resources. *International Journal of Online Pedagogy and Course Design*, 7(3), 17–28. doi:10.4018/IJOPCD.2017070102

Islam, A. A. (2016). Development and Validation of the Technology Adoption and Gratification (TAG) Model in Higher Education: A Cross-Cultural Study Between Malaysia and China. *International Journal of Technology and Human Interaction*, 12(3), 78–105. doi:10.4018/IJTHI.2016070106

Islam, A. Y. (2017). Technology Satisfaction in an Academic Context: Moderating Effect of Gender. In A. Mesquita (Ed.), *Research Paradigms and Contemporary Perspectives on Human-Technology Interaction* (pp. 187–211). Hershey, PA: IGI Global. doi:10.4018/978-1-5225-1868-6.ch009

Jamil, G. L., & Jamil, C. C. (2017). Information and Knowledge Management Perspective Contributions for Fashion Studies: Observing Logistics and Supply Chain Management Processes. In G. Jamil, A. Soares, & C. Pessoa (Eds.), *Handbook of Research on Information Management for Effective Logistics and Supply Chains* (pp. 199–221). Hershey, PA: IGI Global. doi:10.4018/978-1-5225-0973-8.ch011

Jamil, G. L., Jamil, L. C., Vieira, A. A., & Xavier, A. J. (2016). Challenges in Modelling Healthcare Services: A Study Case of Information Architecture Perspectives. In G. Jamil, J. Poças Rascão, F. Ribeiro, & A. Malheiro da Silva (Eds.), *Handbook of Research on Information Architecture and Management in Modern Organizations* (pp. 1–23). Hershey, PA: IGI Global. doi:10.4018/978-1-4666-8637-3.ch001

Related References

Janakova, M. (2018). Big Data and Simulations for the Solution of Controversies in Small Businesses. In M. Khosrow-Pour, D.B.A. (Ed.), Encyclopedia of Information Science and Technology, Fourth Edition (pp. 6907-6915). Hershey, PA: IGI Global. doi:10.4018/978-1-5225-2255-3.ch598

Jha, D. G. (2016). Preparing for Information Technology Driven Changes. In S. Tiwari & L. Nafees (Eds.), *Innovative Management Education Pedagogies for Preparing Next-Generation Leaders* (pp. 258–274). Hershey, PA: IGI Global. doi:10.4018/978-1-4666-9691-4.ch015

Jhawar, A., & Garg, S. K. (2018). Logistics Improvement by Investment in Information Technology Using System Dynamics. In A. Azar & S. Vaidyanathan (Eds.), *Advances in System Dynamics and Control* (pp. 528–567). Hershey, PA: IGI Global. doi:10.4018/978-1-5225-4077-9.ch017

Kalelioğlu, F., Gülbahar, Y., & Doğan, D. (2018). Teaching How to Think Like a Programmer: Emerging Insights. In H. Ozcinar, G. Wong, & H. Ozturk (Eds.), *Teaching Computational Thinking in Primary Education* (pp. 18–35). Hershey, PA: IGI Global. doi:10.4018/978-1-5225-3200-2.ch002

Kamberi, S. (2017). A Girls-Only Online Virtual World Environment and its Implications for Game-Based Learning. In A. Stricker, C. Calongne, B. Truman, & F. Arenas (Eds.), *Integrating an Awareness of Selfhood and Society into Virtual Learning* (pp. 74–95). Hershey, PA: IGI Global. doi:10.4018/978-1-5225-2182-2.ch006

Kamel, S., & Rizk, N. (2017). ICT Strategy Development: From Design to Implementation – Case of Egypt. In C. Howard & K. Hargiss (Eds.), *Strategic Information Systems and Technologies in Modern Organizations* (pp. 239–257). Hershey, PA: IGI Global. doi:10.4018/978-1-5225-1680-4.ch010

Kamel, S. H. (2018). The Potential Role of the Software Industry in Supporting Economic Development. In M. Khosrow-Pour, D.B.A. (Ed.), Encyclopedia of Information Science and Technology, Fourth Edition (pp. 7259-7269). Hershey, PA: IGI Global. doi:10.4018/978-1-5225-2255-3.ch631

Karon, R. (2016). Utilisation of Health Information Systems for Service Delivery in the Namibian Environment. In T. Iyamu & A. Tatnall (Eds.), *Maximizing Healthcare Delivery and Management through Technology Integration* (pp. 169–183). Hershey, PA: IGI Global. doi:10.4018/978-1-4666-9446-0.ch011

Kawata, S. (2018). Computer-Assisted Parallel Program Generation. In M. Khosrow-Pour, D.B.A. (Ed.), Encyclopedia of Information Science and Technology, Fourth Edition (pp. 4583-4593). Hershey, PA: IGI Global. doi:10.4018/978-1-5225-2255-3.ch398

Khanam, S., Siddiqui, J., & Talib, F. (2016). A DEMATEL Approach for Prioritizing the TQM Enablers and IT Resources in the Indian ICT Industry. *International Journal of Applied Management Sciences and Engineering, 3*(1), 11–29. doi:10.4018/IJAMSE.2016010102

Khari, M., Shrivastava, G., Gupta, S., & Gupta, R. (2017). Role of Cyber Security in Today's Scenario. In R. Kumar, P. Pattnaik, & P. Pandey (Eds.), *Detecting and Mitigating Robotic Cyber Security Risks* (pp. 177–191). Hershey, PA: IGI Global. doi:10.4018/978-1-5225-2154-9.ch013

Khouja, M., Rodriguez, I. B., Ben Halima, Y., & Moalla, S. (2018). IT Governance in Higher Education Institutions: A Systematic Literature Review. *International Journal of Human Capital and Information Technology Professionals, 9*(2), 52–67. doi:10.4018/IJHCITP.2018040104

Kim, S., Chang, M., Choi, N., Park, J., & Kim, H. (2016). The Direct and Indirect Effects of Computer Uses on Student Success in Math. *International Journal of Cyber Behavior, Psychology and Learning, 6*(3), 48–64. doi:10.4018/IJCBPL.2016070104

Kiourt, C., Pavlidis, G., Koutsoudis, A., & Kalles, D. (2017). Realistic Simulation of Cultural Heritage. *International Journal of Computational Methods in Heritage Science, 1*(1), 10–40. doi:10.4018/IJCMHS.2017010102

Korikov, A., & Krivtsov, O. (2016). System of People-Computer: On the Way of Creation of Human-Oriented Interface. In V. Mkrttchian, A. Bershadsky, A. Bozhday, M. Kataev, & S. Kataev (Eds.), *Handbook of Research on Estimation and Control Techniques in E-Learning Systems* (pp. 458–470). Hershey, PA: IGI Global. doi:10.4018/978-1-4666-9489-7.ch032

Köse, U. (2017). An Augmented-Reality-Based Intelligent Mobile Application for Open Computer Education. In G. Kurubacak & H. Altinpulluk (Eds.), *Mobile Technologies and Augmented Reality in Open Education* (pp. 154–174). Hershey, PA: IGI Global. doi:10.4018/978-1-5225-2110-5.ch008

Lahmiri, S. (2018). Information Technology Outsourcing Risk Factors and Provider Selection. In M. Gupta, R. Sharman, J. Walp, & P. Mulgund (Eds.), *Information Technology Risk Management and Compliance in Modern Organizations* (pp. 214–228). Hershey, PA: IGI Global. doi:10.4018/978-1-5225-2604-9.ch008

Landriscina, F. (2017). Computer-Supported Imagination: The Interplay Between Computer and Mental Simulation in Understanding Scientific Concepts. In I. Levin & D. Tsybulsky (Eds.), *Digital Tools and Solutions for Inquiry-Based STEM Learning* (pp. 33–60). Hershey, PA: IGI Global. doi:10.4018/978-1-5225-2525-7.ch002

Lau, S. K., Winley, G. K., Leung, N. K., Tsang, N., & Lau, S. Y. (2016). An Exploratory Study of Expectation in IT Skills in a Developing Nation: Vietnam. *Journal of Global Information Management, 24*(1), 1–13. doi:10.4018/JGIM.2016010101

Lavranos, C., Kostagiolas, P., & Papadatos, J. (2016). Information Retrieval Technologies and the "Realities" of Music Information Seeking. In I. Deliyannis, P. Kostagiolas, & C. Banou (Eds.), *Experimental Multimedia Systems for Interactivity and Strategic Innovation* (pp. 102–121). Hershey, PA: IGI Global. doi:10.4018/978-1-4666-8659-5.ch005

Lee, W. W. (2018). Ethical Computing Continues From Problem to Solution. In M. Khosrow-Pour, D.B.A. (Ed.), Encyclopedia of Information Science and Technology, Fourth Edition (pp. 4884-4897). Hershey, PA: IGI Global. doi:10.4018/978-1-5225-2255-3.ch423

Lehto, M. (2016). Cyber Security Education and Research in the Finland's Universities and Universities of Applied Sciences. *International Journal of Cyber Warfare & Terrorism, 6*(2), 15–31. doi:10.4018/IJCWT.2016040102

Lin, C., Jalleh, G., & Huang, Y. (2016). Evaluating and Managing Electronic Commerce and Outsourcing Projects in Hospitals. In A. Dwivedi (Ed.), *Reshaping Medical Practice and Care with Health Information Systems* (pp. 132–172). Hershey, PA: IGI Global. doi:10.4018/978-1-4666-9870-3.ch005

Lin, S., Chen, S., & Chuang, S. (2017). Perceived Innovation and Quick Response Codes in an Online-to-Offline E-Commerce Service Model. *International Journal of E-Adoption, 9*(2), 1–16. doi:10.4018/IJEA.2017070101

Liu, M., Wang, Y., Xu, W., & Liu, L. (2017). Automated Scoring of Chinese Engineering Students' English Essays. *International Journal of Distance Education Technologies*, *15*(1), 52–68. doi:10.4018/IJDET.2017010104

Luciano, E. M., Wiedenhöft, G. C., Macadar, M. A., & Pinheiro dos Santos, F. (2016). Information Technology Governance Adoption: Understanding its Expectations Through the Lens of Organizational Citizenship. *International Journal of IT/Business Alignment and Governance, 7*(2), 22-32. doi:10.4018/IJITBAG.2016070102

Mabe, L. K., & Oladele, O. I. (2017). Application of Information Communication Technologies for Agricultural Development through Extension Services: A Review. In T. Tossy (Ed.), *Information Technology Integration for Socio-Economic Development* (pp. 52–101). Hershey, PA: IGI Global. doi:10.4018/978-1-5225-0539-6.ch003

Manogaran, G., Thota, C., & Lopez, D. (2018). Human-Computer Interaction With Big Data Analytics. In D. Lopez & M. Durai (Eds.), *HCI Challenges and Privacy Preservation in Big Data Security* (pp. 1–22). Hershey, PA: IGI Global. doi:10.4018/978-1-5225-2863-0.ch001

Margolis, J., Goode, J., & Flapan, J. (2017). A Critical Crossroads for Computer Science for All: "Identifying Talent" or "Building Talent," and What Difference Does It Make? In Y. Rankin & J. Thomas (Eds.), *Moving Students of Color from Consumers to Producers of Technology* (pp. 1–23). Hershey, PA: IGI Global. doi:10.4018/978-1-5225-2005-4.ch001

Mbale, J. (2018). Computer Centres Resource Cloud Elasticity-Scalability (CRECES): Copperbelt University Case Study. In S. Aljawarneh & M. Malhotra (Eds.), *Critical Research on Scalability and Security Issues in Virtual Cloud Environments* (pp. 48–70). Hershey, PA: IGI Global. doi:10.4018/978-1-5225-3029-9.ch003

McKee, J. (2018). The Right Information: The Key to Effective Business Planning. In *Business Architectures for Risk Assessment and Strategic Planning: Emerging Research and Opportunities* (pp. 38–52). Hershey, PA: IGI Global. doi:10.4018/978-1-5225-3392-4.ch003

Mensah, I. K., & Mi, J. (2018). Determinants of Intention to Use Local E-Government Services in Ghana: The Perspective of Local Government Workers. *International Journal of Technology Diffusion*, *9*(2), 41–60. doi:10.4018/IJTD.2018040103

Mohamed, J. H. (2018). Scientograph-Based Visualization of Computer Forensics Research Literature. In J. Jeyasekar & P. Saravanan (Eds.), *Innovations in Measuring and Evaluating Scientific Information* (pp. 148–162). Hershey, PA: IGI Global. doi:10.4018/978-1-5225-3457-0.ch010

Moore, R. L., & Johnson, N. (2017). Earning a Seat at the Table: How IT Departments Can Partner in Organizational Change and Innovation. *International Journal of Knowledge-Based Organizations*, 7(2), 1–12. doi:10.4018/IJKBO.2017040101

Mtebe, J. S., & Kissaka, M. M. (2016). Enhancing the Quality of Computer Science Education with MOOCs in Sub-Saharan Africa. In J. Keengwe & G. Onchwari (Eds.), *Handbook of Research on Active Learning and the Flipped Classroom Model in the Digital Age* (pp. 366–377). Hershey, PA: IGI Global. doi:10.4018/978-1-4666-9680-8.ch019

Mukul, M. K., & Bhattaharyya, S. (2017). Brain-Machine Interface: Human-Computer Interaction. In E. Noughabi, B. Raahemi, A. Albadvi, & B. Far (Eds.), *Handbook of Research on Data Science for Effective Healthcare Practice and Administration* (pp. 417–443). Hershey, PA: IGI Global. doi:10.4018/978-1-5225-2515-8.ch018

Na, L. (2017). Library and Information Science Education and Graduate Programs in Academic Libraries. In L. Ruan, Q. Zhu, & Y. Ye (Eds.), *Academic Library Development and Administration in China* (pp. 218–229). Hershey, PA: IGI Global. doi:10.4018/978-1-5225-0550-1.ch013

Nabavi, A., Taghavi-Fard, M. T., Hanafizadeh, P., & Taghva, M. R. (2016). Information Technology Continuance Intention: A Systematic Literature Review. *International Journal of E-Business Research*, 12(1), 58–95. doi:10.4018/IJEBR.2016010104

Nath, R., & Murthy, V. N. (2018). What Accounts for the Differences in Internet Diffusion Rates Around the World? In M. Khosrow-Pour, D.B.A. (Ed.), Encyclopedia of Information Science and Technology, Fourth Edition (pp. 8095-8104). Hershey, PA: IGI Global. doi:10.4018/978-1-5225-2255-3.ch705

Nedelko, Z., & Potocan, V. (2018). The Role of Emerging Information Technologies for Supporting Supply Chain Management. In M. Khosrow-Pour, D.B.A. (Ed.), Encyclopedia of Information Science and Technology, Fourth Edition (pp. 5559-5569). Hershey, PA: IGI Global. doi:10.4018/978-1-5225-2255-3.ch483

Ngafeeson, M. N. (2018). User Resistance to Health Information Technology. In M. Khosrow-Pour, D.B.A. (Ed.), Encyclopedia of Information Science and Technology, Fourth Edition (pp. 3816-3825). Hershey, PA: IGI Global. doi:10.4018/978-1-5225-2255-3.ch331

Nozari, H., Najafi, S. E., Jafari-Eskandari, M., & Aliahmadi, A. (2016). Providing a Model for Virtual Project Management with an Emphasis on IT Projects. In C. Graham (Ed.), *Strategic Management and Leadership for Systems Development in Virtual Spaces* (pp. 43–63). Hershey, PA: IGI Global. doi:10.4018/978-1-4666-9688-4.ch003

Nurdin, N., Stockdale, R., & Scheepers, H. (2016). Influence of Organizational Factors in the Sustainability of E-Government: A Case Study of Local E-Government in Indonesia. In I. Sodhi (Ed.), *Trends, Prospects, and Challenges in Asian E-Governance* (pp. 281–323). Hershey, PA: IGI Global. doi:10.4018/978-1-4666-9536-8.ch014

Odagiri, K. (2017). Introduction of Individual Technology to Constitute the Current Internet. In *Strategic Policy-Based Network Management in Contemporary Organizations* (pp. 20–96). Hershey, PA: IGI Global. doi:10.4018/978-1-68318-003-6.ch003

Okike, E. U. (2018). Computer Science and Prison Education. In I. Biao (Ed.), *Strategic Learning Ideologies in Prison Education Programs* (pp. 246–264). Hershey, PA: IGI Global. doi:10.4018/978-1-5225-2909-5.ch012

Olelewe, C. J., & Nwafor, I. P. (2017). Level of Computer Appreciation Skills Acquired for Sustainable Development by Secondary School Students in Nsukka LGA of Enugu State, Nigeria. In C. Ayo & V. Mbarika (Eds.), *Sustainable ICT Adoption and Integration for Socio-Economic Development* (pp. 214–233). Hershey, PA: IGI Global. doi:10.4018/978-1-5225-2565-3.ch010

Oliveira, M., Maçada, A. C., Curado, C., & Nodari, F. (2017). Infrastructure Profiles and Knowledge Sharing. *International Journal of Technology and Human Interaction*, *13*(3), 1–12. doi:10.4018/IJTHI.2017070101

Otarkhani, A., Shokouhyar, S., & Pour, S. S. (2017). Analyzing the Impact of Governance of Enterprise IT on Hospital Performance: Tehran's (Iran) Hospitals – A Case Study. *International Journal of Healthcare Information Systems and Informatics*, *12*(3), 1–20. doi:10.4018/IJHISI.2017070101

Otunla, A. O., & Amuda, C. O. (2018). Nigerian Undergraduate Students' Computer Competencies and Use of Information Technology Tools and Resources for Study Skills and Habits' Enhancement. In M. Khosrow-Pour, D.B.A. (Ed.), Encyclopedia of Information Science and Technology, Fourth Edition (pp. 2303-2313). Hershey, PA: IGI Global. doi:10.4018/978-1-5225-2255-3.ch200

Related References

Özçınar, H. (2018). A Brief Discussion on Incentives and Barriers to Computational Thinking Education. In H. Ozcinar, G. Wong, & H. Ozturk (Eds.), *Teaching Computational Thinking in Primary Education* (pp. 1–17). Hershey, PA: IGI Global. doi:10.4018/978-1-5225-3200-2.ch001

Pandey, J. M., Garg, S., Mishra, P., & Mishra, B. P. (2017). Computer Based Psychological Interventions: Subject to the Efficacy of Psychological Services. *International Journal of Computers in Clinical Practice, 2*(1), 25–33. doi:10.4018/IJCCP.2017010102

Parry, V. K., & Lind, M. L. (2016). Alignment of Business Strategy and Information Technology Considering Information Technology Governance, Project Portfolio Control, and Risk Management. *International Journal of Information Technology Project Management, 7*(4), 21–37. doi:10.4018/IJITPM.2016100102

Patro, C. (2017). Impulsion of Information Technology on Human Resource Practices. In P. Ordóñez de Pablos (Ed.), *Managerial Strategies and Solutions for Business Success in Asia* (pp. 231–254). Hershey, PA: IGI Global. doi:10.4018/978-1-5225-1886-0.ch013

Patro, C. S., & Raghunath, K. M. (2017). Information Technology Paraphernalia for Supply Chain Management Decisions. In M. Tavana (Ed.), *Enterprise Information Systems and the Digitalization of Business Functions* (pp. 294–320). Hershey, PA: IGI Global. doi:10.4018/978-1-5225-2382-6.ch014

Paul, P. K. (2016). Cloud Computing: An Agent of Promoting Interdisciplinary Sciences, Especially Information Science and I-Schools – Emerging Techno-Educational Scenario. In L. Chao (Ed.), *Handbook of Research on Cloud-Based STEM Education for Improved Learning Outcomes* (pp. 247–258). Hershey, PA: IGI Global. doi:10.4018/978-1-4666-9924-3.ch016

Paul, P. K. (2018). The Context of IST for Solid Information Retrieval and Infrastructure Building: Study of Developing Country. *International Journal of Information Retrieval Research, 8*(1), 86–100. doi:10.4018/IJIRR.2018010106

Paul, P. K., & Chatterjee, D. (2018). iSchools Promoting "Information Science and Technology" (IST) Domain Towards Community, Business, and Society With Contemporary Worldwide Trend and Emerging Potentialities in India. In M. Khosrow-Pour, D.B.A. (Ed.), Encyclopedia of Information Science and Technology, Fourth Edition (pp. 4723-4735). Hershey, PA: IGI Global. doi:10.4018/978-1-5225-2255-3.ch410

Pessoa, C. R., & Marques, M. E. (2017). Information Technology and Communication Management in Supply Chain Management. In G. Jamil, A. Soares, & C. Pessoa (Eds.), *Handbook of Research on Information Management for Effective Logistics and Supply Chains* (pp. 23–33). Hershey, PA: IGI Global. doi:10.4018/978-1-5225-0973-8.ch002

Pineda, R. G. (2016). Where the Interaction Is Not: Reflections on the Philosophy of Human-Computer Interaction. *International Journal of Art, Culture and Design Technologies*, 5(1), 1–12. doi:10.4018/IJACDT.2016010101

Pineda, R. G. (2018). Remediating Interaction: Towards a Philosophy of Human-Computer Relationship. In M. Khosrow-Pour (Ed.), *Enhancing Art, Culture, and Design With Technological Integration* (pp. 75–98). Hershey, PA: IGI Global. doi:10.4018/978-1-5225-5023-5.ch004

Poikela, P., & Vuojärvi, H. (2016). Learning ICT-Mediated Communication through Computer-Based Simulations. In M. Cruz-Cunha, I. Miranda, R. Martinho, & R. Rijo (Eds.), *Encyclopedia of E-Health and Telemedicine* (pp. 674–687). Hershey, PA: IGI Global. doi:10.4018/978-1-4666-9978-6.ch052

Qian, Y. (2017). Computer Simulation in Higher Education: Affordances, Opportunities, and Outcomes. In P. Vu, S. Fredrickson, & C. Moore (Eds.), *Handbook of Research on Innovative Pedagogies and Technologies for Online Learning in Higher Education* (pp. 236–262). Hershey, PA: IGI Global. doi:10.4018/978-1-5225-1851-8.ch011

Radant, O., Colomo-Palacios, R., & Stantchev, V. (2016). Factors for the Management of Scarce Human Resources and Highly Skilled Employees in IT-Departments: A Systematic Review. *Journal of Information Technology Research*, 9(1), 65–82. doi:10.4018/JITR.2016010105

Rahman, N. (2016). Toward Achieving Environmental Sustainability in the Computer Industry. *International Journal of Green Computing*, 7(1), 37–54. doi:10.4018/IJGC.2016010103

Rahman, N. (2017). Lessons from a Successful Data Warehousing Project Management. *International Journal of Information Technology Project Management*, 8(4), 30–45. doi:10.4018/IJITPM.2017100103

Rahman, N. (2018). Environmental Sustainability in the Computer Industry for Competitive Advantage. In M. Khosrow-Pour (Ed.), *Green Computing Strategies for Competitive Advantage and Business Sustainability* (pp. 110–130). Hershey, PA: IGI Global. doi:10.4018/978-1-5225-5017-4.ch006

Rajh, A., & Pavetic, T. (2017). Computer Generated Description as the Required Digital Competence in Archival Profession. *International Journal of Digital Literacy and Digital Competence*, *8*(1), 36–49. doi:10.4018/IJDLDC.2017010103

Raman, A., & Goyal, D. P. (2017). Extending IMPLEMENT Framework for Enterprise Information Systems Implementation to Information System Innovation. In M. Tavana (Ed.), *Enterprise Information Systems and the Digitalization of Business Functions* (pp. 137–177). Hershey, PA: IGI Global. doi:10.4018/978-1-5225-2382-6.ch007

Rao, Y. S., Rauta, A. K., Saini, H., & Panda, T. C. (2017). Mathematical Model for Cyber Attack in Computer Network. *International Journal of Business Data Communications and Networking*, *13*(1), 58–65. doi:10.4018/IJBDCN.2017010105

Rapaport, W. J. (2018). Syntactic Semantics and the Proper Treatment of Computationalism. In M. Danesi (Ed.), *Empirical Research on Semiotics and Visual Rhetoric* (pp. 128–176). Hershey, PA: IGI Global. doi:10.4018/978-1-5225-5622-0. ch007

Raut, R., Priyadarshinee, P., & Jha, M. (2017). Understanding the Mediation Effect of Cloud Computing Adoption in Indian Organization: Integrating TAM-TOE- Risk Model. *International Journal of Service Science, Management, Engineering, and Technology*, *8*(3), 40–59. doi:10.4018/IJSSMET.2017070103

Regan, E. A., & Wang, J. (2016). Realizing the Value of EHR Systems Critical Success Factors. *International Journal of Healthcare Information Systems and Informatics*, *11*(3), 1–18. doi:10.4018/IJHISI.2016070101

Rezaie, S., Mirabedini, S. J., & Abtahi, A. (2018). Designing a Model for Implementation of Business Intelligence in the Banking Industry. *International Journal of Enterprise Information Systems*, *14*(1), 77–103. doi:10.4018/IJEIS.2018010105

Rezende, D. A. (2016). Digital City Projects: Information and Public Services Offered by Chicago (USA) and Curitiba (Brazil). *International Journal of Knowledge Society Research*, *7*(3), 16–30. doi:10.4018/IJKSR.2016070102

Rezende, D. A. (2018). Strategic Digital City Projects: Innovative Information and Public Services Offered by Chicago (USA) and Curitiba (Brazil). In M. Lytras, L. Daniela, & A. Visvizi (Eds.), *Enhancing Knowledge Discovery and Innovation in the Digital Era* (pp. 204–223). Hershey, PA: IGI Global. doi:10.4018/978-1-5225-4191-2.ch012

Riabov, V. V. (2016). Teaching Online Computer-Science Courses in LMS and Cloud Environment. *International Journal of Quality Assurance in Engineering and Technology Education, 5*(4), 12–41. doi:10.4018/IJQAETE.2016100102

Ricordel, V., Wang, J., Da Silva, M. P., & Le Callet, P. (2016). 2D and 3D Visual Attention for Computer Vision: Concepts, Measurement, and Modeling. In R. Pal (Ed.), *Innovative Research in Attention Modeling and Computer Vision Applications* (pp. 1–44). Hershey, PA: IGI Global. doi:10.4018/978-1-4666-8723-3.ch001

Rodriguez, A., Rico-Diaz, A. J., Rabuñal, J. R., & Gestal, M. (2017). Fish Tracking with Computer Vision Techniques: An Application to Vertical Slot Fishways. In M. S., & V. V. (Eds.), Multi-Core Computer Vision and Image Processing for Intelligent Applications (pp. 74-104). Hershey, PA: IGI Global. doi:10.4018/978-1-5225-0889-2.ch003

Romero, J. A. (2018). Sustainable Advantages of Business Value of Information Technology. In M. Khosrow-Pour, D.B.A. (Ed.), Encyclopedia of Information Science and Technology, Fourth Edition (pp. 923-929). Hershey, PA: IGI Global. doi:10.4018/978-1-5225-2255-3.ch079

Romero, J. A. (2018). The Always-On Business Model and Competitive Advantage. In N. Bajgoric (Ed.), *Always-On Enterprise Information Systems for Modern Organizations* (pp. 23–40). Hershey, PA: IGI Global. doi:10.4018/978-1-5225-3704-5.ch002

Rosen, Y. (2018). Computer Agent Technologies in Collaborative Learning and Assessment. In M. Khosrow-Pour, D.B.A. (Ed.), Encyclopedia of Information Science and Technology, Fourth Edition (pp. 2402-2410). Hershey, PA: IGI Global. doi:10.4018/978-1-5225-2255-3.ch209

Rosen, Y., & Mosharraf, M. (2016). Computer Agent Technologies in Collaborative Assessments. In Y. Rosen, S. Ferrara, & M. Mosharraf (Eds.), *Handbook of Research on Technology Tools for Real-World Skill Development* (pp. 319–343). Hershey, PA: IGI Global. doi:10.4018/978-1-4666-9441-5.ch012

Roy, D. (2018). Success Factors of Adoption of Mobile Applications in Rural India: Effect of Service Characteristics on Conceptual Model. In M. Khosrow-Pour (Ed.), *Green Computing Strategies for Competitive Advantage and Business Sustainability* (pp. 211–238). Hershey, PA: IGI Global. doi:10.4018/978-1-5225-5017-4.ch010

Ruffin, T. R. (2016). Health Information Technology and Change. In V. Wang (Ed.), *Handbook of Research on Advancing Health Education through Technology* (pp. 259–285). Hershey, PA: IGI Global. doi:10.4018/978-1-4666-9494-1.ch012

Ruffin, T. R. (2016). Health Information Technology and Quality Management. *International Journal of Information Communication Technologies and Human Development, 8*(4), 56–72. doi:10.4018/IJICTHD.2016100105

Ruffin, T. R., & Hawkins, D. P. (2018). Trends in Health Care Information Technology and Informatics. In M. Khosrow-Pour, D.B.A. (Ed.), Encyclopedia of Information Science and Technology, Fourth Edition (pp. 3805-3815). Hershey, PA: IGI Global. doi:10.4018/978-1-5225-2255-3.ch330

Safari, M. R., & Jiang, Q. (2018). The Theory and Practice of IT Governance Maturity and Strategies Alignment: Evidence From Banking Industry. *Journal of Global Information Management, 26*(2), 127–146. doi:10.4018/JGIM.2018040106

Sahin, H. B., & Anagun, S. S. (2018). Educational Computer Games in Math Teaching: A Learning Culture. In E. Toprak & E. Kumtepe (Eds.), *Supporting Multiculturalism in Open and Distance Learning Spaces* (pp. 249–280). Hershey, PA: IGI Global. doi:10.4018/978-1-5225-3076-3.ch013

Sanna, A., & Valpreda, F. (2017). An Assessment of the Impact of a Collaborative Didactic Approach and Students' Background in Teaching Computer Animation. *International Journal of Information and Communication Technology Education, 13*(4), 1–16. doi:10.4018/IJICTE.2017100101

Savita, K., Dominic, P., & Ramayah, T. (2016). The Drivers, Practices and Outcomes of Green Supply Chain Management: Insights from ISO14001 Manufacturing Firms in Malaysia. *International Journal of Information Systems and Supply Chain Management, 9*(2), 35–60. doi:10.4018/IJISSCM.2016040103

Scott, A., Martin, A., & McAlear, F. (2017). Enhancing Participation in Computer Science among Girls of Color: An Examination of a Preparatory AP Computer Science Intervention. In Y. Rankin & J. Thomas (Eds.), *Moving Students of Color from Consumers to Producers of Technology* (pp. 62–84). Hershey, PA: IGI Global. doi:10.4018/978-1-5225-2005-4.ch004

Shahsavandi, E., Mayah, G., & Rahbari, H. (2016). Impact of E-Government on Transparency and Corruption in Iran. In I. Sodhi (Ed.), *Trends, Prospects, and Challenges in Asian E-Governance* (pp. 75–94). Hershey, PA: IGI Global. doi:10.4018/978-1-4666-9536-8.ch004

Siddoo, V., & Wongsai, N. (2017). Factors Influencing the Adoption of ISO/IEC 29110 in Thai Government Projects: A Case Study. *International Journal of Information Technologies and Systems Approach, 10*(1), 22–44. doi:10.4018/IJITSA.2017010102

Sidorkina, I., & Rybakov, A. (2016). Computer-Aided Design as Carrier of Set Development Changes System in E-Course Engineering. In V. Mkrttchian, A. Bershadsky, A. Bozhday, M. Kataev, & S. Kataev (Eds.), *Handbook of Research on Estimation and Control Techniques in E-Learning Systems* (pp. 500–515). Hershey, PA: IGI Global. doi:10.4018/978-1-4666-9489-7.ch035

Sidorkina, I., & Rybakov, A. (2016). Creating Model of E-Course: As an Object of Computer-Aided Design. In V. Mkrttchian, A. Bershadsky, A. Bozhday, M. Kataev, & S. Kataev (Eds.), *Handbook of Research on Estimation and Control Techniques in E-Learning Systems* (pp. 286–297). Hershey, PA: IGI Global. doi:10.4018/978-1-4666-9489-7.ch019

Simões, A. (2017). Using Game Frameworks to Teach Computer Programming. In R. Alexandre Peixoto de Queirós & M. Pinto (Eds.), *Gamification-Based E-Learning Strategies for Computer Programming Education* (pp. 221–236). Hershey, PA: IGI Global. doi:10.4018/978-1-5225-1034-5.ch010

Sllame, A. M. (2017). Integrating LAB Work With Classes in Computer Network Courses. In H. Alphin Jr, R. Chan, & J. Lavine (Eds.), *The Future of Accessibility in International Higher Education* (pp. 253–275). Hershey, PA: IGI Global. doi:10.4018/978-1-5225-2560-8.ch015

Smirnov, A., Ponomarev, A., Shilov, N., Kashevnik, A., & Teslya, N. (2018). Ontology-Based Human-Computer Cloud for Decision Support: Architecture and Applications in Tourism. *International Journal of Embedded and Real-Time Communication Systems, 9*(1), 1–19. doi:10.4018/IJERTCS.2018010101

Smith-Ditizio, A. A., & Smith, A. D. (2018). Computer Fraud Challenges and Its Legal Implications. In M. Khosrow-Pour, D.B.A. (Ed.), Encyclopedia of Information Science and Technology, Fourth Edition (pp. 4837-4848). Hershey, PA: IGI Global. doi:10.4018/978-1-5225-2255-3.ch419

Sohani, S. S. (2016). Job Shadowing in Information Technology Projects: A Source of Competitive Advantage. *International Journal of Information Technology Project Management, 7*(1), 47–57. doi:10.4018/IJITPM.2016010104

Sosnin, P. (2018). Figuratively Semantic Support of Human-Computer Interactions. In *Experience-Based Human-Computer Interactions: Emerging Research and Opportunities* (pp. 244–272). Hershey, PA: IGI Global. doi:10.4018/978-1-5225-2987-3.ch008

Related References

Spinelli, R., & Benevolo, C. (2016). From Healthcare Services to E-Health Applications: A Delivery System-Based Taxonomy. In A. Dwivedi (Ed.), *Reshaping Medical Practice and Care with Health Information Systems* (pp. 205–245). Hershey, PA: IGI Global. doi:10.4018/978-1-4666-9870-3.ch007

Srinivasan, S. (2016). Overview of Clinical Trial and Pharmacovigilance Process and Areas of Application of Computer System. In P. Chakraborty & A. Nagal (Eds.), *Software Innovations in Clinical Drug Development and Safety* (pp. 1–13). Hershey, PA: IGI Global. doi:10.4018/978-1-4666-8726-4.ch001

Srisawasdi, N. (2016). Motivating Inquiry-Based Learning Through a Combination of Physical and Virtual Computer-Based Laboratory Experiments in High School Science. In M. Urban & D. Falvo (Eds.), *Improving K-12 STEM Education Outcomes through Technological Integration* (pp. 108–134). Hershey, PA: IGI Global. doi:10.4018/978-1-4666-9616-7.ch006

Stavridi, S. V., & Hamada, D. R. (2016). Children and Youth Librarians: Competencies Required in Technology-Based Environment. In J. Yap, M. Perez, M. Ayson, & G. Entico (Eds.), *Special Library Administration, Standardization and Technological Integration* (pp. 25–50). Hershey, PA: IGI Global. doi:10.4018/978-1-4666-9542-9.ch002

Sung, W., Ahn, J., Kai, S. M., Choi, A., & Black, J. B. (2016). Incorporating Touch-Based Tablets into Classroom Activities: Fostering Children's Computational Thinking through iPad Integrated Instruction. In D. Mentor (Ed.), *Handbook of Research on Mobile Learning in Contemporary Classrooms* (pp. 378–406). Hershey, PA: IGI Global. doi:10.4018/978-1-5225-0251-7.ch019

Syväjärvi, A., Leinonen, J., Kivivirta, V., & Kesti, M. (2017). The Latitude of Information Management in Local Government: Views of Local Government Managers. *International Journal of Electronic Government Research, 13*(1), 69–85. doi:10.4018/IJEGR.2017010105

Tanque, M., & Foxwell, H. J. (2018). Big Data and Cloud Computing: A Review of Supply Chain Capabilities and Challenges. In A. Prasad (Ed.), *Exploring the Convergence of Big Data and the Internet of Things* (pp. 1–28). Hershey, PA: IGI Global. doi:10.4018/978-1-5225-2947-7.ch001

Teixeira, A., Gomes, A., & Orvalho, J. G. (2017). Auditory Feedback in a Computer Game for Blind People. In T. Issa, P. Kommers, T. Issa, P. Isaías, & T. Issa (Eds.), *Smart Technology Applications in Business Environments* (pp. 134–158). Hershey, PA: IGI Global. doi:10.4018/978-1-5225-2492-2.ch007

Thompson, N., McGill, T., & Murray, D. (2018). Affect-Sensitive Computer Systems. In M. Khosrow-Pour, D.B.A. (Ed.), Encyclopedia of Information Science and Technology, Fourth Edition (pp. 4124-4135). Hershey, PA: IGI Global. doi:10.4018/978-1-5225-2255-3.ch357

Trad, A., & Kalpić, D. (2016). The E-Business Transformation Framework for E-Commerce Control and Monitoring Pattern. In I. Lee (Ed.), *Encyclopedia of E-Commerce Development, Implementation, and Management* (pp. 754–777). Hershey, PA: IGI Global. doi:10.4018/978-1-4666-9787-4.ch053

Triberti, S., Brivio, E., & Galimberti, C. (2018). On Social Presence: Theories, Methodologies, and Guidelines for the Innovative Contexts of Computer-Mediated Learning. In M. Marmon (Ed.), *Enhancing Social Presence in Online Learning Environments* (pp. 20–41). Hershey, PA: IGI Global. doi:10.4018/978-1-5225-3229-3.ch002

Tripathy, B. K. T. R., S., & Mohanty, R. K. (2018). Memetic Algorithms and Their Applications in Computer Science. In S. Dash, B. Tripathy, & A. Rahman (Eds.), Handbook of Research on Modeling, Analysis, and Application of Nature-Inspired Metaheuristic Algorithms (pp. 73-93). Hershey, PA: IGI Global. doi:10.4018/978-1-5225-2857-9.ch004

Turulja, L., & Bajgoric, N. (2017). Human Resource Management IT and Global Economy Perspective: Global Human Resource Information Systems. In M. Khosrow-Pour (Ed.), *Handbook of Research on Technology Adoption, Social Policy, and Global Integration* (pp. 377–394). Hershey, PA: IGI Global. doi:10.4018/978-1-5225-2668-1.ch018

Unwin, D. W., Sanzogni, L., & Sandhu, K. (2017). Developing and Measuring the Business Case for Health Information Technology. In K. Moahi, K. Bwalya, & P. Sebina (Eds.), *Health Information Systems and the Advancement of Medical Practice in Developing Countries* (pp. 262–290). Hershey, PA: IGI Global. doi:10.4018/978-1-5225-2262-1.ch015

Related References

Vadhanam, B. R. S., M., Sugumaran, V., V., V., & Ramalingam, V. V. (2017). Computer Vision Based Classification on Commercial Videos. In M. S., & V. V. (Eds.), Multi-Core Computer Vision and Image Processing for Intelligent Applications (pp. 105-135). Hershey, PA: IGI Global. doi:10.4018/978-1-5225-0889-2.ch004

Valverde, R., Torres, B., & Motaghi, H. (2018). A Quantum NeuroIS Data Analytics Architecture for the Usability Evaluation of Learning Management Systems. In S. Bhattacharyya (Ed.), *Quantum-Inspired Intelligent Systems for Multimedia Data Analysis* (pp. 277–299). Hershey, PA: IGI Global. doi:10.4018/978-1-5225-5219-2.ch009

Vassilis, E. (2018). Learning and Teaching Methodology: "1:1 Educational Computing. In K. Koutsopoulos, K. Doukas, & Y. Kotsanis (Eds.), *Handbook of Research on Educational Design and Cloud Computing in Modern Classroom Settings* (pp. 122–155). Hershey, PA: IGI Global. doi:10.4018/978-1-5225-3053-4.ch007

Wadhwani, A. K., Wadhwani, S., & Singh, T. (2016). Computer Aided Diagnosis System for Breast Cancer Detection. In Y. Morsi, A. Shukla, & C. Rathore (Eds.), *Optimizing Assistive Technologies for Aging Populations* (pp. 378–395). Hershey, PA: IGI Global. doi:10.4018/978-1-4666-9530-6.ch015

Wang, L., Wu, Y., & Hu, C. (2016). English Teachers' Practice and Perspectives on Using Educational Computer Games in EIL Context. *International Journal of Technology and Human Interaction, 12*(3), 33–46. doi:10.4018/IJTHI.2016070103

Watfa, M. K., Majeed, H., & Salahuddin, T. (2016). Computer Based E-Healthcare Clinical Systems: A Comprehensive Survey. *International Journal of Privacy and Health Information Management, 4*(1), 50–69. doi:10.4018/IJPHIM.2016010104

Weeger, A., & Haase, U. (2016). Taking up Three Challenges to Business-IT Alignment Research by the Use of Activity Theory. *International Journal of IT/ Business Alignment and Governance, 7*(2), 1-21. doi:10.4018/IJITBAG.2016070101

Wexler, B. E. (2017). Computer-Presented and Physical Brain-Training Exercises for School Children: Improving Executive Functions and Learning. In B. Dubbels (Ed.), *Transforming Gaming and Computer Simulation Technologies across Industries* (pp. 206–224). Hershey, PA: IGI Global. doi:10.4018/978-1-5225-1817-4.ch012

Williams, D. M., Gani, M. O., Addo, I. D., Majumder, A. J., Tamma, C. P., Wang, M., ... Chu, C. (2016). Challenges in Developing Applications for Aging Populations. In Y. Morsi, A. Shukla, & C. Rathore (Eds.), *Optimizing Assistive Technologies for Aging Populations* (pp. 1–21). Hershey, PA: IGI Global. doi:10.4018/978-1-4666-9530-6.ch001

Wimble, M., Singh, H., & Phillips, B. (2018). Understanding Cross-Level Interactions of Firm-Level Information Technology and Industry Environment: A Multilevel Model of Business Value. *Information Resources Management Journal*, *31*(1), 1–20. doi:10.4018/IRMJ.2018010101

Wimmer, H., Powell, L., Kilgus, L., & Force, C. (2017). Improving Course Assessment via Web-based Homework. *International Journal of Online Pedagogy and Course Design*, *7*(2), 1–19. doi:10.4018/IJOPCD.2017040101

Wong, Y. L., & Siu, K. W. (2018). Assessing Computer-Aided Design Skills. In M. Khosrow-Pour, D.B.A. (Ed.), Encyclopedia of Information Science and Technology, Fourth Edition (pp. 7382-7391). Hershey, PA: IGI Global. doi:10.4018/978-1-5225-2255-3.ch642

Wongsurawat, W., & Shrestha, V. (2018). Information Technology, Globalization, and Local Conditions: Implications for Entrepreneurs in Southeast Asia. In P. Ordóñez de Pablos (Ed.), *Management Strategies and Technology Fluidity in the Asian Business Sector* (pp. 163–176). Hershey, PA: IGI Global. doi:10.4018/978-1-5225-4056-4.ch010

Yang, Y., Zhu, X., Jin, C., & Li, J. J. (2018). Reforming Classroom Education Through a QQ Group: A Pilot Experiment at a Primary School in Shanghai. In H. Spires (Ed.), *Digital Transformation and Innovation in Chinese Education* (pp. 211–231). Hershey, PA: IGI Global. doi:10.4018/978-1-5225-2924-8.ch012

Yilmaz, R., Sezgin, A., Kurnaz, S., & Arslan, Y. Z. (2018). Object-Oriented Programming in Computer Science. In M. Khosrow-Pour, D.B.A. (Ed.), Encyclopedia of Information Science and Technology, Fourth Edition (pp. 7470-7480). Hershey, PA: IGI Global. doi:10.4018/978-1-5225-2255-3.ch650

Yu, L. (2018). From Teaching Software Engineering Locally and Globally to Devising an Internationalized Computer Science Curriculum. In S. Dikli, B. Etheridge, & R. Rawls (Eds.), *Curriculum Internationalization and the Future of Education* (pp. 293–320). Hershey, PA: IGI Global. doi:10.4018/978-1-5225-2791-6.ch016

Related References

Yuhua, F. (2018). Computer Information Library Clusters. In M. Khosrow-Pour, D.B.A. (Ed.), Encyclopedia of Information Science and Technology, Fourth Edition (pp. 4399-4403). Hershey, PA: IGI Global. doi:10.4018/978-1-5225-2255-3.ch382

Zare, M. A., Taghavi Fard, M. T., & Hanafizadeh, P. (2016). The Assessment of Outsourcing IT Services using DEA Technique: A Study of Application Outsourcing in Research Centers. *International Journal of Operations Research and Information Systems*, 7(1), 45–57. doi:10.4018/IJORIS.2016010104

Zhao, J., Wang, Q., Guo, J., Gao, L., & Yang, F. (2016). An Overview on Passive Image Forensics Technology for Automatic Computer Forgery. *International Journal of Digital Crime and Forensics*, 8(4), 14–25. doi:10.4018/IJDCF.2016100102

Zimeras, S. (2016). Computer Virus Models and Analysis in M-Health IT Systems: Computer Virus Models. In A. Moumtzoglou (Ed.), *M-Health Innovations for Patient-Centered Care* (pp. 284–297). Hershey, PA: IGI Global. doi:10.4018/978-1-4666-9861-1.ch014

Zlatanovska, K. (2016). Hacking and Hacktivism as an Information Communication System Threat. In M. Hadji-Janev & M. Bogdanoski (Eds.), *Handbook of Research on Civil Society and National Security in the Era of Cyber Warfare* (pp. 68–101). Hershey, PA: IGI Global. doi:10.4018/978-1-4666-8793-6.ch004

About the Contributors

Riaz Ahmed Shaikh is currently an Associate Professor at the Computer Science Department of the King Abdulaziz University, KSA. He received a Ph.D. degree from the Computer Engineering Department of Kyung Hee University, Korea in 2009; a MS degree in IT from the National University of Sciences and Technology, Pakistan in 2005; and a BSc degree in Computer Engineering from the Sir Syed University of Engineering and Technology, Pakistan, in 2003 respectively. He wrote 50+ research articles published in peer-reviewed conferences and journals. He is a reviewer of various international journals, e.g., IEEE Transactions on Parallel & Distributed Systems, IEEE Computer Journal, Elsevier International Journal of Systems, Control & Communications, Elsevier Mathematical & Computer Modeling, Transactions on Emerging Telecommunications Technologies and many more. His research interests include privacy, security, trust management, risk estimation, sensor networks, Vehicular networks, and IoT. For more information please visit http://sites.google.com/site/riaz289.

* * *

Abdullahi Baffa was the Dean Faculty of Computer Science and Information Technology, Bayero University Kano. He received a Bachelor and Master's degree in Mathematics from Bayero University Kano, and a PhD from University of Birmingham. His current research is on IoT, next-generation networks and artificial intelligence.

Ismail Butun received B.Sc. and M.Sc. degrees in Electrical and Electronics Engineering from Hacettepe University in 2003 and 2006, respectively. He received another M.Sc. and a Ph.D. degree in Electrical Engineering from the University of South Florida in 2009 and 2013, respectively. In the years 2014 and 2015, he worked as an Assistant Professor for the Department of Mechatronics Engineering at Bursa Technical University. In 2016, he worked as a Post-Doctoral Fellow for Department of Electrical and Computer Engineering at University of Delaware. In

2017, he was affiliated with the Department of Computer Engineering at Abdullah Gul University as an Assistant Professor. Since June 2017, he has been working for the Department of Information System and Technology at Mid Sweden University as a Post-Doctoral Fellow. His research interests include computer networks, wireless communications, cryptography, network security, and intrusion detection.

Jairo A. Gutiérrez is the Deputy Head of the School of Engineering, Computer and Mathematical Sciences at Auckland University of Technology in New Zealand. He received a Systems and Computing Engineering degree from Universidad de Los Andes in Colombia, a Master's degree in Computer Science from Texas A &M University, and a Ph.D. in Information Systems from the University of Auckland. His current research is on viable business models for IT-enabled enterprises, next-generation networks and security issues in wireless networks.

Malka N. Halgamuge is a Researcher in the Department of Electrical and Electronic Engineering of the University of Melbourne, and she has also obtained her Ph.D. from the same department in 2007. She is passionate about research and teaching university students (Data Science, Business Intelligence using Big Data, Internet of Things (IoT), Sensor Network, Digital Health, Telemedicine, Bioelectromagnetics, and Hyperthermia). She has published more than 90 peer-reviewed technical articles attracting over 944 Google Scholar Citations with h-index = 16, and her Research Gate RG Score is 33.96. She is currently supervising a PhD student at the University of Melbourne, and 4 PhD students completed their theses in 2013, 2015 and 2018 with her as the principal supervisor. She successfully sought 7 short-term research fellowships at premier Universities in the World. She excels in major commercial work related to the investigation of electromagnetic radiation hazards safety assessment and the provision of extensive technical reports. This work allows her to help organizations and individuals to promote precautionary approaches to the use of technology. This gave her access to contribute in Media (Newspaper Articles and interviews on Radio and Television).

S. Raheel Hassan has done Ph.D. from Université de Franche-Comté (UFC) in Network Security. He worked as an Assistant Professor in the Department of Computer Systems Engineering at Quaid-e-Awam University of Engineering Science and Technology in Pakistan for more than five years. He has recently joined Faculty of Computing and Information Technology (FCIT) and associated with the Department of Computer Science at King Abdulaziz University in Jeddah, Saudi Arabia.

Muhammad Aminu Lawal received his B.ENG in Electrical and Computer Engineering from the Federal University of Technology Minna, Nigeria in 2006 and MSc in Computer Science from Universiti Putra Malaysia in 2014. He is currently pursuing a Ph.D. in Computer Science at King Abdulaziz University, Jeddah. His research interest includes Network Security and Privacy in Smart City Environment.

Jayapandian N. is a PhD assistant professor at the Christ University, Department of Computer Science and Engineering, Bangalore, India. His research interest includes information security, cloud computing, and grid computing. Dr. Jayapandian holds a Bachelor of Technology degree in Information Technology from IRTT, Anna University and Master degree in Computer Science and Engineering from KEC, Anna University. He is published various research article in reputed international journals. He is an active reviewer of reputed international journals. He has participated in numerous national and international conferences and has made a remarkable contribution to cloud data security field and publishing several articles.

Patrik Österberg received his M.Sc. degree in Electrical Engineering from Mid Sweden University, Sundsvall, Sweden, in 2000, the degree of Licentiate of Technology in Teleinformatics from the Royal Institute of Technology, Stockholm, Sweden, in 2005, and the Ph.D. degree in Computer and System Science from Mid Sweden University in 2008. During 2007, he worked as a development engineer at Acreo AB in Hudiksvall, Sweden, and from 2008 to 2010, he was employed as researcher at Interactive TV Arena KB in Gävle, Sweden. Since 2008, he is an Assistant Professor at Mid Sweden University and from 2013, he is also the head of the Department of Information System and Technology.

James S. Purkis is a lecturer at Charles Sturt University Study Centres Melbourne and is currently completing a PhD in educational technology from Charles Sturt University Wagga Wagga. He has a broad range of teaching experience working in primary, secondary and tertiary both in Australia and overseas in the fields of English, History, ESL and Behavioural Science. His recent publications include conference papers and presentations at the Australian Consortium for Social and Political Research Inc (ACSPRI) and the Australian Association for Research in Education (AARE). He has also worked as a research assistant in developing online learning platforms and modules for tertiary courses.

Junaid Mohammad Qurashi is currently pursuing his Ph.D. in King Abdul Aziz University. He completed his Masters from International Islamic University Malaysia in 2016, with dissertation titled "Investigation of Mobile Health Applications Usage among University Students: IIUM Case Study." His Bachelors was in Computer Science from Kashmir University.

Mamata Rath is currently Assistant Professor at C.V. Raman College of Engineering, Bhubaneswar, India. Her research interests include Wireless Networks, Internet of Things (IoT), Computer Security, Smart Applications, Real Time Systems, E-commerce & ERP. She has around 30 number of good quality research publications in the area of Mobile Ad-hoc networks, Real time Applications, Smart Applications for Smart City and Internet of Things in good journals with high impact and Internationally reputed Conferences. She has been part of few good quality International Conferences and acted as reviewer in few selected International Journals.

Gaganjot Kaur Saini received her Bachelor's of Technology degree in Computer Science and Engineering at Punjab Technical University, Punjab, India in 2010, after she worked as an Assistant Professor at Guru Nanak Institute of Technology for 3 years and Lord Krishna Polytechnic College for 10 months and Masters of Information Technology degree from Charles Sturt University, Melbourne, in 2017. Her main research includes cyber Security issues and their proposed solutions.

Jana Shafi pursed Bachelor of Computer Science from MDU, India in 2010, And Master of Technology from MDU, India in 2012. She is currently working as lecturer in Computer Science Department, College of Arts and Science, Prince Sattam Bin Abdul Aziz University, KSA since 2014. Her research interest includes Soft Computing, Social Networks and Internet of things (IoT). She has published many papers in international journals and has presented papers at international conferences. She has been serving as member of review committee for the Asian Journal of Computer Science Engineering (AJCSE).

Vijey Thayananthan is an Associate Professor at Computer Science Department in the King Abdulaziz University, Jeddah, KSA. He obtained his Ph.D. in Engineering (Digital communication engineering) from Department of Communication Systems, University of Lancaster, UK, in 1998. Further, he had worked as Postdoctoral research Fellow in Department of Computer Science, Glasgow/Sterling University and Communication division, Department of Electrical and Electronic Engineering, Strathclyde university, UK for 2 years. Since 2000, he had been working as a

Research engineer and senior algorithm development engineer in Advantech Ltd, Southampton University Science Park, UK and Amfax Ltd, UK respectively. His research interests include wireless communication algorithm design and mobile communication analysis, security management of communication network and big data, computer security and wireless sensor network. He has been a full-time member of the Institution of Engineering Technology (IET), UK since 2005.

Aminu Bello Usman is a Senior Lecture in Computer Science: Networks and Security, School of Art, Design and Computer Science, York St John University, UK. He received his PhD in Network Security at AUT, New Zealand, a Master's degree in Computer Network and Security at Middlesex University, London and a Bachelor in computer Science at Bayero University, kano Nigeria. His current research is on Secure Wireless Mobile Networks using trust mechanisms, IoT and Cloud Computing Security.

Amtul Waheed received Bachelor of Computer Application from Osmania University, Hyderabad, India in 2003, and Master of Computer Application from Osmania University in year 2006 and Master of Technology from Jawaharlal Nehru Technological University. She is currently working as lecturer in Computer Science Department in College of Arts and Science, Prince Sattam Bin Abdul Aziz University, Saudi Arabia since 2011. Her research interest includes Web Services, Social Networks and Internet of things (IoT). She has published many papers in international journals and has presented papers at international conferences. She has been serving as member of review committee for the Asian Journal of Computer Science Engineering (AJCSE).

Index

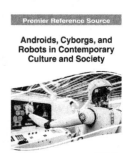

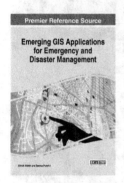

Printed in the United States
By Bookmasters